Sustainable Development Goals Series

The **Sustainable Development Goals Series** is Springer Nature's inaugural cross-imprint book series that addresses and supports the United Nations' seventeen Sustainable Development Goals. The series fosters comprehensive research focused on these global targets and endeavours to address some of society's greatest grand challenges. The SDGs are inherently multidisciplinary, and they bring people working across different fields together and working towards a common goal. In this spirit, the Sustainable Development Goals series is the first at Springer Nature to publish books under both the Springer and Palgrave Macmillan imprints, bringing the strengths of our imprints together.

The Sustainable Development Goals Series is organized into eighteen subseries: one subseries based around each of the seventeen respective Sustainable Development Goals, and an eighteenth subseries, "Connecting the Goals", which serves as a home for volumes addressing multiple goals or studying the SDGs as a whole. Each subseries is guided by an expert Subseries Advisor with years or decades of experience studying and addressing core components of their respective Goal.

The SDG Series has a remit as broad as the SDGs themselves, and contributions are welcome from scientists, academics, policymakers, and researchers working in fields related to any of the seventeen goals. If you are interested in contributing a monograph or curated volume to the series, please contact the Publishers: Zachary Romano [Springer; zachary.romano@springer.com] and Rachael Ballard [Palgrave Macmillan; rachael.ballard@palgrave.com].

Stellina Jolly · Nafees Ahmad ·
Matthew Scott
Editors

Climate-Related Human Mobility in Asia and the Pacific

Interdisciplinary Rights-Based Approaches

Springer

Editors
Stellina Jolly
Faculty of Legal Studies
South Asian University
New Delhi, Delhi, India

Nafees Ahmad
Faculty of Legal Studies
South Asian University
New Delhi, Delhi, India

Matthew Scott
Senior Researcher and Leader of Human
Rights and Environment Thematic Area
Raoul Wallenberg Institute of Human
Rights and Humanitarian Law
Lund, Sweden

ISSN 2523-3084 ISSN 2523-3092 (electronic)
Sustainable Development Goals Series
ISBN 978-981-97-3233-3 ISBN 978-981-97-3234-0 (eBook)
https://doi.org/10.1007/978-981-97-3234-0

This Springer imprint is published by the registered company Springer Nature Singapore Pte Ltd.
The registered company address is: 152 Beach Road, #21-01/04 Gateway East, Singapore 189721, Singapore

Foreword

It is an honour to write a Foreword for this important and timely volume on climate mobility in the Asia-Pacific region. This collection brings together a range of perspectives from academics from across the region to examine how displacement, migration, and planned relocation are being experienced in the Asia-Pacific, and possible policy responses. Through a human rights lens, it draws on expertise from a wide range of disciplinary areas, including law, political ecology, social work, disaster management, climate change adaptation, and artificial intelligence.

The Asia-Pacific is the region most affected by disaster displacement, with an annual average of almost 19 million internal disaster displacements since 2010—nearly 80 per cent of the global total.[1] As in other parts of the world, the cascading and intertwined effects of conflict, inequality, poverty, human rights violations, disasters, and climate change amplify existing risks and vulnerabilities, which in turn affect people's ability to remain in their homes. Increased displacement—whether internal or cross-border, short-term or long-term—may create instability within countries and across the region, contributing to cultural tensions, resource scarcity, and, in some cases, even civil unrest.

As experts from the Asian Development Bank and the Internal Displacement Monitoring Centre have explained: 'While displacement often demands a humanitarian response, it is first and foremost a development issue. It disrupts and erodes the development gains of affected communities and can have an impact on their longer-term stability and resilience to future shocks.'[2] Disasters in the Asia-Pacific already cost several hundred billion dollars annually—and this does not include the secondary economic impacts of displacement, let alone the psychological, social, and cultural impacts. As the chapters in this book show, displacement, migration, and planned relocation need to be considered within a broader social and economic context. Policy responses must be nuanced and appropriate to local conditions.

[1]Internal Displacement Monitoring Centre, *Disaster Displacement in Asia and the Pacific* (2022) 20.

[2]'Foreword' in ibid, 12.

Recognising that addressing displacement and other forms of climate mobility are a key part of planning for disasters is an important step in building resilience and enabling recovery. Indeed, the evidence shows that when people receive prompt and adequate assistance in the aftermath of a disaster, they are more likely to stay and rebuild than to move on in search of work to survive. By contrast, a lack of durable solutions may push people to keep moving onwards.

The past decade has seen enormous advancements in understandings of and responses to climate mobility, but much work remains to be done. In particular, more fine-grained analyses—such as those in this volume—are needed to ensure that responses are contextually appropriate, well-attuned, and will have local buy-in. That is also why the voices of affected communities must be front and centre. The region is already home to a number of world-leading practices on climate mobility, largely by necessity. A number of countries in the Pacific, for instance, have adopted national guidelines on internal displacement and planned relocations in the context of climate change, and a regional framework on climate mobility is also under development [update as late as possible: intention was for it to be adopted by the PIF in November 2023].

The present volume brings diverse new perspectives to bear through qualitative interviews, case studies, and critiques of existing legal and policy frameworks. It is a significant and meaningful contribution to the literature on climate mobility, with the capacity for real-world impact.

November 2023

Jane McAdam
Professor of Law, Director
of the Andrew & Renata Kaldor Centre
for International Refugee
Law University of New South Wales (UNSW)
Sydney, Australia

Acknowledgements

The concept of this edited volume book emanated out of the deliberations and suggestions from the workshop on Climate Change, Human Mobility, and Human Rights in the Asia Pacific organised by Raoul Wallenberg Institute of Human Rights and Humanitarian Law (RWI) on 25–27 October 2022 at Bangkok, Thailand. We want to acknowledge the intellectual contribution of the participants and presenters in shaping the initial conceptualisation of the book. It was suggested that the issue of climate mobility needs to be evaluated from the lived experience of the home-grown researchers from the region, and that formed the backbone of this research project. We also recognise that many of them could not be part of this academic collaboration, but their thinking profoundly impacted conceiving this project.

After this initial stage of fruition, many assisted in putting together the current version of the edited volume. We would like to place on record our heartfelt gratitude to the authorities at South Asian University for supporting and encouraging the academic collaboration between RWI and Faculty of Legal Studies, South Asian University (FLS-SAU), which made the fruition of the book possible.

The editors sincerely thank Prof. Jane McAdam for the valuable Foreword. Her contribution has enriched the value of the book. The editors would like to thank all the contributors who shared their ideas and cooperated during the entire manuscript preparation process. We must acknowledge that the numerous meetings and interactions we had with the contributors were of mutual learning.

The editors wish to place on record the suggestions and comments of the reviewers for this edited collection. We would also like to thank the entire Springer team, especially Ms. Nupoor Singh, for her encouraging response to our project, the constant cooperation, and for ensuring an accelerated publication process.

Our special thanks to Mr. Danang Aditya Nizar, the programme officer at RWI Regional Asia Pacific Office, who has become our friend in this academic journey. Our thanks to Danang for ensuring smooth communication with the editors and authors, always with a smiling face. We would also like to thank Mr. Jason Squire, the director of RWI Regional Asia Pacific Office, for his continuous support throughout the process.

We would like to acknowledge the generous financial support from the Swedish International Development Cooperation Agency (Sida), which made the open access of the volume possible for wider accessibility. We take this opportunity to acknowledge the research assistance rendered by Ms. Gursimran Kaur Bakshi and Md. Iftesham, our LLM students at FLS-SAU. We also want to thank Ms. Shilpa George for her help with the standardisation of language and copyediting. In the end, we immensely thank our family for their enduring support.

February 2024

Stellina Jolly
Nafees Ahmad
Matthew Scott

Contents

Editors and Contributors

About the Editors

Dr. Stellina Jolly is a Lead Editor and an Associate Professor at the Faculty of Legal Studies, South Asian University (SAU). She is a Visiting Senior Research Associate with the Centre for Emerging Countries in Private international law at the University of Johannesburg. Dr. Jolly is a Fulbright Scholar with the University of San Francisco and a recipient of the International Visitors Leadership Program (IVLP). She researches international environmental law and conflict of laws. In environmental law, her research papers have explored the north-south dimensions of environmental justice. Her book, *Climate Refugees in South Asia* published by Springer, explored the legal and policy framework of climate refugees in South Asia; she has also elaborately written on the environmental justice and governance concerns and intersectionalities of climate change, water, gender, disaster management, and sustainable development goals focusing on South Asian jurisdictions. Her recent researches also focus on the emerging global trend of the rights of nature. In her research, she advocates an approach based on just sustainability, hybrid law mechanism, and environmental justice to tide over the ecological crisis. Dr. Jolly has published in several leading international journals like the *Transnational Environmental Law*, *Washington Journal of Environmental Law and Policy*, *Texas Environmental Law Journal*, and *Chinese Journal of Environmental Law*, to name a few. Hart published her book on *Indian Private International law*. She has also co-edited a book collection published by Springer on *Private International Law of South Asian Jurisdictions*. She is a member of the *IUCN World Commission on Environmental Law and part of the Network of Environmental Law Champions*, *Asian Development Bank*. She has undertaken projects and consultancies with various organizations, including the *Rajiv Gandhi Foundation, European Union, International Renewable Energy Agency* (IRENA), and *Asia Europe Foundation*. She was awarded an educational grant on Civil Society Law from International Centre for Not-for-Profit Law (ICNL) and USAID.

Dr. Nafees Ahmad is an Editor, and an Associate Professor at the Faculty of Legal Studies, South Asian University (SAU)-New Delhi. He holds a doctorate in International Refugee Law and Human Rights. His scholarship focuses on RAMS (Refugees, Asylum-seekers, Migrants, Stateless) and the role of Artificial Intelligence (AI) in their protection, Global Forced Displacement, Global Circumstantial Migration (GCM) Governance, and Climate Refugees in South Asia. He also addresses the International Politics of Asylum, Refugee Policy Paradigms, Invisible Frames of Asylum, Disconnects of Durable Solutions, and SAARC connects and contexts of refugee protection. He conceived and introduced a new SAARC-specific Program in 2011 called *Comparative Constitutional Law of SAARC Nations* (CCLSAARCN) at the LLM level. He publishes inter-alia in the *International Handbook on Disaster Research* (Springer 2023), *The Asian Yearbook of Human Rights and Humanitarian Law* (Brill 2023), *Groningen Journal of International Law* (The Netherlands), *Harvard International Law Journal* (Harvard) *Asia-Pacific Journal of Human Rights and Law* (Brill), *International Journal of Environment and Waste Management* indexed in SCOPUS, *International Journal on Minority and Group Rights*, *Kings' Student Law Review* (KCL-London), *ISIL Year Book on International Humanitarian Law and Refugee Law*, *ELCOP Year Book of Human Rights* and *NUJS International Journal of Legal Studies and Research* (IJLSR) etc. Dr. Ahmad has co-authored a book on *Climate Refugees in South Asia* (Springer 2019). He is a member of Editorial Advisory Board of *Iranian Journal of International and Comparative Law*. Dr. Ahmad has been a Resource Person and External Reviewer for the Ministry of Law; Government of India-sponsored Research Project on "*Judicial Reforms since June 2016*" at the Indian Institute of Management (IIM-Kashipur) and accomplished an ICSSR Project on "*The Municipal Solid Waste Management in Delhi: A Socio-Legal Study of Okhla Lanfill*." Further, he has been a visiting professor at the Indian Society of International Law (ISIL), Jamia Milia Islamia-New Delhi, and Judicial Academies of various states of India.

Dr. Matthew Scott is an Editor and co-leads the Human Rights and the Environment thematic area at the Raoul Wallenberg Institute of Human Rights and Humanitarian Law in Lund, Sweden. His scholarship integrates social science perspectives with international legal standards to promote context-sensitive, human rights-based law, policy, and practice relating to disaster risk reduction and climate change adaptation. Dr. Scotts' primary area of expertise concerns migration and displacement in the context of disasters and climate change, on which he published a monograph entitled *Climate Change, Disasters and the Refugee Convention* (CUP 2020), an edited volume entitled *Climate Change, Disasters and Internal Displacement in Asia and the Pacific: A Human Rights-Based Approach* (Routledge 2021), and a range of book chapters and academic articles in, amongst others, the *International Journal of Refugee Law*, the *Nordic Journal of International Law* and the *Yearbook of International Disaster Law*. The edited volume was a major output of a regional thematic study he coordinated as part of RWI's Asia-Pacific programme on human rights and sustainable development. Dr. Scott works on the role of local authorities in addressing climate- and disaster-related migration and displacement and how human rights law can contribute to building resilience to pandemic risk. He holds a Ph.D. in Public International Law from Lund University (2018) and a MA in Social Anthropology of Development from SOAS (1998). He practiced immigration and asylum law before entering academia. He is a member of the advisory committee of the Platform on Disaster Displacement, a member of the editorial board of the *Yearbook of International Disaster Law*, and a founding member of the *Nordic Network on Climate-Related Displacement and Mobility*. At Lund University, he convenes the LLM course on human rights law, the environment, and climate change and lectures on international refugee and human rights law at the Faculty of Law. He also lectures on the M.Sc. programme in Disaster Risk Reduction and Climate Change Adaptation at the Department of Risk Management and Societal Safety. He is also actively engaged in international collaboration initiatives and currently contributing technical expertise on human rights-based approaches to disaster risk reduction across eight countries in Asia in collaboration with the Asia Disaster Preparedness Center, Stockholm Environment Institute, and the Swedish Civil Contingencies Agency.

Contributors

Dinushika M. Yapa Abeywardhana Department of Sociology, University of Peradeniya, Peradeniya, Sri Lanka

Nafees Ahmad Faculty of Legal Studies, South Asian University, New Delhi, India

Inayah Bastin Al Hakim Masters in Disaster Management, Universitas Gadjah Mada, Yogyakarta, Indonesia

Ly Quoc Dang Mekong Delta Development Research Institute, Can Tho University, Can Tho, Viet Nam

Ishari Gunarathna Department of Sociology, University of Peradeniya, Peradeniya, Sri Lanka

Stellina Jolly South Asian University, New Delhi, India

Reni Juwitasari Disaster Resilience and Environmental Sustainability (DRES) Program, Asian Research Center for International Development (ARCID), School of Social Innovation, Mae Fah Luang University, Chiang Rai, Thailand

Om N. Katel Department of Environment & Climate Studies, College of Natural Resources, Royal University of Bhutan, Punakha, Bhutan

Md. Abdul Awal Khan Independent University, Dhaka, Bangladesh

Dora Kuir-Ayius School of Humanities and Social Sciences, The University of Papua New Guinea, Port Moresby, Papua New Guinea

Nazia Khan Lina Department of Geography and Environment, Jagannath University, Dhaka, Bangladesh

Md. Abdul Malak Department of Geography and Environment, Jagannath University, Dhaka, Bangladesh

Feby Aulia Marsida Masters in Disaster Management, Universitas Gadjah Mada, Yogyakarta, Indonesia

Liam Moore International Politics and Policy, James Cook University, Townsville, Australia

Anooja Nair Department of Food Science & Technology, College of Natural Resources, Royal University of Bhutan, Punakha, Bhutan

Danang Aditya Nizar Raoul Wallenberg Institute of Human Rights and Humanitarian Law, Jakarta, Indonesia

Lovelyn Otoiasi Environment Department, Faculty of Agriculture, Fisheries and Forestry, Solomon Islands National University, Honiara, Solomon Islands

Priyadarshani Premarathne Department of Sociology, University of Peradeniya, Peradeniya, Sri Lanka

Cahyadi Ramadhan Masters in Disaster Management, Universitas Gadjah Mada, Yogyakarta, Indonesia

James Reinaldo Rumpia Center for Public Policy and Human Rights Studies, Faculty of Law, Lampung University, Bandar Lampung, Indonesia

Dina Ruslanjari Masters in Disaster Management, Universitas Gadjah Mada, Yogyakarta, Indonesia

Matthew Scott South Asian University, New Delhi, India

Chogyel Wangmo Department of Environment & Climate Studies, College of Natural Resources, Royal University of Bhutan, Punakha, Bhutan

Ugyen Yangchen Department of Sustainable Development, College of Natural Resources, Royal University of Bhutan, Punakha, Bhutan

Introduction

Stellina Jolly, Nafees Ahmad, and Matthew Scott

This edited volume brings together academics from Indonesia, Thailand, Sri Lanka, Vietnam, Bangladesh, Bhutan, India, the Solomon Islands, Papua New Guinea, Australia and the Philippines, to examine lived experiences of and actual as well as potential responses to the varieties of migration, displacement, planned relocation as well as voluntary and involuntary immobility (collectively termed human mobility) that take place in the context of disasters and climate change that increasingly impact the lives of people living across Asia and the Pacific. A product of the editorial collaboration between the South Asian University and the Raoul Wallenberg Institute for Human Rights and Humanitarian Law (RWI), the volume represents the collective efforts of members of the Asia Pacific Academic Network on Disaster Displacement. This academic network was formed in 2018 as part of a regional thematic study on human rights and disaster displacement. Since then, the network has expanded its membership and activities, including collaboration on publications, submissions to international consultative processes, conferences and collaboration on design and implementation of research as well as more practitioner-focused initiatives.[1]

This volume explores multiple dimensions of climate mobility, examines the extent to which it threatens, but which can also support the realisation of, fundamental human rights in countries in Asia and the Pacific, and engages with ideas about how to overcome observed challenges while recognising that 'solutions' can only be achieved in the local context. The objective is to provide evidence-informed insights that can support the work of multiple actors, including civil society, regional and national government, the private sector and international agencies.

The collection of contributions provides expert, vernacular insights into the complexities of disaster- and climate-related human mobility in the region.[2] This volume offers unique and discernible research treatment from home-grown scholars who are closely experiencing and enduring the ground realities of climate change impact and resultant human mobility. At

S. Jolly (✉) · N. Ahmad · M. Scott
South Asian University, New Delhi, India
e-mail: stellinajolly@sau.ac.in

N. Ahmad
e-mail: drnafeesahmad@sau.ac.in

M. Scott
e-mail: matthew.scott@rwi.lu.se

[1] 'Climate Change, Disasters and Internal Displacement in Asia and Pacific: A Human Rights-Based Approach' (*Raoul Wallenberg Institute)* https://rwi.lu.se/disaster-displacement/ accessed 25 November 2023; For more information on the APANDD network, see https://rwi.lu.se/disaster-displacement/.

[2] Abbass et al. [1].

S. Jolly et al. (eds.), *Climate-Related Human Mobility in Asia and the Pacific*,
Sustainable Development Goals Series, https://doi.org/10.1007/978-981-97-3234-0_1

a time when international and regional agendas are calling for more action to address the phenomenon, this volume expands the knowledge base. It is crystallised with the required visibility for stakeholders to re-calibrate and re-formulate the laws, policies and programmes in context for addressing Asia and the Pacific-related climate crisis in their respective jurisdictions, locally and regionally. It stimulates interdisciplinary engagement in law, political ecology, social work, disaster management, climate change adaptation and artificial intelligence. These contributions are combined under the universalising language of human rights. This introduction sets the scene and outlines the key themes that will be addressed across the volume.

The RWI conducted a bibliometric study on literature on human rights, human mobility and climate change in Asia and the Pacific,[3] using the PRISMA methodology for systematic review and transparency. The study found that while research on human mobility is growing, a significant number of academic outputs are produced by authors from the Global North, such as the USA, Australia and the UK. China and India are the second and fifth most productive countries, respectively. The study suggests increasing the representation of scholars from the region who can write about their experiences and perspectives. The study reveals a significant focus on displacement (71% of the literature), with less attention given to planned relocation (18%), immobility (7%) and migration (4%). The current volume addresses these findings by highlighting the role of local researchers in understanding human mobility within the region.

1 Climate Change Impacts in Asia and the Pacific

The Asia–Pacific region hosts 4.3 billion inhabitants that represent more than 50% of the population of the world.[4] Droughts, floods, tropical cyclones and other weather-related climate hazards are threatening the lives and livelihoods of millions of people throughout the region, displacing a large number of people.[5] Extreme weather and climate events are having an impact on agriculture arability, food security and water availability. One notable instance of climate change-related impacts was Pakistan's historic floods, where the most substantial losses were in the housing, agriculture and livestock and transportation and communications sectors, at USD 5.6 billion, USD 3.7 billion and USD 3.3 billion, respectively.[6] Due to climate change in low-lying areas, typhoons and river flooding threaten a growing number of people in India,[7] Bangladesh,[8] the Philippines[9] and across the Asia and the Pacific region.

Many factors are significant in assessing Asia–Pacific's vulnerability and exposure to climate change. Firstly, the location is a key factor in determining exposure to climate change. Asia's river deltas and coastal regions are among the world's most densely populated regions and climate hotspots.[10] The majority of Pacific Island nations, some of them including low-lying atolls, are also quite exposed.[11] Another critical issue is rapid urbanisation and

[3] Raoul Wallenberg Institute [2].

[4] Asian Development Bank, *Key Indicators for Asia and the Pacific* (2021).

[5] 'Over 57 Million Affected by Climate Disasters Across Asia Pacific in 2021' (*International Federation of Red Cross and Red Crescent Societies,* 15 December 2021) https://www.ifrc.org/press-release/over-57-million-affected-climate-disasters-across-asia-pacific-2021 accessed 25 November 2023.

[6] 'Pakistan: Flood Damages and Economic Losses Over USD 30 billion and Reconstruction Needs Over USD 16 billion-New Assessment' (*The World Bank,* 28 October 2022) https://www.worldbank.org/en/news/press-release/2022/10/28/pakistan-flood-damages-and-economic-losses-over-usd-30-billion-and-reconstruction-needs-over-usd-16-billion-new-assessment accessed 15 June 2023.

[7] Roy et al. [3].

[8] Zami [4].

[9] Delano [5].

[10] Parry et al. [6].

[11] 'A Region at Risk: The Human Dimensions of Climate Change in Asia and the Pacific' (*Asian Development Bank,* 2017) 1, 84; Siddiqui and Sahay [7].

population growth. The rising population pressure in the Asia–Pacific, where the urban population is projected to exceed 50% in the next ten years with a significant coastal concentration, increases exposure and vulnerability.[12]

Vulnerability to climate change is not solely a matter of where a person lives but also depends on how they live. Many Asia–Pacific nations share common characteristics, such as their dependency on the agricultural and natural resource industries, densely populated coastal areas, inadequate institutional framework and the fact that a sizeable portion of the population lives in poverty.[13] The population of the Asia–Pacific is especially susceptible to climate effects because it is also home to the vast majority of the world's poor.[14] Though many countries in the region have undergone economic transformation and are developed,[15] poverty remains widespread, with about 1.8 billion people living on less than $2 per day.[16] Unchecked global warming is already significantly undermining prior gains in economic growth and advancements, endangering the region's future. Without coordinated action, the region could see an additional 7.5 million people fall into poverty as a result of climate impacts by 2030.[17] This figure should be considered alongside the considerable development losses resulting from the COVID-19 pandemic and increases in the cost of living, which has pushed more than 67 million more people into extreme poverty since the outbreak of the pandemic in 2020.[18] The deteriorating effects of climate change on the economy, society, culture and environment have negative repercussions for the effective promotion and defence of human rights and the accomplishment of the Sustainable Development Goals (SDGs) outlined in the 2030 Agenda.[19]

According to the Pacific Islands Forum's 2018 Boe Declaration on Regional Security, 'Climate change remains the single greatest threat to the livelihoods, security, and wellbeing of the peoples of the Pacific.'[20] One of the climate change impacts is climate-related human mobility.

2 Climate-Related Human Mobility in Asia and the Pacific

This volume explores variations of human mobility in specific national and sub-national contexts across this vast region. It contributes to existing literature by presenting the diversity of the phenomenon in a collection where common challenges and potential responses can be discerned.

The First IPCC Assessment Report (FAR), published in 1990, emphasised the significance of climate change as a problem needing global cooperation. According to the first IPCC assessment in 1990, the most severe effects of climate change may be on human mobility when major disasters such as coastal floods and shoreline erosion, force millions of people to flee their

12 'Urban Development' (*The World Bank*, 03 April 2023) https://www.worldbank.org/en/topic/urbandevelopment/overview accessed 26 November 2023.

13 'Multidimensional Poverty Analysis for the Asia Pacific Region 2022' (*Embassy of Sweden and Development Cooperation Section Bangkok*, 2022) 1, 15.

14 'The Economics of Climate Change in the Asia-Pacific Region' (*United Nations Economic and Social Commission for Asia and the Pacific*)1, 5–6.

15 'Human Development Report 2010, The Real Wealth of Nations: Pathways to Human Development' (*United Nations Development Programme*, 2010).

16 'Addressing Climate Change and Migration in Asia and the Pacific' (*Asian Development Bank*, 2012) 1,2.

17 'Climate and Development in East Asia and Pacific Region' (*The World Bank*, 08 November 2023) https://www.worldbank.org/en/region/eap/brief/climate-and-development-in-east-asia-and-pacific-region accessed 15 November 2023.

18 Asian Development Bank, Key indicators for Asia and the Pacific 2023 (2023). https://www.adb.org/publications/key-indicators-asia-and-pacific-2023 accessed 24 January 2024.

19 UN General Assembly [8].

20 'Boe Declaration on Regional Security' (*Pacific Islands Forum*, 2018) https://www.forumsec.org/2018/09/05/boe-declaration-on-regional-security/ accessed 26 November 2023.

homes.[21] The aforementioned finding was developed and nuanced in later IPCC assessments that also includes the Sixth Assessment Report (AR6), which also covered several scenarios of climate mobility for the Asia–Pacific and other regions.[22] By this point in time, all varieties of human mobility are extensively discussed in this state of the art reports.[23]

Regarding climate-induced displacement, several projections are made, but the accuracy is limited due to this complicated web of components. What is clear is that every year, an average of 25 to 30 million new displacements occur in the context of sudden-onset disasters, particularly floods and storms.[24] Data concerning displacement in the context of slower-onset hazard events like droughts and sea level rise are much more challenging to collect owing to the lack of a temporally clear trigger of movement. Planned relocation is only beginning to receive attention.[25] Although predicting future movements has been strongly discouraged by a broad spectrum of academic authorities owing to the multiple drivers and varieties of human agency,[26] the IPCC reports that paint increasingly urgent pictures of current and foreseeable climate impacts strongly point towards an increase in all forms of climate mobility, including immobility, in the coming years. More attention to the issue is clearly called for. Although there is a growing body of research on climate-related mobility in Asia and the Pacific,[27] these studies only begin to reveal the varieties of movement arising in the context of interconnected social and environmental dynamics across the myriad social-ecological systems characterising the region.

Even with the growth in armed conflict, the majority of IDPs move in the context of climate-related disasters. Between 2008 and 2020, the Asia–Pacific region surpassed all other regions in the world in terms of climate-related displacement,[28] accounting for 80% of all new cases. In 2021, the biggest per-country numbers of displaced individuals were recorded in China, the Philippines, India and Vietnam.[29] There were 32.6 million movements in 2022 compared to 2021, reflecting a whopping 40% increase. Pakistan, the Philippines, China, India and Nigeria were the top five nations where there were new disaster-related displacements.[30]

Although displacement receives attention in this volume, as a point of departure, we note the importance of elevating the term climate-related human mobility owing to the inadequacy of the terms migration and displacement to reflect observed experiences. These terms, which typically reflect different ends of a spectrum between forced and voluntary movement, can mislead in situations where multiple forms of movement are taking place, as observed in the case studies in this volume. For instance, in their study of the mobility of elderly people in the context of recurrent cyclones in Bangladesh, Malak and Lina describe how some people have moved to urban centres, while others have remained. Those who moved reflect on the adverse and cumulative impact of cyclones, starting with Cyclone Aila in 2009, but also on the benefits associated with improved access to services in towns. Similarly, Nizar and Dang draw attention to what they describe as 'micro-level mobility dynamics' reflected in how people move within areas impacted by (increasingly

[21] 'Climate Change: The IPCC 1990 and 1992 Assessments': IPCC First Assessment Report Overview and Policymakers Summaries and 1992 IPCC Supplement (*Intergovernmental Panel on Climate Change,* 1992) 1,103.

[22] IPCC [9].

[23] See Scott [10].

[24] 'Global Report on International Displacement 2023: Internal displacement and food security' (*Internal Displacement Monitoring Centre,* 2023)1, 68.

[25] Bower et al. [11].

[26] 'Foresight: Migration and Global Environmental Change, Final Project Report' (*The Government Office for Science, London* 2011) 1, 11.

[27] Raoul Wallenberg Institute [12].

[28] 'Global Internal Displacement Database' (*Internal Displacement Monitoring Centre*) https://www.internal-displacement.org/database/displacement-data accessed 26 November 2023.

[29] Ibid.

[30] DW [13].

frequent) flooding events, without necessarily being described as 'displaced.' Katel et al. describe rural to urban 'migration,' but demonstrate the significant impact of a changing climate on agricultural livelihoods as a significant factor compelling people to move.

Further, the phenomenon of planned relocation warrants substantially more attention, not least owing to the serious human rights issues that arise in this context. The term is defined as:

> … a planned process in which persons or groups of persons move or are assisted to move away from their homes or places of temporary residence, are settled in a new location, and provided with the conditions for rebuilding their lives. Planned Relocation is carried out under the authority of the State, takes place within national borders, and is undertaken to protect people from risks and impacts related to disasters and environmental change, including the effects of climate change. Such Planned Relocation may be carried out at the individual, household, and/or community levels.[31]

Several texts in the volume, such as Abeywardhana's chapter on relocations in Sri Lanka, also include the term 'resettlement.' One way to distinguish resettlement from relocation is the process of resettling relocated individuals or groups into new locations while maintaining significant socio-political features at the resettlement destination, as well as restoring or creating new livelihoods in these sites. In conceptualising the planned relocation dimension of human mobility in this volume, we follow the description provided in the Brookings, Georgetown and UNHCR guidance as 'a community is physically moved to another location and resettled there.'[32] Further, drawing on Section V of the 1998 UN Guiding Principles on Internal Displacement, we understand resettlement as including the character of permanence as one of the three durable solutions to displacement. However, whereas resettlement in this sense can only take place after a person has been displaced, relocation has a pre-emptive character.

[31] Ferris et al. [14].

[32] Ibid.

Although there has been more focus on relocating people due to climate change events in recent years, there is still a great deal of ambiguity over how populations will respond to long-term climate change, and existing case studies show widespread adverse outcomes.[33] Consequently, planned relocation should not be undertaken lightly, must follow international good practice and ought generally to be pursued as an option of last resort.

Thus the volume adopts the term human mobility to address the varieties of movement that take place in the context of disasters and climate change. We also use the term immobility where the lack of movement takes place in a context where movement is somehow a relevant consideration in light of lived experiences in a particular location. In this context, for instance, Reni Juwitisari in this volume describes decision-making by older farmers in northern Thailand to remain in situ notwithstanding drought-related challenges to their agricultural livelihoods, even as some members of the younger generation engage in circular forms of labour migration to complement agricultural livelihoods.

3 Intersectionality

As other social and economic elements operate in tandem and will be the deciding factor in individual and community mobilities, climate change amplifies existing pressures. The thresholds of exposure, vulnerability and capacity all impact the mobility of individuals and groups of people in climate change situations. Therefore, climate mobility cannot be viewed as monolithic, and the experience, as well as measures to address the phenomenon, varies according to a variety of governance, environmental and social factors at different regional and national scales. Consequently, it is important to explore mobility dynamics in national and local contexts to enhance the understanding of drivers, formal

[33] Bower et al. [11].

and informal governance mechanisms, impacts and other characteristics necessary to catalyse transformation.[34]

Climate change makes it harder for people in intersecting situations of gendered vulnerability to adapt to and deal with the social, political and economic obstacles they confront.[35] Due to their expertise in coping with and mitigating the effects of climate change at the home and community levels, their meaningful participation in disaster risk reduction and climate action initiatives is essential to reducing these disparities. Some of the contributions to this volume engage intersectionality to a certain extent. For instance, Nizar and Dang explore varieties of mobility experienced by women in different socio-economic groups living in an urban context in Vietnam. Gunarathna and Premarathne describe varieties of intersecting vulnerabilities associated with gender, age and disability in relation to disaster-related evacuations and emergency response in Sri Lanka.

Another dimension can be associated with the vulnerability faced by Indigenous Peoples. Climate change exacerbates the difficulties already confronted by Indigenous Peoples, including political and economic marginalisation, loss of land and resources, human rights abuses and violations, discrimination and unemployment.[36] In this volume, Rumpia explores indigenous experiences of immobility and resistance to development-induced displacement in the Philippines.

4 International Law and Climate Mobility

The volume does not address movement of people across international borders. Consequently, notwithstanding the prevalence of the term 'climate refugee,' international refugee law is not relevant to the case studies because it requires people to cross an international border before it can be engaged. At the same time, we recognise that the increasing scale of displacement and other forms of human mobility already include a cross-border dimension, even though few legal mechanisms exist to regulate this kind of movement,[37] and no reliable statistics are available to reflect how many people are moving across borders in this context, either in Asia–Pacific or in other parts of the world. Similarly, the international climate regime anchored in the United Nations Framework Convention on Climate Change and the Paris Agreement has not adequately addressed the issue of cross-border climate mobility,[38] although initiatives linked to this process, including the Warsaw International Mechanism and the Platform on Disaster Displacement, have contributed to notable advances at the regional and sub-regional level in Africa, the Americas and the Pacific, with less progress evident in Asia.[39] The field may be expanding here as climate change adaptation, which is where human mobility has normally been discussed, is now institutionally complemented by a formal UNFCCC mechanism on climate change-related loss and damage. Moore's contribution to this volume advances a loss and damage perspective on processes of planned relocation in Fiji and Tuvalu. Human mobility is expressly envisaged as relating to loss and damage, as reflected in the UNFCCC 2023 global stocktake:

> The Conference of the Parties serving as the meeting of the Parties to the Paris Agreement…
>
> 131. Calls on Parties and relevant institutions to improve coherence and synergies between efforts pertaining to disaster risk reduction, humanitarian assistance, rehabilitation, recovery and reconstruction, and displacement, planned relocation and migration, in the context of climate change impacts, as well as actions to address slow onset events, in order to make progress in averting, minimizing and addressing loss and damage associated with climate change impacts in a coherent and effective manner[40];

[34] Nicholls et al. [15]; Mayer [16].

[35] Duerto-Valero and Kaul [17].

[36] Abate and Kronk [18].

[37] Ahmad [19].

[38] Jolly and Ahmad [20].

[39] Scott [21].

[40] UNFCCC [22].

While waiting, and advocating, for regional and international mechanisms to emerge, international human rights law applies to all persons all of the time, irrespective of whether they have crossed an international border or not.[41] The human rights dimensions of climate-related human mobility are extensive, and adopting a human rights-based approach (HRBA) to this phenomenon provides a framework for rich description as well as policy-relevant recommendations. Whereas by now the 1998 UN Guiding Principles, themselves grounded in international human rights law, are well-established as providing a useful tool for all actors concerned with climate-related displacement,[42] a human rights-based approach is also relevant to addressing migration, planned relocation and immobility as well.

Human rights are universal legal safeguards that defend individuals, communities and nations from acts and omissions that limit their fundamental freedoms and rights. Human rights law obliges governments (in general) and other duty-bearers to respect, advance, preserve and uphold all human rights.[43] Every individual's inherent value and dignity is the cornerstone of human rights. They are equal, indivisible, inalienable, interrelated and interdependent.[44]

It is now beyond question that the full enjoyment of human rights is negatively impacted by climate change brought on by human activities. Climate change substantially affects many human rights, including the rights to life, self-determination, development, food, health, water, sanitation and housing.[45] People who are marginalised on a social, economic, cultural, political, institutional or other level are particularly sensitive to climate change as well as to various adaptation and mitigation measures.[46] The catalogue of rights and related obligations created by international law is the foundation for the HRBA to climate change. The Universal Declaration of Human Rights (UDHR), International Covenant on Civil and Political Rights (ICCPR) and International Covenant on Economic, Social, and Cultural Rights (ICESCR), collectively known as the 'International Bill of Rights,'[47] are the most critical components of these approaches. In the present context, human rights are invoked in the Paris Agreement on climate change, the Sendai Framework for Disaster Risk Reduction and Agenda 2030 on sustainable development. The normative depth generated by 75 years of work at international, national and sub-national levels to progressively realise the rights of all persons without discrimination provides guidance relevant to the implementation of these agendas, including in the context of climate-related human mobility.

The human rights-based approach highlights those who are the most marginalised, excluded or subjected to discrimination. To make sure that interventions reach the most marginalised sections of the community,[48] it is frequently necessary to analyse gender stereotypes, various forms of discrimination and power disparities. In accordance with the human rights framework, efforts to reduce the effects of climate change and prepare for them must be directed by pertinent human rights norms and principles, such as the rights to participation and information, transparency, accountability, equity and non-discrimination.[49] Several well-formulated norms that constitute the core policies of human rights include:

- Advancing human rights in the context of climate change mitigation, adaptation and measures to address loss and damage is considered the ultimate goal.
- People-centric approach is the core element of tackling climate change.

[41] Jaswal and Jolly [23].

[42] Scott [24].

[43] Krajewski [25].

[44] Duffy [26].

[45] Das et al. [27]; Jolly [28].

[46] Human Security' in Field et al [29].

[47] Flowers [30].

[48] Muigua [31].

[49] Horne et al. [32].

- Participation serves as a means to an end, but also an end in itself.
- The climate activities are locally controlled, and the programmes give special attention to excluded and marginalised communities.
- A crucial component of the human rights framework is the idea of vulnerability. The legal commitment of the States to respect, protect and fulfil human rights is supported by the need to identify and remedy vulnerability in addition to the need to uphold human dignity.

Human rights strategies require socially inclusive climate adaptation measures to build resilience and reduce underlying vulnerabilities contributing to harmful forms of human mobility. An HRBA also attempts to ensure that mobility is based on choice. However, when movement is unavoidable, this will also require plans and programmes to support safe, voluntary and dignified movement as a positive choice and, if necessary, through participatory planning to relocate to safer settlement areas with full respect for human rights.[50] Simply expressed, the human rights framework must be a part of the entire solution cycle to the climate-related mobility issue.

In the last ten years, several regional initiatives and partnerships addressing climate issues have evolved in response to these significant global trends. However, how each region approaches and conceptualises such cooperation varies. Therefore, it is crucial to support ongoing efforts at all levels to address the negative forms of climate-related human mobility on the enjoyment of human rights in light of scientific data and assessments and in a well-integrated manner that advances progress towards the pragmatic implementation of the Sustainable Development Goals, the Paris Agreement and the United Nations Framework Convention on Climate Change, as well as towards the promotion, protection, realisation and enjoyment of all human rights.

Academic and research collaboration is an integral part of the solution-seeking. In this context, the present book attempts to bring together home-grown research from scholars in Asia and the Pacific and highlight the challenges and opportunities of climate change mobility in Asia and the Pacific from an interdisciplinary perspective with a special focus on an HRBA.

5 Local Perspectives

The edited volume examines the multiple dimensions of climate-related human mobility from countries in Asia and the Pacific, intending to advocate and highlight the appropriateness of the HRBA. Given the multiple dimensions of climate mobility, evaluating the issue from a single-domain perspective will have limited coverage. Hence the instant book frames the issues of climate mobility cumulatively through an interdisciplinary perspective that encompasses other subjects like migration, gender, social disability and human rights. The book showcases the research abilities and subject matter expertise of scholars from Asia and the Pacific on climate-related human mobility in the region. This is done to bring forth the lived experiences and local solutions of the people of the region. Various chapters in the instant book, while covering the policy and theoretical dimensions of mobility, also evaluate the issue through empirical studies. The objective is to provide evidence-informed insights to build strategies that are resistant to the impacts of climate change and equipped to ensure the basic human rights of all. As the context is significant in understanding the nuances of mobility, the findings of empirical research help scholars and policymakers understand how environmental factors interact with socio-political and economic factors to influence people's movements. They also highlight promising future research directions and raise critical

[50] 'A Toolbox: Planning Relocations to Protect People from Disasters and Environmental Changes' (*Brookings Institute, Georgetown University, United Nations High Commissioner for Refugees and Institute for the Study of International Migration*, 2015) 1, 53.

questions for global policymaking. In this manner, the book examines the multi-dimensional issues of climate mobility in Asia and the Pacific through an interdisciplinary perspective.

The scholars from the region have adopted the human rights-based approach and the interdisciplinary approaches consistent with the international human rights law framework in their contributions that make the unique focus on the impacts of climate change-related human mobility, justified in this region. However, there is no consensus on terminologies among the scholars of this part of the world. It is qualitative or quantitative and depends on the authors' research treatment, understanding and intellectual preferences. The volume's primary focus is not to critique the climate mobility from the Global North–South dynamics lens but to instead highlight the lived experience and ground realities of climate change impacts and consequent mobilities and immobilities happening in the Asia–Pacific from the perspective of vulnerable populations. The expertise and understanding gathered through the studies are crucial in evolving a durable solution based on the HRBA approach, which is inclusive and not merely a top-down model.

6 Book Structure

The volume is divided into two part. Part 1 focuses on mobilities and immobilities most closely associated with migration and displacement. Most case studies also examine lived experiences from the perspective of different groups within society, such as older persons, Indigenous Peoples or persons with disabilities. Part 2 explores mobilities related to planned relocation and resettlement, as well as proximate forms of mobility such as development-related displacement. The section will also see the responsibility of the State and non-State actors in exploring planned relocation/resettlement.

This Chapter—The introductory chapter offers an overview of the book and provides a conceptual background. It brings out the issue of climate change and mobility, focusing on the Asia–Pacific. It explains the focus of the book, the structure and arrangement of the book.

Part 1 of the book titled Dimensions of Migration and Displacement: Experiences, Policies, Laws and State Practice contains eight chapters.

Chapter "Tropical Cyclone and the Mobility of Older Persons: Insights from Coastal Bangladesh" by Md. Abdul Malak, and Nazia Khan Lina, explores the mobility of older individuals in Bangladesh in the context of recurring tropical cyclones, highlighting the significant impact on coastal populations. It highlights the challenges faced by these individuals, including economic, social and cultural losses. The study also reveals that these mobility choices often make them more storm-resistant, contributing to our understanding of their adaptation strategies.

Chapter "To Move or Not to Move? Drought-Related Internal Migration and Voluntary Immobility Among the Hmong Forest Community in Northern Thailand" by Reni Juwitasari, explains the drought in Northern Thailand, particularly near Myanmar and Laos, that is affecting the Hmong ethnic minority group in Chiang Rai. The drought has led to a significant water shortage, affecting agriculture, forests, income, food security and primarily circular labour migration. The study, examining the social effects of climate change in the Hmong community's home district of Wiang Kaen, reveals that factors such as limited educational opportunities, job readiness and land ownership status influence migration patterns. The older generation, using indigenous knowledge, has shown resilience in managing forests and agriculture, while the younger generation seeks financial stability, yet neither group reflects any interest in moving away from their predominantly rural livelihoods, reflecting a form of voluntary immobility and highlighting the importance of supporting such communities to strengthen resilience in situ.

Chapter "Sensitisation of Disaster Relief Operations towards Persons with Disabilities" by Ishari Gunarathna and Priyadarshani Premarathne, explains that persons with disabilities often face marginalisation in society, especially after disasters. The findings highlight the need for pre-emptive mapping, adequate training for evacuation, cultural awareness and increased participation of social workers in disaster management initiatives.

Chapter "Rising Waters, Stagnant Paths: Gendered Experiences of Flooding and Restricted Mobility in Can Tho City, Viet Nam" by Danang Aditya Nizar and Ly Quoc Dang, explains the integration of climate change into the regulations of Vietnam recognising the link between mobility, gender equality and climate change. However, the authors point out the challenges faced in implementing national commitments, as seen in Can Tho City's urban flooding. The authors argue for consideration of micro-level mobility dynamics and gender principles in climate initiatives, emphasising the need for greater consideration of this aspect in the debate on climate change-related mobility.

Chapter "Establishing a Human Rights-Based Approach to Climate Change-Induced Internal Displacement in the Regime of Bangladesh: Challenges and Way Forward" by Md Abdul Awal Khan, deals with the impact of climate change leading to internal displacement in Bangladesh. The author highlights the challenges faced by Bangladesh in linking human rights law with climate change legislation. The chapter reviews the current Bangladeshi domestic law and policy framework, highlighting inadequacies in preventative and preparedness systems, and suggests long-term remedies to protect communities displaced due to climate change.

Chapter "Climate Change, Agriculture and Internal Human Mobility in the Bhutan Himalayas" by Om Katel, Anooja Nair, Ugyen Yangchen and Chogyel Wangmo, explains that Bhutan is among the most vulnerable nations to climate change, despite not contributing to it. The country's geographical location and the effect of climate change have a detrimental impact on sectors like forestry, agriculture and natural resource management. Climate change also increases the risk of disasters, impacting biodiversity and community livelihoods leading to primarily rural–urban labour migration.

Chapter "Climate Change-Induced Disaster Displacement and Law in India: Positioning the Operationalization of Artificial Intelligence for Protecting Human Rights" by Nafees Ahmad, explains that climate change-induced disaster displacements (CiDD) in India and other countries are causing human rights concerns. AI applications can significantly impact both CiDD and low-emission, climate-resilient development. The UN Climate Change Initiative on AI for Climate Action suggests that the crisis needs to be reimagined and operationalised to defend human rights under a legally enforceable international Human Rights Protection Framework (HRPF). The author highlights that India could develop AI-driven automatic decision-making systems to defend human rights in CiDD scenarios. To ensure AI technologies operationalise human rights protection and contribute to the accomplishment of Sustainable Development Goals, humanitarian actors must be proactive and inclusive in creating AI apps, policies and accountability systems that protect human rights.

Chapter "Humanitarian Aid Distribution in the Context of Human Rights-Based Approach Among Vulnerable Communities: Flash Floods and Climate Change in North Luwu, Indonesia" by Dina Ruslanjari, Cahyadi Ramadhan, Inayah Bastin Al Hakim and Feby Aulia Marsida, deals with the outcome of a catastrophic flash flood that occurred in 2020 in South Sulawesi Province, Indonesia's North Luwu Regency, leaving 106 persons missing and 38 confirmed deaths. Using a qualitative methodology, the study explores the comprehensiveness of laws and policies in Indonesia, finding a high level of satisfaction among people assisted in this multi-level, multi-stakeholder initiative.

Part 2, titled Planned-Relocation, Resettlement, and State Responsibility, contains six chapters.

Chapter "Climate Change-Related Displacement: Inter-Island and Rural–Urban Migration in the Solomon Islands: Options for a Viable National Resettlement Plan" by Lovelyn K.M. Otoiasi, deals with climate change that is causing a surge in forced migration, particularly from low-lying atolls in the Solomon Islands. This chapter examines the intervening aspects of climate change-related migration and rural–urban migration, suggesting institutional, legal and financial mechanisms for the nation's resettlement programme. It aims to improve the understanding of the social, economic and cultural characteristics of the mobile population, promoting wise decisions regarding climate change-related migration and resettlement.

Chapter "Impact of Climate Change and Accessing Services in Papua New Guinea" by Dora Kuir-Ayius, examines the issues of government service delivery, focusing on communities from the Carteret and Manam Islands who have been relocated in the context of disasters and climate change. The chapter highlights the centrality of land ownership to identity and livelihoods and considers how people who relocate and do not secure land rights face serious adverse human rights impacts because state authorities are unable to ensure the right to an adequate standard of living for those unable to support themselves on the land.

Chapter "State Responsibilities and International Obligations in Responding to Climate Mobilities: What Should International Assistance Look Like?" by Liam Moore, highlights that Fiji and Vanuatu have developed innovative policies, recognising the rights of people forced to relocate and the responsibilities of states to uphold these rights. However, there are emerging implementation challenges, and the international community has a responsibility to support these states in managing climate mobility and relocation. Foreign assistance should be directed locally and customised to local circumstances, supporting local actors and increasing their potential.

Chapter "Marginalisation of Adat Communities: Intersectionality of Land Grabbing, Human Rights, Climate Adaptation and Human Mobility in Indonesia" by James Reinaldo Rumpia, discusses the marginalisation of Indigenous Peoples particularly the Adat Community, in Indonesia. It highlights the impact of land grabs, human mobility, climate adaptation and human rights violations. The Sikep and Rendu people, despite their traditional farming practices, face land grabbing and migration due to the cement industry and dam projects. The situation is aggravated by the impact of climate change. The lack of the role of indigenous law and participation in environmental policy further exacerbates these vulnerabilities.

Chapter "Climate Change and Socio-economic Issues in Sri Lanka: An Assessment of Landslide-Induced Relocations in Aranayake" by Dinushika M. Yapa Abeywardhana, examines the socio-economic effects of landslides in Aranayake, Sri Lanka, which led to the displacement of communities. The relocations increased social and economic vulnerabilities, jeopardising their well-being. The study suggests that for effective policy formation in Sri Lanka, community requirements must be identified and monitored before and after the relocation, ensuring that the community's rights to freedom, development and an adequate standard of living are upheld.

Chapter "Conclusion"—The concluding chapter explores the findings and learnings gathered from the chapters highlighting the impact of climate change on global migration and the need for an inclusive policy response founded on human rights and reflecting the local reality. Although highlighting the varieties of human mobility taking place across the region, the volume also identifies core similarities relating to differential exposure and vulnerability based on characteristics like age, gender and class, limitations of law and policy to tackle the complex development challenge of climate-related human mobility and the consequences of failing to take the dimensions reflected in a human rights-based approach into account, when taking action to address disaster risk and the adverse impacts of climate change. With this volume we offer ground-level insights into lived experiences, while framing issues in relation to international standards and global policy processes. These two levels need to speak more to one another, and the APANDD network provides a unique forum for facilitating such an exchange.

References

1. Abbass K, Qasim MZ, Song H, Murshed M, Mahmood H, Younis I (2022) A review of the global climate change impacts, adaptation, and sustainable mitigation measures. Environ Sci Pollut Res 29:42539–42559
2. Raoul Wallenberg Institute (2023) A bibliometric analysis of research at the nexus of climate change, human mobility, and human rights. The Raoul Wallenberg Institute of Human Rights and Humanitarian Law, pp 2–92
3. Roy P, Pal SC, Chakraborty R, Chowdhuri I, Saha A, Shit M (2023) Is the spelling correct? Please check 'effects of climate change and sea-level rise on coastal habitat: vulnerability assessment, adaptation strategies and policy recommendations.' J Environ Manage 330:117187
4. Zami MT (2023) Bangladesh confronts growing threat of warming-driven floods. Context. https://www.context.news/climate-risks/bangladesh-confronts-growing-threat-of-warming-driven-floods. Accessed 26 Nov 2023
5. Delano JW (2022) Sinking land and rising seas threaten manila bay's coastal communities. Inside Climate News. https://insideclimatenews.org/news/18112022/sinking-land-and-rising-seas-threaten-manila-bays-coastal-communities/. Accessed 11 Nov 2023
6. Parry ML et al (eds) (2007) Climate change 2007: impacts, adaptation and vulnerability. In: Working group II contribution to the fourth assessment report of the intergovernmental panel on climate change. Cambridge University Press, pp 1, 59
7. Siddiqui AR, Sahay A (2022) Climate change, disaster and adaptations. Springer
8. UN General Assembly (2015) Transforming our world : the 2030 agenda for sustainable development, 21 Oct 2015, A/RES/70/1. Available at: https://www.refworld.org/docid/57b6e3e44.html. Accessed 27 Nov 2023
9. IPCC (2022) Climate change 2022: impacts, adaptation and vulnerability. In: Pörtner H-O, Roberts DC, Tignor M, Poloczanska ES, Mintenbeck K, Alegría A, Craig M, Langsdorf S, Löschke S, Möller V, Okem A, Rama B (eds) Contribution of working group II to the sixth assessment report of the intergovernmental panel on climate change. Cambridge University Press
10. Scott M (2022) How the new IPCC report addresses migration and human rights. Raoul Wallenberg Institute. https://rwi.lu.se/blog/how-the-new-ipcc-report-addresses-migration-and-human-rights/. Accessed 24 Jan 2024
11. Bower E et al (2023) Enabling pathways for sustainable livelihoods in planned relocation. Nat Clim Chang 13:919
12. Raoul Wallenberg Institute (2023) A bibliometric analysis of research at the nexus of climate change, human mobility, and human rights. Raoul Wallenberg Institute, Jakarta. https://rwi.lu.se/wp-content/uploads/2023/06/64828d9890605746195582.pdf. Accessed 22 Jan 2024
13. DW (2023) Number of internally displaced people worldwide hit a record high in 2022. In: The Wire. https://thewire.in/world/number-of-internally-displaced-people-worldwide-hit-a-record-high-in-2022. Accessed 12 Nov 2023
14. Ferris E, Riera J, Weerasinghe S (2015) Guidance on protecting people from disasters and environmental change through planned relocation. Georgetown University and UNHCR, Brookings, p 5
15. Nicholls RJ et al. (2007) Coastal system and low-lying areas. In: Parry ML et al (eds) Climate change 2007: impacts, adaption and vulnerability. Working group II contribution to the fourth assessment report of the intergovernmental panel on climate change. Cambridge University Press
16. Mayer B (2016) The concept of climate migration: advocacy and its prospects. Elgar Studies in Climate Law
17. Duerto-Valero S, Kaul S (2023) Gendered impact of climate change: empirical evidence from Asia. UN Women & Women Count, p 3
18. Abate RS, Kronk EA (2013) Climate change and indigenous peoples: the search for legal remedies. J Environ Law 25(2):1,5
19. Ahmad N (2023) Disaster displacement and international refugee law: locating legal protections in the context of climate change migration (eds). In: International handbook of disaster research, Springer
20. Jolly S, Ahmad N (2015) Climate refugees under international climate law: towards addressing the protection gaps and exploring the legal alternatives for climate justice. In: ISIL year book of international humanitarian and refugee law, pp 216–248
21. Scott M (2024) Law and human mobility. In: Bartolini G et al (eds) Yearbook of international disaster law, vol 5 2022. Brill
22. UNFCCC (2023) Outcome of the first global stocktake (Decision-/CMA.5). https://unfccc.int/sites/default/files/resource/cma5_auv_4_gst.pdf. Accessed 22 Jan 2024
23. Jaswal PS, Jolly S (2013) Climate refugees: challenges and opportunities for international law. J Indian Law Instit 55(1):45–58
24. Scott M (2021) The role of national law and policy in addressing displacement in the context of disasters and climate change in Asia and the Pacific. In: Scott M, Salamanca A (eds) Climate change, disasters and internal displacement in Asia and the Pacific: a human rights-based approach. Routledge
25. Krajewski M (2023) Mandatory human rights due diligence laws: blurring the lines between state duty to protect and corporate responsibility to respect? Nordic J Human Rights 265–278

26. Duffy A (2023) Universality: recognising the right to have rights. Intercult Human Rights Law Rev 18:77–110
27. Malakar KD, Kumar M, Anand S, Kuzur G (2023) Climate change, justice, and rights. In: Climate change and socio-ecological transformation. advances in geographical environmental sciences. Springer
28. Jolly S (2020) UN human rights committee on climate displacement: legal reflections. Hong Kong J Law Public Affairs 3:44–47
29. Field CB et al (eds) (2015) Climate change 2014 impact, adaptation, and vulnerability, Part A: global and sectoral aspects contribution of working group II to the fifth assessment report of the intergovernmental panel on climate change. Cambridge University Press, pp 755–791
30. Flowers (2023) Human rights glossary. University of Minnesota Human Rights Resource Centre. http://hrlibrary.umn.edu/edumat/hreduseries/hereandnow/Part-5/6_glossary.htm. Accessed 02 Dec 2023
31. Muigua K (2023) Entrenching a human rights based approach to sustainable development. J Conflict Manag Sustain Dev 10(4):166
32. Horne K, Tigre MA, Gerrard MB (2023) Status report on principles of international and human rights law relevant to climate change. https://scholarship.law.columbia.edu/faculty_scholarship/3924/. Accessed 03 Dec 2023

Dr. Stellina Jolly is a Lead Editor and a Senior Associate Professor at the Faculty of Legal Studies, South Asian University (SAU). She is a Visiting Senior Research Associate with the Centre for Emerging Countries in Private international law at the University of Johannesburg. Dr. Jolly is a Fulbright Scholar with the University of San Francisco and a recipient of the International Visitors Leadership Program (IVLP). She researches international environmental law and conflict of laws. In environmental law, her research papers have explored the north-south dimensions of environmental justice. Her book, 'Climate Refugees in South Asia' published by Springer, explored the legal and policy framework of climate refugees in South Asia; she has also elaborately written on the environmental justice and governance concerns and intersectionalities of climate change, water, gender, disaster management, and sustainable development goals focusing on South Asian jurisdictions. Her recent researches also focus on the emerging global trend of the rights of nature. In her research, she advocates an approach based on just sustainability, hybrid law mechanism, and environmental justice to tide over the ecological crisis. Dr. Jolly has published in several leading international journals like the Transnational Environmental Law, Washington Journal of Environmental Law and Policy, Texas Environmental Law Journal, and Chinese Journal of Environmental Law, to name a few. Hart published her book on Indian Private International law. She has also co-edited a book collection published by Springer on Private International Law of South Asian Jurisdictions. She is a member of the IUCN World Commission on Environmental Law and part of the Network of Environmental Law Champions, Asian Development Bank. She has undertaken projects and consultancies with various organizations, including the Rajiv Gandhi Foundation, European Union, International Renewable Energy Agency (IRENA), and Asia Europe Foundation. She was awarded an educational grant on Civil Society Law from International Centre for Not-for-Profit Law (ICNL) and USAID.

Dr. Nafees Ahmad is an Editor, and an Associate Professor at the Faculty of Legal Studies, South Asian University (SAU)-New Delhi. He holds a doctorate in International Refugee Law and Human Rights. His scholarship focuses on RAMS (Refugees, Asylum-seekers, Migrants, Stateless) and the role of Artificial Intelligence (AI) in their protection, Global Forced Displacement, Global Circumstantial Migration (GCM) Governance, and Climate Refugees in South Asia. He also addresses the International Politics of Asylum, Refugee Policy Paradigms, Invisible Frames of Asylum, Disconnects of Durable Solutions, and SAARC connects and contexts of refugee protection. He conceived and introduced a new SAARC-specific Program in 2011 called Comparative Constitutional Law of SAARC Nations (CCLSAARCN) at the LLM level. He publishes inter-alia in the International Handbook on Disaster Research (Springer 2023), The Asian Yearbook of Human Rights and Humanitarian Law (Brill 2023), Groningen Journal of International Law (The Netherlands), Harvard International Law Journal (Harvard) Asia-Pacific Journal of Human Rights and Law (Brill), International Journal of Environment and Waste Management indexed in SCOPUS, International Journal on Minority and Group Rights, Kings' Student Law Review (KCL-London), ISIL Year Book on International Humanitarian Law and Refugee Law, ELCOP Year Book of Human Rights and NUJS International Journal of Legal Studies and Research (IJLSR) etc. Dr. Ahmad has co-authored a book on Climate Refugees in South Asia (Springer 2019). He is a member of Editorial Advisory Board of Iranian Journal of International and Comparative Law. Dr. Ahmad has been a Resource Person and External Reviewer for the Ministry of Law; Government of India-sponsored Research Project on "Judicial Reforms since June 2016" at the Indian Institute of Management (IIM-Kashipur) and accomplished an ICSSR Project on "The Muncipal Solid Waste Management in Delhi: A Socio-Legal Study of Okhla Lanfill." Further, he has been a visiting professor at the Indian Society of International Law (ISIL), Jamia Milia Islamia-New Delhi, and Judicial Academies of various states of India.Dr. Mathew Scott is an Editor and co-leads the Human Rights and the Environment thematic area at the Raoul Wallenberg Institute of

Human Rights and Humanitarian Law in Lund, Sweden. His scholarship integrates social science perspectives with international legal standards to promote context-sensitive, human rights-based law, policy, and practice relating to disaster risk reduction and climate change adaptation. Dr. Scotts' primary area of expertise concerns migration and displacement in the context of disasters and climate change, on which he published a monograph entitled Climate Change, Disasters and the Refugee Convention (CUP 2020), an edited volume entitled Climate Change, Disasters and Internal Displacement in Asia and the Pacific: A Human Rights-Based Approach (Routledge 2021), and a range of book chapters and academic articles in, amongst others, the International Journal of Refugee Law, the Nordic Journal of International Law and the Yearbook of International Disaster Law. The edited volume was a major output of a regional thematic study he coordinated as part of RWI's Asia-Pacific programme on human rights and sustainable development. Dr. Scott works on the role of local authorities in addressing climate- and disaster-related migration and displacement and how human rights law can contribute to building resilience to pandemic risk. He holds a PhD in Public International Law from Lund University (2018) and a MA in Social Anthropology of Development from SOAS (1998). He practiced immigration and asylum law before entering academia. He is a member of the advisory committee of the Platform on Disaster Displacement, a member of the editorial board of the Yearbook of International Disaster Law, and a founding member of the Nordic Network on Climate-Related Displacement and Mobility. At Lund University, he convenes the LLM course on human rights law, the environment, and climate change and lectures on international refugee and human rights law at the Faculty of Law. He also lectures on the M.Sc. programme in Disaster Risk Reduction and Climate Change Adaptation at the Department of Risk Management and Societal Safety. He is also actively engaged in international collaboration initiatives and currently contributing technical expertise on human rights-based approaches to disaster risk reduction across eight countries in Asia in collaboration with the Asia Disaster Preparedness Center, Stockholm Environment Institute, and the Swedish Civil Contingencies Agency.

Dimensions of Displacement: Experiences, Policies, Laws and State Practice

Tropical Cyclones and the Mobility of Older Persons: Insights from Coastal Bangladesh

Md. Abdul Malak and Nazia Khan Lina

Abstract

Cyclones occur almost annually in Bangladesh, causing significant harm to people living on the coast. It is argued that less attention has been paid to older persons in the development and migration-related literature. In response, this study explores how cyclone impacts influence the decision-making capabilities of older persons with respect to mobility as an adaptation strategy. The study area is in Gabura (a coastal union) as a place of origin, but migrant participants for interviews were also selected from the coastal municipality of Shyamnagar and the district city of Satkhira, as well as the mega-capital city of Dhaka. The data was collected through qualitative research, including in-depth interviews and focus group discussions. The study results show that older adults face challenges during cyclones, which can lead to migration. Older persons face various economic, social, and cultural issues, including the loss of property and assets, disruption of social networks, and loss of cultural practices due to climate change-generated hazards. The results suggest that mobility to nearby towns or big cities is a common response of older persons to climate change. It is also found that such mobility decisions of older persons improved their resilience to cyclones. Overall, this study advances the knowledge concerning mobility among the elderly in the face of tropical cyclones and other coastal environmental stressors. The findings can be useful in developing policies and programmes that include older persons and address their specific needs and vulnerabilities in the context of climate change.

1 Introduction

South Asia is affected by tropical cyclones that form over nearby waters, particularly the Bay of Bengal.[1] Cyclones have long been an issue for the people of Bangladesh because they immediately impact individuals who live along the coast and indirectly harm the entire country's economy, agriculture, and standard of living.[2] Cyclones are most frequently observed around the beginning of the summer and at the end of the rainy season and affect Bangladesh's coastal districts almost annually. At least 15 significant

M. A. Malak (✉) · N. K. Lina
Department of Geography and Environment, Jagannath University, Dhaka, Bangladesh
e-mail: amalak@geography.jnu.ac.bd

[1] Mondal et al. [1]; Quader et al. [2].

[2] Hoque et al. [3]; McMichael et al. [4].

S. Jolly et al. (eds.), *Climate-Related Human Mobility in Asia and the Pacific*, Sustainable Development Goals Series, https://doi.org/10.1007/978-981-97-3234-0_2

cyclones, in the Bay of Bengal (including West Bengal in India), have struck Bangladesh's coastal regions in the past 50 years, killing 1.54 million people.[3] The 1970 Bhola cyclone alone was the deadliest tropical cyclone ever recorded, claiming between 300,000 and 500,000 lives.[4] In addition to causing significant casualties, these disasters profoundly impact countless other people, destroying their possessions and impairing their livelihoods. Apart from the direct impact, cyclones and other environmental stressors have some far-reaching repercussions, including increase of violence against women and socially marginalised groups, escalating social conflict over natural resources, price hikes, psychological stresses, etc.[5]

It should be noted that the disproportionate impact of disasters on older individuals is not unique to Bangladesh. Even in wealthy nations,[6] older persons are particularly vulnerable to problems including increased disease risk,[7] higher susceptibility to harsh weather (both hot and cold), and increased numbers of individuals who are either moved or staying behind.[8] In Bangladesh, 'staying behind' refers to situations when elderly people voluntarily stay or/and are left behind as younger adults migrate to different locations for to secure their livelihood.[9] Rural areas often have an above-average senior population (60+ age) due to the seasonal and permanent out-migration of working-age persons.[10]

In addition, deaths from disasters are disproportionately more likely to affect older persons.[11] Threats from environmental stressors frequently damage the capacity of older persons to adapt and cope later in life.[12] The interplay of several characteristics related to exposure, vulnerability, and coping abilities makes older persons more susceptible to these environmental stressors.[13] In some cases, their ability to deal with challenges associated with climate change is hampered by social helplessness, limited access to resources, and precarious health circumstances.

In Bangladesh, socio-economically marginalised women, children, and people with disabilities are among the various groups whose members often experience vulnerability to climate change. NGOs, funders, and the government routinely provide specialised care and support to people in these groups.[14] However, existing studies indicate that older individuals are frequently forgotten in climate change adaptation plans.[15] Climate change adaptation programmes rarely consider older people's needs.[16] According to estimates, Bangladesh has 9.28% of its population consisting of senior citizens (60+) in 2022, which will account for 22% of the total population by 2050.[17] So, to deal with the senior population and the changes they will inevitably bring about to the system, we must adjust the current situation and re-focus on managing our systems and resources.[18]

A growing body of research addresses the nexus between[19] human mobility and climate change adaptation, but globally, little attention has been paid to the mobility of older persons. More broadly, Bastia has argued that migration and development research have overlooked older adults.[20] Considering this, the chapter describes the cyclone-induced impact and mobility nexus

[3] Mondal and others (n 1).

[4] Ibid.

[5] Desai and Mandal [5]; Malak et al. [6]; Wachholz [7]; Whittenbury [8].

[6] Walkling and Haworth [9].

[7] Malak and others (n 5).

[8] Furlong et al. [10].

[9] Ibid.

[10] Mobarak and Akram [11]; Tripathy Furlong and others (n 8).

[11] Walkling and Haworth (n 6); Kawasaki [12].

[12] Golam Rabbani et al. [13].

[13] Ibid.

[14] Malak and others (n 5).

[15] Haq et al. [14].

[16] Ibid.

[17] BBS [15].

[18] Kwan and Walsh [16].

[19] Adger and Adams [17]; Bardsley and Hugo [18]; Maharjan et al. [19]; Mortreux and Barnett [20].

[20] Bastia et al. [21].

in relation to older persons in Bangladesh. The chapter is divided into several sections. After the introduction, Sect. 2 provides a critical review of the literature related to the vulnerability of older persons' to climate change and their mobility in this context. Section 3 describes the relevant legal and policy framework in Bangladesh. Then, Sect. 4 describes the study context and methods followed in conducting this research. The results and analysis of the first-hand accounts are examined and developed in Sect. 5. Finally, the last section concludes the chapter with a future direction.

2 Climate Change, Older People's Vulnerability, and Mobility

The understanding of vulnerability is diverse across disciplines. Academics and practitioners in different fields of study, understandably, also use different definitions and meanings of vulnerability and concepts related to vulnerability to serve their own goals.Therefore, for the purposes of this study, vulnerability refers to the characteristics of an individual or group in terms of their capability to foresee, cope with, resist, and recover from the effects of natural hazards.[21] Additionally, it is acknowledged that experiences of vulnerability differ from person to person and community to community, no matter how they may be explored from the outside. For instance, age plays an important role in determining an individual's ability to withstand and recover from climate-induced damage.

According to the World Health Organization two billion individuals will be 60 or older by 2050[22] and 80% of them will live in developing countries.[23] Climate change seriously threatens ageing populations, particularly in less developed and developing countries.[24] These nations frequently have the lowest ability to adapt to changing weather patterns and rising temperatures, making them the most susceptible to the effects of climate change.[25] The disaster vulnerability experience of older persons is distinct from young adults[26] owing to their diminished mobility, physiological changes, and limited resource access, impairing their adaptation capacity.[27] Older persons who live on the coast are more likely to be vulnerable to cyclones because of their age-related low mobility, lack of information, and their financial dependence on their children.[28]

The vulnerability of older persons is relative and depends on their socio-economic capacity in some cases. Compared to older individuals who were not impoverished, persons who are poor and older tend to be more vulnerable to climate change.[29] For instance, displacement is one of the consequences of climate change, and a socio-economically marginalised older person is more susceptible to uprooting and relocation than an older person who is economically solvent and lives with young family members and neighbours. Destitute older persons are more likely to suffer from the long-term effects of displacement, such as ill health and social isolation.[30] Older persons are more inclined to move if they have prior storm experience, a higher level of education, and access to transportation.[31] It is argued that older persons with lower incomes and educational levels are more susceptible to the effects of extreme weather events and are more likely to migrate as a form of adaptation.[32]

Several variables, including economic, societal, and environmental ones, can compel older

[21] Blaikie et al. [22].

[22] WHO [23].

[23] Ibid.

[24] UNESCAP [24].

[25] Ibid.

[26] Malak and others (n 5).

[27] Gamble et al. [25].

[28] Malak and others (n 5).

[29] Shanahan et al. [26].

[30] Ibid.

[31] Bukvic et al. [27].

[32] Koubi et al. [28].

individuals to migrate. Internal and international climate change-related migration can occur permanently or temporarily, due to environmental stressors.[33] It is estimated that climate change may increase displacement, and people who are poor with fewer resources may migrate more in response to extreme weather events, particularly in low-income developing countries.[34] In addition to that, in response to climate change, migration may also be an effective adaptation strategy.[35] Human mobility in climate change is also considered a crucial development concern under Sustainable Development Goals. Goals 10.7 and 13 encourage nations to enable orderly, safe, regular, and responsible migration and mobility of people and require rapid action to combat climate change and its repercussions by increasing mitigation, adaptation, and disaster risk reduction strategies.[36] Global environmental change puts significant pressure on the decision-making regarding human mobility.[37] Human mobility is crucial in maintaining livelihoods, managing risk, responding to social and environmental change, and supporting hopes and goals.[38]

Environmental deterioration and social and economic constraints brought on by climate change may drive elderly people to migrate against their will.[39] Statistics show that older persons are often disproportionately represented among people displaced during a hurricane in the USA. Fussell et al. argue that older persons in the USA are more likely to move due to climate change effects like rising temperatures and sea levels.[40] In Bangladesh too, older persons were more likely to move due to the impacts of climate change, such as cyclones, flooding, and drought.[41] In fact older persons were found to be predisposed to relocate to city areas following cyclones, since they had insurmountable economic hardships in their place of origin in countries like Bangladesh and Fiji.[42]

Climate change adaptation refers to certain strategies of responding and adjusting to the consequences of climate change in order to lessen its adverse effects on natural and societal systems.[43] Such strategies are categorised as transformative, anticipatory, and reactive adaptation strategies.[44] Adjusting natural or human systems in response to present or anticipated climatic pressures moderates harm or exploits advantageous chances.[45] Various actions, such as modifications to infrastructure, land-use techniques, social behaviour, and human mobility, can be seen as adaptation.[46]

McMichael et al. argue for a climate change adaptation plan that considers particular requirements of senior citizens.[47] Improved healthcare access, social support, community resilience, and greater climate change awareness are some solutions that studies have found to be useful in minimising the susceptibility of ageing populations to climate change.[48] For instance, over the last 25 years, the WHO has developed and scaled up its programme on climate change and health to support and guide implementation of protective measures to reduce the vulnerability of older persons.[49] Strategies for adaptation are crucial to minimising the effects of climate change on the elderly.

However, these adaptation strategies need to be socially responsive as ageing people rely on their social networks to help them cope with the effects of climate change. People (e.g.

[33] Bardsley and Hugo (n 17).

[34] IPCC [29].

[35] Ibid.

[36] Galez-Davis et al. [30].

[37] Neil Adger et al. [31].

[38] Mobarak and Akram (n 10); IDRC [32]; Malak [33].

[39] Bardsley and Hugo (n 17).

[40] Fussell et al. [34].

[41] McMichael, Schwerdtle and Ayeb-Karlsson (n 2).

[42] Ibid.

[43] Mortreux and Barnett (n 17).

[44] Neil Adger et al. [35].

[45] IPCC (n 34).

[46] Ibid.

[47] McMichael, Schwerdtle and Ayeb-Karlsson (n 2).

[48] Antal and Bhutani [36].

[49] WHO [37].

neighbours, relatives, volunteers) connected via social networks can offer practical and emotional support amid severe weather.[50] Older persons may benefit from community-based adaptation techniques such as early warning systems, communal gardens, and social networks. By taking these steps, older persons can reduce their exposure to climate change vulnerability, preserve their independence, and maintain social connections.[51] They may be less likely to engage in actions associated with adaptation to climate change due to things like financial constraints, a lack of understanding, and limited access to information.[52] It is advised to develop outreach and education initiatives that help them deal with climate change.[53] One strategy to attain these aims is by implementing age-friendly policies that consider their unique needs while preparing for climate change adaptation.[54] Such measures include creating inexpensive housing, accessible transportation systems, social support networks, and encouraging physical activity.[55] In this context, the following section evaluates whether the legal and policy framework of climate change in Bangladesh accommodates the concerns of older persons.

3 Legal and Policy Framework on Climate Change, Human Mobility, and Older Persons

Several national frameworks and policy guidelines are available to address Bangladesh's climate change and disaster risk management. These documents include the Standing Orders on Disaster (SOD), the Disaster Management Act 2012 (DMA), the National Adaptation Plan (2023), the National Strategy on the Management of Disaster and Climate-Induced Internal Displacement (NSMDCIID), and so forth. This section critically reviews how these documents incorporate the vulnerability of older persons and their adaptation and mobility in the face of climate change.

The Standing Orders on Disaster (SOD), introduced in 1997 (later modified in 1999, 2010, and 2019), was Bangladesh's initial Disaster Risk Management (DRM) regulatory framework. The SOD ensures that all stakeholders are included and emphasises the involvement of the elderly and people with disabilities at all stages of disaster risk management in the spirit of 'Leaving No One Behind.' It guides rescue teams in transporting older persons and people with disabilities to a safe shelter following instructions. The evacuation plan stated in SOD provides priority to humanitarian assistance, addresses the issues of separate accommodation provisions in shelters for elderly people, as well as separate toilets for older persons and persons with disabilities, develops and implements disaster risk reduction and emergency response plans by following social inclusion principles and ensuring the participation of the elderly.[56]

The Government of Bangladesh has passed the Disaster Management Act (Act No. 34 of 2012) which describes the legal obligations of the various disaster risk reduction bodies and ministries. The Disaster Management Act 2012 includes various communities, which may open an opportunity to prepare any disaster risk reduction guideline, particularly for older persons. Bangladesh implemented the Sendai Framework as part of its ongoing DRM commitment, and as part of this, the National Disaster Management Policy of 2015 has been introduced, aiming to offer a national understanding of disaster risk governance.[57]

Similarly, another recently approved and published document is the National Adaptation Plan (NAP) 2023. Ensuring socially inclusive adaptation, including for the elderly and people with disabilities, is an essential guiding principle

[50] Sari Kovats et al. [38].

[51] ESCAP (n 24).

[52] Pillemer et al. [39].

[53] Ibid.

[54] McMichael, Schwerdtle and Ayeb-Karlsson (n 2).

[55] ESCAP (n 24).

[56] DDM [40].

[57] MoDMR [41].

of the NAP. The formulation process ensured an inclusive approach and fortified engagement with and the participation of older persons, persons with disabilities, minority ethnic communities, and other marginalised groups in every stage, especially during the rigorous consultative process to identify adaptation needs and prioritise adaptation options for reducing the disproportionate impacts of climate change.[58]

The National Strategy on the Management of Disaster and Climate-Induced Internal Displacement (NSMDCIID) directs authorities to recognise that people in situations of vulnerability, such as older persons have special needs that must be given due attention. It ensures that the specific needs of particular groups of CIIDPs, such as older persons, can be adequately addressed and that their rights are appropriately protected. It also ensures measures for preventing and responding to the protection risks older persons face.[59]

To sum up, it is evident from Bangladesh's climate change and disaster-related legal and policy documents that special attention is given to socio-economically marginalised people, including older persons. Therefore, there exists a legal obligation and policy priority to focus on older persons, including persons with disabilities and women. The following section explores the study context and methods, the findings and discussion, addressing the question of how this legal and policy framework is reflected in practice.

4 Study Context and Methods

Bangladesh's coastal part is the most prone to climate-induced disaster. The region is constantly affected by cyclones, tidal waves, and floods, making the people living there vulnerable and exposed to hazards. The Gabura Union, the lowest administrative unit of Shymanagar sub-district, Satkhira District (Fig. 1), is a southwest coastal island cut off from the mainland by rivers. Over 15,000 people live in the union, which comprises 12 settlement villages, with most of them working in agriculture and fishing.[60] The susceptibility of the Gabura Union to climate change and severe weather events like cyclones, salinity intrusion, and storm surges has drawn attention in recent years. Cyclones *A ila* (2009), *Amphan* (2020), and *Yash* (2021) are the three significant cyclones that have visited this area over the last 15 years. The local population, many of whom depend on agriculture and fishing for a living, has suffered greatly because of these hazards, causing severe damage to the area's infrastructure, including roads, homes, and schools.[61]

4.1 Data Collection and Methods

This research employed a qualitative approach for data collection, including semi-structured interviews and focus group discussions (FGDs). The sampling method used for the study is the purposive sampling method. This method is adopted as the target population needed to be selected from a specific trait or experience related to the research. Table 1 shows the participant category for interviews. Both old and young adult participants including male and female groups were interviewed. The group included people who had migrated, as well as those who had not migrated, both of who were interviewed. Since the focus of the study was mainly Cyclone *Aila* and the associated mobility concerns of older persons, the participants included, some elderly in the 80+ age bracket, and some participants who are 75+ (they were 60+ at the time). The FGDs were conducted in two different villages with males and females, separately.

The interviews were conducted in different villages in the Gabura Union, including *Lokkhikhali, Chokbara, Chandimukha, and*

[58] MoEFCC [42].

[59] MoDMR [43].

[60] Afrin [44].

[61] Ibid.

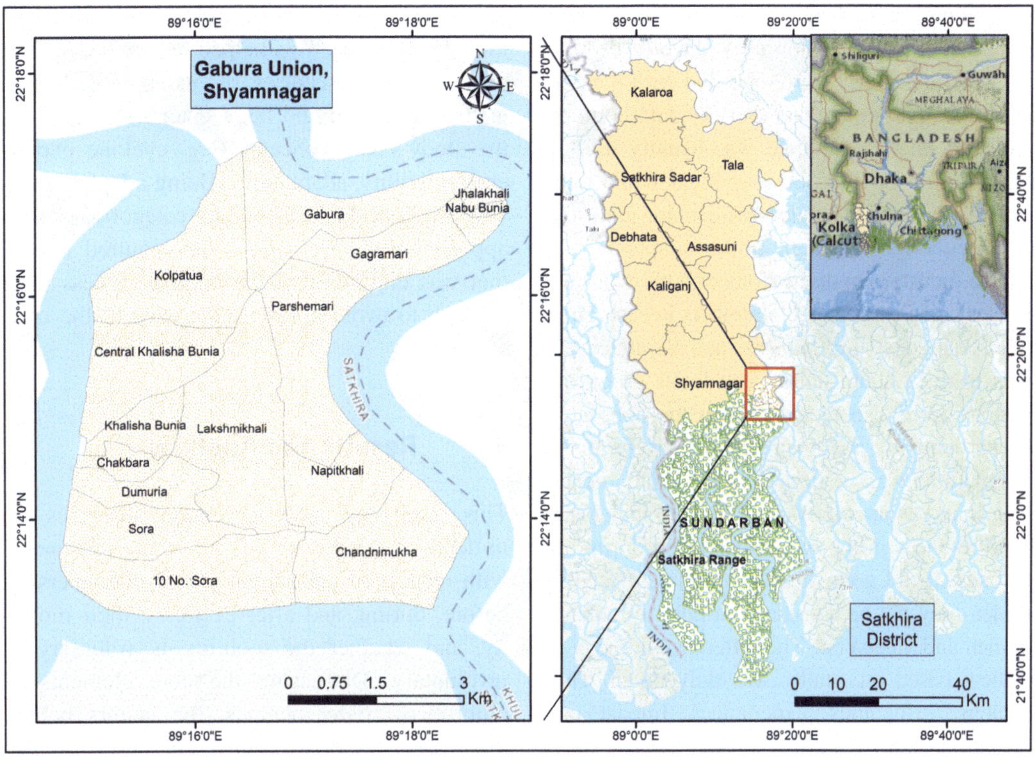

Fig. 1 Gabura Union, the study context. *Source* Authors

Table 1 Participants for interviews

Participant categories	Age	Male	Female	Total
Migrant older persons who are currently living in Shyamnagar municipality town, Satkhira district city, and Dhaka Mega-city	60+	4	1	5
Older persons who are currently living in Gabura Union	60+	3	2	5
Non-migrated young adults	30+	3	2	5
	Total no. of participants			15

Kholishabunia. The people who had migrated were also interviewed at Dhaka, Satkhira, and Shyamnagar. Several young adults were selected from the Gabura Union. Participants were contacted before an interview, and some local gatekeepers (local people) helped us reach the participants. This made it easier to find them, and the interviews were conducted faster. Before taking the interviews and conducting the FGDs, consent was taken from each participant. A guide including a set of keywords based on the theme was prepared for conducting successful interviews and FGDs. The questions were open-ended, so the participants could answer them and share their experiences elaborately. Interviews were a mix of informal and formal questions depending on the situation and context. Ten interviews were taken from the Gabura Union as a place of origin, and the other five interviews were conducted with the migrant participants (who had migrated from Gabura) outside the Gabura Union as a place of destination. The interviews were recorded with the consent of the participants so that the data could be evaluated later. A notepad was usually used to take notes and write follow-up ideas while interviewing participants. Interviewees are identified with pseudonyms throughout this chapter.

Two focus group discussions of both males and females consisting of eight members in each group were conducted in *Kholpetua* and *Napitkhali* villages in the Gabura union. The focus group discussions were also recorded with the consent of the participants, and their answers were noted. Information was collected from both gender groups, and an attempt was to made to understand the differences in the difficulties faced during the cyclones. It did seem that the problems faced by women may not be considered a point of concern for men and vice-versa. The difference in the point of view based on gender was duly noted. Their opinions, difficulties, experiences, impact, and adaptation strategies differed. All research documentations were digitally preserved for future reference with access only to the researcher so as to protect the identities of the people involved.

4.2 Data Coding and Analysis

The coding and analysis of empirical data involved several stages. Audio-recorded interviews were transcribed, edited, and checked to ensure they were clearer in language, sentence structure, and so on for the understanding of readers. Qualitative data was then coded in a deductive way and analysed through the NVivo software and thematic descriptive analysis. Nodes such as socio-demographic, experience to cyclone, mobility, and adaptation were set up.

After creating the first coding structure, the first step was to read the transcriptions to highlight and categorise every passage related to one of several categories (such as socio-demographic, experience, mobility, and adaptation). The text was tagged in both categories when connected to more than one category (or node). Each text related to the main themes was given a new code but was not fitted into the pre-determined categories. This helped reduce the neglect of important categories by using a deductive technique to find and build categories based on the study's key concepts (e.g. cyclone experience, mobility, adaptation). Being adaptable and allowing for additional coding categories as they appeared was an essential and required feature that was embraced as the coding process went through to avoid a strict adherence to the code structure.

5 Results and Analysis

This result and analysis section illustrates the in-depth perceptions of all the informants' with respect to the experience of older persons before, during, and after cyclones, their mobility, and whether the mobility is voluntary or involuntary. Discussing the core elements of mobility of older persons, the factors behind their voluntary or involuntary mobility are also deliberated. The conversation also includes the experiences of the interviewees with respect to the *Aila* (2009), the *Amphan* (2020), and other cyclonic situations. Among the cyclones, the *Aila* was the most devastating and severely impacted the lives of people in Gabura. The older persons stated how cyclones affected their life and influenced their decision-making regarding mobility as an adaptation strategy, the benefits after migration, and the changes in their livelihood.

5.1 Older Persons and Cyclone-Related Mobility

Climate-related mobility can take many forms, such as intentional relocation and movement away from climate-risk areas, displacement following disasters, and temporary labour migration to diversify livelihoods.[62] Older persons who choose to relocate are usually confronted

[62] McMichael, Schwerdtle and Ayeb-Karlsson (n 2).

with various economic, social, and cultural issues, including the loss of property and assets, eviction from traditional lands and resources, disruption of social networks, and loss of cultural practices.[63]

Participants reported about the mobility pattern of ageing people and that some older persons migrated for a short period, such as for a cyclone period or a few months to a year after the cyclone. On the other hand, some people migrate permanently from Gabura to get a better livelihood and facilities and mainly to escape the devastation that Cyclone *Aila* brought. The other cyclones, for instance, Cyclones *Bulbul* (2019) and *Amphan* (2020) were less impactful for the people in Gabura. They did not suffer much loss from these recent cyclones because of several issues (e.g. not so intense landfall, fastest early warning dissemination, quick evacuation to shelter centres, and increased number of shelter centres) which indicates an improvement in the early warning system and increase in the number of shelters and facilities as a result of the increased attention to disaster risk management. These cyclones seem to have impacted their lives a lot lesser than the impact of Cyclone *Aila* wherein the participants faced a long-term impact. Various factors that emerge from the FGD and interviews of the participants are argued upon to decide whether people would or would not migrate. These factors include water crisis, changes in income and livelihood and access to health and communication, and remoteness of physical location. The following section describes people's experiences with cyclones, the early warning system, and the evacuation process.

5.1.1 Early Warning and Evacuation Experience of Older Persons During Cyclones

Cyclones like the *Aila* (2009) caused widespread destruction and devastated the regular life of the elderly in the Gabura Union of Bangladesh. The informants we spoke to reported of how they saw the fierceness of the cyclone in front of their very eyes. They had no inkling regarding the impending cyclone, and everything was very calm in the beginning. In a flash, water began entering from all sides, flooding every place. Findings from the FGDs and interviews also suggested that it was so unexpected, that they were unable to think of what to do. An elderly participant stated the condition:

> It was sudden. We were eating. We just left our rice and curry in the pan. We left it all there and just ran out of here. They [the volunteers] announced and instructed us to come out as soon as possible. We could not wrap anything up as we didn't have time. The water was coming from everywhere, destroying everything on its way. Such a helpless situation.

Similarly, the first-hand account also suggested that while they were uninformed, some were reluctant to acknowledge the cyclone. When asked about the early warning, the participants said that they did not get any early warning during the Cyclone *Aila*. An older adult participant stated regarding the *Aila*:

> People lost their lives due to flooding during *Aila*. Everyone said nothing serious would happen, and gradually it would stop, like before. So, they weren't conscious.

Although the recent cyclones (e.g. *Bulbul*, *Amphan*) had a weaker landfall than the *Aila*, participants reported that they got a well-dessiminated early warning. Also the older persons got the early warning with the help of volunteers, relatives, and neighbours. Moreover, participants reported that due to the recent development of mobile networks and the affordability of mobile phones, older adults were kept informed by their relatives and other family members from the urban areas about the recent cyclone landfall. However, participants reported that where the cyclone was devastating, people were required to move to a safer place once a warning was issued, and this is still a challenge for older persons as they are required to quickly move to shelters, and it is such challenges that influence an older person's decision-making in migrating to an urban area.

The older participants stated that there were many people in the cyclone shelter during the

[63] Ibid.

recent cyclones (e.g. Cyclone *Bulbul,* Cyclone *Amphan*), and it was difficult to maintain provisions, particularly for older people, due to inadequate resources. For example, the cyclone shelter (built before 2000) and the school which act as a cyclone shelter have no separate toilet for males and females. Hence, there was no separate toilet for male and female older persons as required by the Standing Order on Disaster (SOD). Moreover, all age groups of people who went to the cyclone shelter stayed together. This was largely because of scarcity of separate spaces for different age groups. The lack of separate accommodation provisions in shelters for older people was also against the provisions of the SOD, which emphasised that disaster risk reduction and emergency response plans must be based on social inclusion principles by ensuring the participation of older persons.

During the recent cyclones such as the *Bulbul* and the *Amphan*, the older persons moved to different places to get a safe place as part of the evacuation. However, the participants reported that the older persons were late in moving from the place as they needed assistance. They were mostly helped by their near and dear ones, such as their children, relatives, neighbours, and volunteers. The interview participants stated that they were moved by boat or other means. The oldest were carried on shoulders or on the lap. A young adult participant said that "*they (older persons) were taken on lap, shoulder. Those who had vans, they took older people on vans who could not walk. Also, they were taken on swings.*" Therefore, it is evident that some older persons are facing evacuation difficulties, making them vulnerable in the face of cyclones. Along with the evacuation challenges, older persons also faced with the problem of saving their assets, protecting their health and dealing with cultural shock. The scenario also points to the inability of the government to play an active role in prioritising the relocation of older persons in disasters as mandated by the SOD which insists on priority to be given for transporting older persons to safe shelters. The next section illustrates the findings with respect to the property loss health issues and cultural experiences of older persons during a cyclone.

5.1.2 Loss of Assets, Health Concerns, and Cultural Experience of Older Persons with Respect to Cyclones

The empirical study done suggests that older persons lost most of their assets to the cyclone, including houses, trees, cultivable land, and a significant portion of their belongings. They could not take anything or could only take a handful of things with them. During Cyclone *Aila*, it was reported that the water level rose to the roof of some of their houses, and some of the houses even got demolished by the force of the water. They tried to take their essential documents and some valuable items. However, they were unable to save the rest. When inquired about this, an older adult stated that:

> The house got damaged. The goats I had got lost in the flood. The chickens and ducks got lost as well. The house is still damaged, and we could not fully fix it. There were dead bodies of livestock everywhere that day.

Consequently, older people faced damage to their food and trees which used to grow in that area. After Cyclone *Aila*, the soil became salty due to all the salt water on the land, which took around three years to dissipate, rendering the soil unfit to grow any fruits or vegetables. The older persons also lost their cattle, hens, and ducks along with vegetables and trees. An older interviewee stated about their grief:

> Nothing could be saved. How can trees be saved in the salty water? There were 7-8 mango trees that used to give fruits. They got destroyed.

In the same vein, Cyclone *Aila* left them financially vulnerable. Some people had nothing left, as everything was swept away with the water. The water level rose so much that in some places, it was hard to even feel the ground due to the depth of the water. They lost their furniture, clothes, utensils, and other household materials. Although some claimed it was inadequate, they did get immediate help from the government and received food, shelter, medicines, drinking water, and cash from the government and non-government organisations. Even after so many years, they continue to receive reliefs, for instance, sanitary items, water tanks to preserve rainwater. The next section illustrates the difficulties they faced related to drinking and cooking water and the loss of land property.

5.1.3 Water Crisis and Loss of Agricultural Land

The area is susceptible to cyclones, tidal surges, and other natural hazards, and the empirical accounts reported that the region's few freshwater resources were repeatedly and often contaminated by cyclone events, turning them unhealthy for human consumption or use. Additionally, the embankment was broken due to the cyclone (*Aila*), which left the area waterlogged for a long time, and it took 4–5 years to fix the embankment. So people have had to face salinity issues even 3–5 years after the Cyclone *Aila* which damaged agriculture land and sweet water sources for a long time. Moreover, saltwater intrusion salinised both the surface and underground water, turning agricultural land into fallow land for a long time. A young participant opined that:

> Our agricultural land was inundated for a long time, and it took nearly three years to water free. Even the biggest issue was drinking water. It was very scarce at that time.

The above statement suggests that the water had become salty and undrinkable and water scarcity was very prominent in the region. Participants from both FGDs and interviews stated that they did not have any reliable source of drinking water. The ponds from which they used to collect drinking water were also affected by the salty water when the area got flooded, and they were forced to rely on water from external sources for the first 2–3 years after Cyclone *Aila*. These limited drinking water resources were not easily accessible, and drinking water had to be collected from distant places. Therefore, it was quite difficult for older persons to get drinking water. They often sought help from their relatives and neighbours. The government and different NGOs also provided them with drinking water. While the government is taking steps and trying to solve this water crisis by providing rainwater harvesting tanks and building pond sand filtering (PSF) mechanisms and desalination plants, it is a fact that they continue to face a drinking water crisis. Participants reported that there is no option to consider age-specific requirements to install the PSF and desalination plants. The location for establishing the PSF and desalination plants (either by NGOs or the government) depends on available space and optimum distance from villages because many people rely on one PSF or desalination plant. However, there remains an option to consider the needs of older persons in establishing or providing a rainwater harvesting mechanism.

5.1.4 Remoteness and Difficulties in Access to Health Services and Transport

It is found that many older persons have age-related problems or chronic illnesses. The findings suggest that during the Cyclones *Aila* and *Amphan*, the mental and physical health of older persons deteriorated due to the stress and pressure. Findings from the FGDs and interviews suggest that older persons had suffered physical injuries due to strong winds, tidal bores, and structural damage from cyclones. Falls, bruises, cuts, fractures, and head injuries are all too prevalent for the oldest persons in the 80+ bracket. Moreover, some older persons had a mental breakdown seeing the destruction caused by the cyclones. Participants also reported that they were affected by waterborne diseases. Some of them continuously and repetitively suffered from diarrhoea, cholera, fevers, and stomach aches.

Another factor is the difficulties in access to health, transport, and the issue of remoteness. The participants stated that the cyclones caused a colossal loss of community resources, including schools, hospitals, and road networks and other infrastructure. Older persons thus needed to go a long way for any treatment. The nearest hospital is situated in Shyamnagar town (sub-district town), and rivers on all four sides surround the area. There is no connecting road with the outer regions of Gabura. Participants also stated that a trip to the nearest sub-district hospital would take two hours, using three different modes of transport (e.g. boat, bike/easy bike, and bus). One of the older participants interviewed reported about the roads saying:

> There are small and narrow roads. We face difficulties during movement. There were few bricks made road only. Only motorcycles are used here. Nevertheless, it is worst in the rainy season. It becomes muddy. These roads are not older people friendly.

When asked about the facilities for primary treatment for older persons, participants argued that only community clinics were the nearest available medical facilities, which only provide basic health facilities and are only open during the daytime. Unprescribed medicines from medicine shops are the alternate way to tackle diseases such as fever, cold, and *amashoe* (a waterborne disease). The next section discusses the changes in income and livelihoods of older persons in Gabura due to the cyclone, which has also played a critical role in their mobility.

5.1.5 Changes in Income and Livelihoods Due to Cyclones

Participants explained that older persons (in the age bracket of 60–70 years) also decide to migrate when they and the young adult family members lose their income and have to change their livelihoods. The findings suggested that agriculture and fishing are the two sectors of Gabura's economy most affected by climate change. Reduced fish stocks, saltwater intrusion, and reduced access to natural resources have all contributed to falling earnings. The first-hand account suggested that the amount of damage was immense during the cyclones, leading to economic losses. Even older adult participants argued that active older adults (60–70 years) lost their income. This vast economic loss also led to poverty. An older interviewee said:

> The most damage was dealt by *Aila* in 2009. We had organised household and fishery; everything was destroyed during *Aila*. Everything was very calm and peaceful till 9 am. Everything was well organised in our house, pond, fishery, paddy in the field, and crops in our yard. At 9.30 am, everything got flooded in about 2 hours. *Aila* went over my croplands. After *Aila*, the salt water has ruined the ground's fertility, and nothing grows there anymore.

After the devastating damage, the young adult earning members have therefore been forced to relocate to urban regions in search of alternate means of subsistence. Along with the livelihood loss, restriction in access to natural resources (e.g. restriction to the mangrove forest for fishing, wood collection, honey harvesting, leaf collection) and economic disparity to get social safety net programme have led to food insecurity. Therefore, as older persons depend on young adults in many cases (e.g. earning, care, daily meal preparation), they decide to migrate with other family members. Moreover, the participants stated that they moved from their place of origin in order to find better working opportunities. They are trying to permanently migrate from the place by selling their properties. Some sold their lands cheaply to relatives and neighbours who did not want to leave Gabura. On the other hand, some could not sell their land, so they moved to other regions, leaving their water-trapped lands behind. Moreover, participants reported that elite people and political leaders from outside Gabura also bought such land at low prices for shrimp farming.

Finally, the empirics suggest that several factors have made life difficult for older persons in Gabura, especially after the Cyclone *Aila* (2009). Cyclone *Aila* initiated several difficulties which are discussed above, and later, other cyclones and economic and political factors multiplied the challenges, which increased the problems for older people, making it harder for them to remain in this Gabura Union. It is also important to mention that cyclones are often associated with the sudden onset of displacement, but recurrent cyclones can result in long-term impacts affecting mobility. Mobility plays a critical role in response to cyclones and other disasters and as an adaptation strategy for some older persons. The next section illustrates how mobility can be an adaptation strategy for the elderly population in the face of climate change impacts.

5.2 Mobility as an Adaptation Strategy

Mobility is multi-faceted and an element of adaptation to climate change. The result of the empirical study suggests that tropical cyclones, tidal surges, salinity intrusion, and land degradation are some of the examples of climate-related disasters predicted to cause displacement. According to the participants, the Gabura community have suffered a lot due to these catastrophes, and some of them continue to fight against cyclones. In contrast, others have migrated to other places to reduce their distress, and older migrant participants revealed the fact that they migrated from Gabura to save their lives and now feel relieved after migration. After migrating to Shyamnagar, an older adult had the following to say about the relocated place, "*A lot better place we are living in! This environment is a lot better than Gabura. This area is less vulnerable than Gabura. The damages caused by cyclones are comparatively lesser.*"

The participants also reported that as a consequence of migrating to another place, ageing people were getting better healthcare facilities along with quicker access to doctors and hospital facilities. As such older adults do not need to worry about their treatments as the transportation costs to visit a doctor had been reduced, and they did not need to wait long or travel to Shyamnagar and Satkhira towns in order to receive treatment. In the same vein, the informants also reported that income opportunities in Gabura had greatly reduced. This had become a survival issue, and active older adults (60–70 years) could not survive on such a poor income. After migration, however they were able to secure employment, and their income increased, which ultimately contributed to their family's well-being. An older participant stated about better working opportunities after migration saying, "*I get to work every day there. They do pay every day.*" The findings reveal the role of mobility in climate change adaptation and its impact on vulnerable communities such as the Gabura community who are deeply affected by climate-related disasters that can lead to displacement. It does appear that after migrating to a new place, older persons have been able to secure better access to healthcare and employment that pays them on a regular basis. This has, in turn, helped them get financially solvent and adapt to the cyclone. The mobility dynamics of older persons are twofold. In the place of origin, older persons faced difficulties with disaster impact, which were also triggered by socio-political and economic factors influencing their decision to stay there. On the other hand, the lure of several urban facilities indirectly pulls them to migrate from rural areas. The following section discusses the interplay between climate change impact, human mobility, and the adaptation strategy.

5.3 Discussion

This research provides an in-depth understanding of the impact that cyclones have on the lives of older persons and its bearing on their decisions to migrate. It also reveals their vulnerability which in turn influences their decision to migrate to another place as an adaptation strategy. The research placed older persons at the centre of the study, thereby allowing us to fill research gaps caused due to the exclusion of such age groups from the migration and development literature. The empirical evidence gathered from the study provides critical points to the climate change adaptation and migration literature.

The study reveals that older persons are vulnerable to cyclones due to their age and physical constraints.[64] Older adults are usually not self-sufficient and need the help of the people around them.[65] The findings suggest that the experiences of older persons in Gabura before, during, and after cyclones are far more deplorable when compared to other age groups. It was elderly individuals in Gabura who lost their

[64] Malak and others (n 5); Quader, Khan and Kervyn (n 1).

[65] Walkling and Haworth (n 6).

homes, arable land, trees, and many possessions during the cyclones.[66] These older persons could not recoup because of the substantial loss of assets and significant economic losses that plunged some into abject poverty. Although the recent cyclones (e.g. *Bulbul* and *Amphan*) have had a less severe impact in the area, due to their weaker landfall, the people still suffered a lot due to the severe destruction caused by Cyclone *Aila* (2009). These unexpected cyclones have made life very unpredictable and therefore at risk, and though older persons are familiar with these environmental challenges, it has taken them longer to return to normalcy post the shock and stress of cyclones.

Similarly, the study also shows that even though older persons are attached to their roots and their motherland, they have had to migrate despite not really wanting to make the move.[67] Due to the slow-onset of changes like desertification, sea level rise, and extreme weather occurrences like cyclones, floods, droughts, and storms, it has been difficult for older people to support themselves and their families in their place of origin.[68] Many older adults have been forced to relocate to urban areas due to poverty and environmental problems such as riverbank erosion, cyclones, and floods.[69] However, in addition to pre-existing migratory pathways and economic, political, historical, demographic, social, and environmental elements that interact with climate threats and migration decisions are also influenced by pre-existing movement patterns.[70] One of the main strategies employed by households to decrease the consequences of cyclones was moving to cities where cyclones impacted very little or not at all. Some of them moved to places where there were alternative working opportunities and strong social networks. In fact this mobility has become their adaptation strategy to save themselves from different cyclones.

Migration is thus an adaptation strategy for young adults from socio-economically marginalised strata.[71] Displacement is anticipated from extreme weather, tropical cyclones, rising sea levels, and land degradation.[72] The findings of this research echoe the same understanding even for older persons in climate-vulnerable households, challenging narratives that older people tend to stay behind. The results suggest that the elderly are using mobility as an adaptive technique to protect themselves from the impact of cyclones. The findings reveal that while older persons have struggled with disasters, many moved to secure urban places to lessen their worries and distress.

6 Conclusion

The study explored how climate change-induced vulnerability influences people's decisions regarding their choice to stay or migrate outside of their area. The main focus was on the vulnerability of older persons during cyclones due to their age and mobility restrictions. The study revealed the various factors that influence the decision of older persons to migrate, such as lack of water, access to healthcare and communication facilities, remoteness, changes in income and livelihood, and stresses the significance of taking the observations and points of view of older persons into consideration when designing ways to adapt. The findings shed light on the mobility of the elderly in Gabura and illuminated their experiences before, during, and after cyclones. The mobility-related findings of the study highlighted the vulnerability of the elderly to cyclones because of their advanced age. The study emphasises how crucial it is to consider their viewpoints and experiences when creating adaptation strategies to deal with the effects of climate change on the

[66] Alam and Collins [45].

[67] Adams [46].

[68] Malak and others (n 5).

[69] Adger, de Campos and Mortreux (n 39); Malak and others (n 5); Hossain et al. [47].

[70] Courtoy [48].

[71] Adger and Adams (n 17).

[72] McMichael, Schwerdtle and Ayeb-Karlsson (n 2).

elderly. Moreover, the research provides valuable insights into the complex decision-making processes of persons in response to cyclones. Overall, this study advances the knowledge about elderly mobility in the face of tropical cyclones and other coastal environmental stressors. The findings are useful in developing policies and programmes that include older persons and address their specific needs and vulnerabilities in the context of climate change. Moreover, the first-hand accounts may equip the authorities to bring about necessary changes in the Standing Order on Disaster (SOD), the Disaster Management Act (2012), National Adaptation Plan (2023), climate financing, and climate justice programme.

The study recommends that the future focus of the authorities should be on developing rules and policies as well as adaptation techniques that consider the perspectives and needs of older persons, in order to mitigate climate change impacts on them. For instance, the government is implementing a housing project for climate-related displaced people, and this will be an opportunity to include older persons, and therefore it may consider building age-sensitive housing under this scheme. Thus, further work on governance, institutions, and social networks could significantly improve the understanding of what needs to be done to enhance future adaptive capacities.

References

1. Mondal et al (2022) Spatio-temporal behaviours of tropical cyclones over the Bay of Bengal basin in last five decades. Trop Cyclone Res Rev 11(1)
2. Quader MA, Khan AU, Kervyn M (2017) Assessing risks from cyclones for human lives and livelihoods in the coastal region of Bangladesh. Int J Environ Res Public Health 14:831
3. Hoque MA et al (2021) Cyclone vulnerability assessment of the western coast of Bangladesh. Geomatics Nat Hazards Risk 12:198
4. McMichael C, Schwerdtle PN, Ayeb-Karlsson S (2023) Waiting for the wave, but missing the tide: case studies of climate-related (Im) mobility and health. J Migr Health 7:10014
5. Desai BH, Mandal M (2021) Role of climate change in exacerbating sexual and gender-based violence against women: a new challenge for international law. Environ Policy Law 51:137
6. Malak MA et al (2020) We are feeling older than our age: vulnerability and adaptive strategies of aging people to cyclones in coastal Bangladesh. Int J Disaster Risk Reduct 48:101595
7. Wachholz S (2003) 'At risk': climate change and its bearing on women's vulnerability to male violence. In: Beirne P, South N (eds) Issues in green criminology. Willan
8. Whittenbury K (2012) Climate change, women's health, wellbeing and experiences of gender based violence in Australia. In: Alston M, Whittenbury K (eds) Research, action and policy: addressing the gendered impacts of climate change. Springer, pp 207–221
9. Walkling B, Haworth BT (2020) Flood risk perceptions and coping capacities among the retired population, with implications for risk communication: a study of residents in a North Wales Coastal Town, UK. Int J Disaster Risk Reduct 51:101793
10. Furlong BT et al (2022) Gendered (Im) mobility: emotional decisions of staying in the context of climate risks in Bangladesh. Reg Environ Change 22:123
11. Mobarak AM, Akram AA (2016) Seasonal migration to increase incomes of poor households in Bangladesh. Copenhagen Consensus Center and BRAC, Bangladesh Priorities Project
12. Kawasaki L (2014) Disasters and the frail elderly. In: Cefalu CA (ed) Disaster preparedness for seniors: a comprehensive guide for healthcare professionals. Springer
13. Golam Rabbani MM, Cotton M, Friend R (2022) Climate change and non-migration—exploring the role of place relations in rural and coastal Bangladesh. Popul Environ 44:99
14. Haq G, Brown D, Hards S (2010) Older people and climate change: the case for better engagement. Stockholm Environment Institute Project Report
15. BBS (2022) Bangladesh population and housing census. Ministry of Planning, Government of Bangladesh
16. Kwan C, Walsh CA (2017) Seniors' disaster resilience: a scoping review of the literature. Int J Disaster Risk Reduct 25:259
17. Adger WN, Adams H (2013) Migration as an adaptation strategy to environmental change. World Social Science Report Changing Global Environments, pp 261–264
18. Bardsley DK, Hugo GJ (2010) Migration and climate change: examining thresholds of change to guide effective adaptation decision-making. Popul Environ 32:238
19. Maharjan A et al (2020) Migration and household adaptation in climate-sensitive hotspots in South Asia. Curr Clim Change Rep 6:1
20. Mortreux C, Barnett J (2009) Climate change, migration and adaptation in Funafuti, Tuvalu. Glob Environ Change 19:105

21. Bastia T, Lulle A, King R (2022) Migration and development: the overlooked roles of older people and ageing. Prog Hum Geogr 46:1009
22. Blaikie P et al (2014) At risk: natural hazards, people's vulnerability and disasters, Routledge
23. WHO (2015) World report on ageing and health. World Health Organization
24. UNESCAP (2022) Climate change and population ageing in Asia-Pacific Region: status, challenges and opportunities. United Nations
25. Gamble JL et al (2013) Climate change and older Americans: state of the science. Environ Health Perspect 121:15
26. Shanahan DF et al (2015) The health benefits of urban nature: how much do we need? Bioscience 65:476
27. Bukvic A et al (2018) Aging in flood-prone coastal areas: discerning the health and well-being risk for older residents. Int J Environ Res Public Health 15:2900
28. Koubi V et al. (2022) Climate events and the role of adaptive capacity for (Im-)mobility. Popul Environ 1
29. IPCC (2014) Climate change 2014: impacts, adaptation, and vulnerability. Part A: global and sectoral aspects. Contribution of working group II to the fifth assessment report of the intergovernmental panel on climate change. Intergovernmental Panel on Climate Change
30. Galez-Davis C et al (2022) Goal 16/8/5: leveraging administrative data to fight human trafficking. In: Vidal EM, Laczko F (eds) Migration and the SDGs: measuring progress—an edited volume. International Organization for Migration. https://publications.iom.int/books/migration-and-sdgs-measuring-progress-edited-volume. Accessed 25 Oct 2022
31. Neil Adger W, de Campos RS, Mortreux C (2018) Mobility, displacement and migration, and their interactions with vulnerability and adaptation to environmental risks. In: Routledge handbook of environmental displacement and migration. Routledge
32. IDRC (2021) Understanding mobility in the context of climate change. https://idrc-crdi.ca/en/research-in-action/understanding-mobility-context-climate-change. Accessed 25 Dec 2022
33. Malak MA (2022) "Resilient livelihoods?" Insights from Tanguar Haor in Bangladesh. University of Wollongong Thesis Collection. https://ro.uow.edu.au/theses1/1389/. Accessed 29 Oct 2022
34. Fussell E, Sastry N, VanLandingham M (2010) Race, socioeconomic status, and return migration to New Orleans after Hurricane Katrina. Popul Environ 31:20
35. Neil Adger W, Arnell NW, Tompkins EL (2005) Successful adaptation to climate change across scales. Glob Environ Chang 15:77
36. Antal H, Bhutani S (2022) Identifying linkages between climate change, urbanisation, and population ageing for understanding vulnerability and risk to older people: a review. Age Int 1
37. WHO (2014) Strengthening health resilience to climate change. World Health Organization. chrome-extension://efaidnbmnnnibpcajpcglclefindmkaj/https://www.afro.who.int/sites/default/files/2017-06/conference_briefing_1_healthresilience_27aug.pdf. Accessed 29 Oct 2022
38. Sari Kovats R, Hajat S, Wilkinson P (2004) Contrasting patterns of mortality and hospital admissions during hot weather and heat waves in Greater London, UK. Occup Environ Med 61:893
39. Pillemer K, Cope MT, Nolte J (2021) Older people and action on climate change: a powerful but underutilized resource. HelpAge International
40. DDM (2019) Standing order on disaster (SOD). Ministry of Disaster Management and Relief. http://www.ddm.gov.bd/site/view/policies. Accessed 30 Sept 2023
41. MoDMR (2015) National disaster management policy 2015. Ministry of Disaster Management and Relief. http://203.76.123.197/egls/documents/458/search-detail. Accessed 29 Oct 2023
42. MoEFCC (2023) National adaptation plan of Bangladesh. Ministry of Environment, Forest and Climate Change. https://moef.gov.bd/site/news/110358e4-670c-4058-b20f-e7ab6c89feb9/National-Adaptation-Plan-of-bd-2023-2050. Accessed 12 Jun 2023
43. MoDMR (2021) National strategy on internal displacement management. Ministry of Disaster Management and Relief. https://modmr.portal.gov.bd/sites/default/files/files/modmr.portal.gov.bd/publications/d05820e8_b72f_43a1_8cfe_4b4bd34a2560/NSIDM%20Publication(Size%206.5%20X%209.5%20in).pdf. Accessed 17 Oct 2022
44. Afrin S (2021) Disaster induced water crisis in gabura. Dhaka Tribune, Dhaka, Bangladesh. https://www.dhakatribune.com/science-technology-environment/climate-change/248381/disaster-induced-water-crisis-in-gabura. Accessed 20 Apr 2023
45. Alam E, Collins AE (2010) Cyclone disaster vulnerability and response experiences in Coastal Bangladesh. Disasters 34:931
46. Adams H (2016) Why populations persist: mobility, place attachment and climate change. Popul Environ 37:429
47. Hossain NJ et al (2023) Livelihoods under pressure: insights from riverine community in Bangladesh. Asian Geogr 1
48. Courtoy M (2022) "To leave is to die": states' use of mobility in anticipation of land uninhabitability. German Law J 23:992

Dr. Md. Abdul Malak is a human geographer with over ten years of research and teaching experience in climate change adaptation, resilience, livelihoods, migration, refugees, the political economy of environmental change, and critical review in Bangladesh and overseas. He has obtained a PhD in Human Geography from the

School of Geography and Sustainable Communities at the University of Wollongong. He also received B.Sc. and M.S. degrees in Geography and Environment from Jahangirnagar University, Bangladesh. He has a long experience in handling qualitative and Quantitative data in academia and development research. A wide range of professional experience in conducting field-based research, the ability to critically assess problems, and experience in communicating research results through a number of channels, such as publishing in academic journals/books, writing reports, and oral presentations, made Dr. Malak an expert in areas where he works, as mentioned above. He has published a total of 20 journal articles and book chapters. Currently, he is working as an Associate Professor of Geography and Environment at the Jagannath University, Bangladesh.Nazia Khan Lina is working as a research assistant at the Department of Geography and Environment, Jagannath University. Her research focuses on Human Mobility, Climate Change Adaptation, and Livelihoods. She has obtained her MS and BSc degrees from the Department of Geography and Environment at Jagannath University, Bangladesh.

To Move or Not to Move? Drought-Related Internal Migration and Voluntary Immobility Among the Hmong Forest Community in Chiang Rai, Northern Thailand

Reni Juwitasari

Abstract

Over four decades, from the 1980s to the 2020s, Thailand has experienced a severe and prolonged drought, primarily attributed to the effects of climate change. Consequently, this drought has led to a significant and pervasive shortage of water resources across the country. Drought has substantial environmental, social, and economic ramifications on farmers, manifesting as diminished agricultural and forest productivity, depleted water reserves, heightened vulnerability to wildfires, and consequent losses in income, food security, well-being, and population displacement. The region of Northern Thailand has seen the emergence of drought as a distinct slow-onset calamity in the country. Chiang Rai is characterised by its extensive forested areas and agricultural landscapes, which provide residence for at least ten ethnic minority groups from neighbouring countries, including Hmong. The survival of these ethnic minority people is dependent on the agricultural production they rely upon. Therefore, climate change has emerged as a catalyst for forced migration. This study examines the social implications of climate-related drought in the Wiang Kaen district, which the Hmong ethnic group inhabits. The analysis focuses on three main areas: (1) the impact of drought on the community's forced migration patterns; (2) the disparities in perceptions of human mobility between older and younger generations in response to the drought; and (3) the factors influencing well-being, such as food security and adaptive capacity, in the context of climate-related drought. The study reveals that climate-induced migration has become one of the climate change adaptation efforts, especially for economic reasons. The internal migration has patterned at the moving from rural to urban areas in joining the informal sector. Although this effort is positively adopted, the disparity in migration perspective is distinctly reflected between older and younger generations, with particular emphasis on food security and availability. Food sufficiency for the household is considered the benchmark of well-being for the Hmong elderly per se, while the younger generation weightens the currency for sustainable well-being. In addition to climate change adaptation, the older generation has demonstrated their ability to cope with climate-induced drought by using traditional or local knowledge in managing forests and

R. Juwitasari (✉)
Disaster Resilience and Environmental Sustainability (DRES) Program, Asian Research Center for International Development (ARCID), School of Social Innovation, Mae Fah Luang University, Chiang Rai, Thailand
e-mail: reni.juw@mfu.ac.th

S. Jolly et al. (eds.), *Climate-Related Human Mobility in Asia and the Pacific*, Sustainable Development Goals Series, https://doi.org/10.1007/978-981-97-3234-0_3

agriculture, acknowledged as a kind of resilience rather than resorting to migration, while the younger generation chooses migration as an adaptation strategy. Therefore, this study acknowledges the significance of climate-induced human mobility as a pressing worldwide issue that poses climate change adaptation for sustainable development, although it has been differently perceived by different generations. Hence, the formulation of policies based on climate change adaptation and resilience by both generations is being undertaken.

1 Introduction

Thailand is located in Southeast Asia and is bordered by countries including Cambodia, Laos, Malaysia, and Myanmar. Thailand has seen a notable increase in temperatures ranging from 0.10 to 0.18 °C every decade over the period spanning from 1979 to 2005.[1] A reduction in both the frequency and amount of precipitation and an increase in the severity of precipitation events also accompanied an increase in annual mean temperatures.[2] The changes in the spatial arrangement, occurrence rate, and strength of precipitation events have significantly influenced the volume and characteristics of water reserves in the hydrological basins of Thailand. It is estimated that by the century's conclusion, there will be an increase in average temperatures and an increased probability of heat waves, particularly in the southern regions of Thailand.[3] It is also anticipated that Thailand will see an escalation in extreme events, including prolonged floods and drought, landslides, and intense storm surges.

Drought is a very destructive phenomenon in the area. Over almost three decades, from 1978 to 2007, the Royal Irrigation Department (RID) of Thailand and the Thai Meteorological Department (TMD) observed a consistent decrease in yearly precipitation, ranging from 900 to 1700 mm. The duration of the dry season has extended beyond its previous length, resulting in significant challenges for individuals during periods of severe drought.[4] In 2010, several regions in Thailand declared that they were experiencing severe disaster due to shortages of water. Because the tropical rainy season ended earlier than usual in November 2009, together with global warming and the El Nino phenomenon, Thailand experienced unusually hot weather and a lack of rainfall at the beginning of 2010.[5] After the 2010 drought coupled with the adverse floods, the Department of Disaster Prevention and Mitigation in Thailand documented that 46 out of 76 provinces in the nation had experienced drought in 2014. Among these provinces, the northern area, particularly Chiang Rai Province, was afflicted. The mean monthly precipitation during the rainy season in the reference period is around 196 mm, but the precipitation during the rainy season in the most recent year was roughly 192 mm. Consequently, the anticipated variation in rainfall in the northern area is predicted to be within a narrow range of 100–200 mm compared to the baseline rainfall data.[6] Thais residing in the northern region, namely in Chiang Rai, had significant difficulties as they were unable to engage in fishing, which serves as a crucial means of revenue and good source of protein. This predicament arose due to the river experiencing exceptionally low water levels, impacting the communities' access to drinking water, agricultural needs, and livestock.[7]

Consequently, drought is likely to harm crops by reducing the amount of land available for cultivation, eventually affecting the yield of crops, resulting in a loss of revenue, and other

[1] The Economics of Climate Change in Southeast Asia: A Regional Review (Asian Development Bank 2009) 22.

[2] Mark [1].

[3] 'Climate Risk Country Profile: Thailand' (World Bank Group 2021) 13.

[4] Homdee et al. [2].

[5] Garbero and Muttarak [3].

[6] Seeboonruang [4].

[7] Marks [1].

social problems. The phenomenon of climate change-induced migration has had a significant impact on those affected by natural disasters. The process of internal migration from rural regions to metropolitan areas, particularly Bangkok, has had a significant impact.[8] As mentioned by Burrows and Kinney in 2016, drought has been associated with a rise in both temporary movement (as a means to broaden sources of income) and permanent migration, leading to an escalation in permanent resettlement.[9] Moreover, the potential for unequal distribution of the consequences among different social groups, particularly those economically disadvantaged and marginalised, is very high due to variations in adaptation policies and plans.[10] The concept of migration in relation to climate change has been reconceptualised as the promotion of social resilience.[11] Hence, migration is acknowledged to have beneficial impacts.

According to the Ministry of the Interior, 64 out of 77 provinces, including Chiang Rai, experienced severe drought in 2010.[12] In Chiang Rai, droughts occur frequently, resulting in low water quality.[13] This climate-related catastrophe is anticipated to have the most significant impact on the agricultural sector because the agricultural sector serves as the primary source of income and employment for a considerable portion of the populace while also making a substantial contribution to the country's overall gross domestic product (GDP). In the rural areas of Thailand's northern region, especially Chiang Rai, small-scale family plantations have become the centre of economic activity. Drought and heightened temperature are projected to have a detrimental impact on agricultural output. Consequently, the decline of agricultural production in rural areas in Chiang Rai resulting from drought could significantly undermine food security and exacerbate the socio-economic circumstances of impoverished rural populations, forcing some to migrate, even if others may experience forms of climate-related immobility.[14]

Migration[15] is a frequently used approach for adjusting to the fluctuating patterns of agricultural production, land constraints, and economic instability in the light of climate change.[16] Before the manifestation of climatic effects, individuals resorted to migration, especially from the rural regions, to adapt to the changing livelihood patterns. However, in the light of the emergence of climate change, there has been a notable rise in migration, driven by the need for mitigation and adaptation and social, political, and economic considerations. The act of migration has the potential to alleviate challenges faced by traditional agricultural communities, such as the decline in crop production due to prolonged and intensified periods of drought.[17] When mitigation strategies are well planned and executed, migration may be a viable approach for safeguarding livelihoods and fostering resilience.

The migration of vulnerable families lacking access to land and facing abject poverty often represents a critical means of ensuring their survival when faced with disasters like droughts. However, it is to be noted that such migratory patterns might undermine their resilience. For example, individuals who migrate during droughts or famine may provide little or negligible remittances, yet migration concurrently diminishes the labour pool available for domestic food production.[18] Conversely, migration can undermine the communities where individuals move from and those who receive them.[19] In addition, it has been shown that migrants or individuals from vulnerable groups can

[8] Ibid.

[9] Burrows and Kinney [5].

[10] Lebel et al. [6].

[11] De Guttry et al. [7].

[12] UPI [8].

[13] Lebel (n 10) 932.

[14] Calzadilla et al. [9].

[15] Zander et al. [10].

[16] Ibid.

[17] Curran and Meijer-Irons [11].

[18] Warner and Afifi [12].

[19] Burrow and Kinney (n 9) 1.

experience limited prospects for local adaptation or the ability to relocate to mitigate risks.[20] Individuals who find themselves in perilous circumstances and express a desire to migrate but lack the means to do so are a matter of specific concern, as they may face a heightened vulnerability to poverty and mortality.

This chapter elaborates on the case study conducted in Wiang Kaen district, located in Chiang Rai Province, which serves as the residence of the Hmong ethnic community. The first three parts examine the interconnection between climate change, forest communities, the agriculture sector, migration, and the Sustainable Development Goals. The initial section examines the intersection of forest communities, agriculture, and climate change to gain a comprehensive understanding of the regional circumstances. The second section examines the role of climate change, particularly drought, as a significant factor influencing migratory patterns. The third section explores the significance of migration as a tool for enhancing societal resilience. The fourth section pertains to the methodology employed in the research, whereas the fifth section comprises an analysis of the social impacts in Chiang Rai. The analysis primarily centres on three key aspects: (1) the impact of climate-related drought on the occurrence of forced migration within the community; (2) the disparities in perceptions regarding human mobility resulting from the drought, between older and younger generations; and (3) the factors influencing well-being, such as food security and adaptive capacity, in the context of climate-related drought.

2 Climate Change Issues, Adaptation, and Mobility Pattern in Thailand

Thailand has already experienced the impacts of global climate change. Thailand's annual temperature has risen by approximately one degree Celsius from 26 to 27° between 1981 and 2007, and it is predicted to increase by 1.2–1.9 °C by 2050.[21] In addition, the number of rainy days and the level of precipitation in Thailand has decreased over the last fifty years, and it has led to prolonged drought, especially in the Mekong River.[22] Along with the drought and the increase in temperature, Thailand has suffered from coastal erosion and sea level rise caused by changes in sediment loads, land subsidence, and loss of mangroves.[23]

Northern Thailand, especially Chiang Rai, is recognised as one of the eight hotspot provinces in the Lower Mekong Basin region experiencing significant changes in temperature, precipitation, and flooding, including smoke and haze problems caused by wildfires and open-space burning, and this has resulted in environmental problems that have negatively affected the regional climate, the environment and socio-economic development, and in particular, the health of the region's population.[24] Chiang Rai stands out as a province that is vulnerable to climate change.[25] It is due to an analysis of environmental degradation and ecological collapse caused by the alleged impacts of anthropogenic activities, e.g. deforestation and urbanisation.[26] Therefore, the adaptation has to be enhanced, and its planning needs to be addressed autonomously, driven by how actions can reduce risks, environmental degradation, and climate change.

Besides, Chiang Rai is home to at least 60% of the seventeen ethnic groups in Thailand,[27] such as Akha, Bisu, Dara-ang, Hmong, Isan, Karen, Khmu, Lahu, Lisu, Lwa, Mien, Tai-Yuan, Tai-Lue, Tai Khun, Tai Ya, Ta Yai, and Tai Yong, and is possibly the area with the highest ethnic group population inhabitants in the

[20] Neil Adger et al. [13].

[21] Danny Mark (n 2) 232.

[22] Ibid.

[23] Lebel [14].

[24] Sirimongkonlertkul et al. [15].

[25] Carew-Reid and Taylor [16].

[26] Forsyth and Evans [17].

[27] Hu [18].

northern region.[28] These ethnic groups with cultural diversities have driven Chiang Rai's tourism industry to promote the slogan 'City of Art and Diversity.'[29] These ethnic groups have experienced migration in history due to several reasons, such as civil war, shifting cultivation, land settlement, drug problems, and deforestation.[30] In addition, there are six main ethnic groups residing in highland Thailand, namely Akha, Lahu, Lisu, Mien, Karen, and Hmong.[31] Specifically, the Hmong reside primarily in the Wiang Kaen and Mae Suai districts of the Chiang Rai province.[32] This ethnic group of Hmong migrated to Thailand searching for new Swiddens, most notably for opium poppy growing—which quickly depleted the soil—and spread among the Hmong in the mid-nineteenth century.[33] However the Hmong community has ameliorated and acculturated their native culture through the generations. While the elder Hmong are still maintaining native customs and traditions, for instance, in family matters, agriculture practices, and so forth, the young generations are slowly acculturating to Western society, which chooses dominant intra-personal matters over their native identity, including migration to obtain more opportunities and material capital.[34] The native culture of the Hmong remains patriarchal, emphasising the power of males to provide security, take care of finances for the family, farming and hunting, while female Hmong take on the role of caretaking and cleaning, with young females possessing skills in sewing and garment making.[35] Therefore, the most common reason for the Hmong community migration refers to well-being and factors, specifically agriculture and economic factors.

Additionally, climate change becomes one of the major reasons for migration due to its impacts on social, economic, political, and demographic variables, influencing the quality of life and geographical distribution of human populations.[36] Sakdapolrak et al. [25][37] indicate that both existing mobility patterns and post-disaster migration can be necessary and deemed a suitable adaptive reaction for individuals and households. For instance, 'migration as adaptation' views migration as a means for families to diversify their income[38] and distribute risk. It is particularly so for small farmers who rely on rainfed agriculture,[39] where environmental problems are common and a sparse period is characteristic.[40] Warner [24] highlights that the convergence of two important issues, namely human mobility and climate change, is posing significant challenges to governments in managing cross-border migration and refugees.[41] While there has been much research on climate-induced migration caused by flooding in Southeast Asia conducted by scholars like Carl Middleton et al., this work specifically examines a case study on migration resulting from drought. In 2017, Carl Middleton et al. said that common migration follows a pattern of spatial mobility, including rural–urban, rural–rural, and transnational movements, which result in economic, environmental, and socio-political disruptions.[42] In 2013, Lebel explained that migration can be either permanent or temporary, depending on the feasibility or desirability of adaptation, as well as the potential for the environment to recover, among other factors. This suggests that migration could serve as a significant strategy for adapting to climate change and that internal migration within a country is

[28] Tongdhamachart and Alwi [19].

[29] Jantakad [20].

[30] Tongdhamachart and Alwi (n 28) 3.

[31] Ibid. 5.

[32] Oupra [21].

[33] Culas and Michaud [22].

[34] Tatman [23].

[35] Ibid.

[36] Warner [24].

[37] Solomon and Sheldon [26].

[38] Klepp [27].

[39] Deshingkar [28].

[40] Bryan et al. [29].

[41] Warner (n 36) 389.

[42] Middleton et al. [30].

likely to surpass migration across international borders.[43] In addition, migration may be seen as a possible reaction to the negative effects of activities conducted outside of a specific location. This migration can be seen as a deliberate action taken by individuals, referred to as 'slow-forced displacement,' as shown by the building of a dam in southern Laos, resulting in increased vulnerability.[44]

Additionally, migration can be interpreted from a vulnerability perspective. It facilitates a deeper understanding of the associated relationships by contextualising the environment-migration nexus: household members are exposed to many inter-personal and ecological factors, including stresses related to climate change. Migration in this context could indicate that the local ability to adapt has failed as a means of adaptation, or a development strategy that reduces susceptibility and increases resilience. Migration as an adapting strategy reflects insights from this line of research by situating (environmental) migration within the larger context of rural existence,[45] recognising migration as not (solely) an emergency but a regular aspect of how people make a living[46] and theoretically locating migration within the structure of vulnerability and resilience.[47] As a result, human mobility has been influenced not merely by economic factors but also climate change.

3 Forest Community and Agriculture in Thailand

Forests cover thirty percent of the entire geographical surface of the planet. Ten countries account for two-thirds of the world's forest area, while 57 countries have less than 10% of their land area covered by forests.[48] In Thailand, seventy-two percent (72%) of the extant forest community resides in North and Northeast Thailand, including in Chiang Rai, where natural forest species are found. Moreover, most impoverished people are also reported to live in these two regions.[49] The forest is often the fundamental pillar of their heritage and culture. The term 'forest community' is used comprehensively in the context of the chapter to include elements such as initiatives, institutions, and procedures that aim to enhance the involvement of the local people in the regulation and management of forest resources. The forest community encompasses a range of dimensions, including social, economic, and conservation aspects, evident in various activities. These activities include traditional management of culturally significant sacred sites, the operation of small-scale forest-based enterprises, the implementation of forestry out-grower schemes, the establishment of company-community partnerships, and the adoption of decentralised and devolved forest management approaches.[50]

Preserving forests is recognised as an important strategy for protecting biodiversity and the human race. Local populations heavily depend on forests as a primary provider of sustenance, including food, fuel, fodder, and building materials. Locatelli et al. in 2011 mentioned that forests, sometimes called 'forests for people's adaptation,' contribute significantly to the communities' and societies' adaptation to climate change.[51]

There is a growing acknowledgement that well-managed ecosystems have the potential to assist civilisations in adapting to existing

[43] Lebel [31].

[44] Middleton [32].

[45] Khoa et al. [33].

[46] Tacoli [34].

[47] Klasen et al. [35].

[48] Adaptation to Climate Change in Agriculture, Forestry, and Fisheries: Perspective, Framework and Priorities, (FAO 2007).

[49] Asia–Pacific Forestry Sector Outlook Study II: (Working Paper Series, (FAO), 2009) https://www.fao.org/3/am617e/am617e00.pdf.

[50] People and Forests in a Time of Rapid Change: Strengthening Capacities for Community Forestry to Respond- RECOFTC Strategic Plan 2008–2013, (RECOFTC 2008).

[51] Locatelli et al. [36].

climate threats as well as future climate change via the provision of diverse ecosystem services. Forest ecosystems play a crucial role in facilitating adaptation efforts by offering valuable local ecosystem services that effectively mitigate susceptibility of societies to the impacts of climate change.[52] For instance, mangroves serve as a protective barrier for coastal regions, safeguarding them from potentially intensified storms and waves resulting from climate change and increased sea levels. Forest products serve as a crucial support for local populations when climatic occurrences adversely affect crops. The preservation and effective governance of ecosystems and their associated services can provide many socio-ecological advantages and foster enduring strategies for adapting to climate change in the long run.[53]

On the other hand, it is essential to note that existing forests, including community forests, are expected to encounter direct and indirect impacts of climatic change through changes in temperature, droughts, etc.[54] The earliest indicators of climate change impacts on forests include the decline of forests on former permafrost soils in Canada and Alaska and the decline of cloud forests in the tropics. Furthermore, there is a global increase in the frequency and severity of forest fires, a shift in the timing of germination, and a rise in insect and disease outbreaks.[55] These climatic impacts on forests would seriously compromise the value of goods and services provided by the forests. Such a scenario becomes heavily burdensome for ethnic minorities, whose livelihood and food security are already affected by reduced agricultural output due to droughts. However, in stark contrast to the agricultural sector, choices pertaining to the management of forests, such as the choice of tree species, possess a long-lasting and irrevocable impact that may last for many decades or even centuries. Conversely, a certain amount of time will be necessary to determine suitable seed sources for modified climatic circumstances.[56]

For instance, the northern region of Thailand (Chiang Rai) is now experiencing a prolonged period of increase in the cultivation of monocropped maize. Local peasants and small-scale farmers mostly carry out this practice to meet the demands of the global animal feed industry, which has a significant presence in Thailand. The prevalence of maize monoculture in agricultural landscapes and its economic significance for farmers has raised several social and environmental issues. These include farmer indebtedness, food insecurity, forest destruction, and seasonal burning. One instance is the attribution of severe particulate matter haze pollution in urban areas due to the actions of peasants burning their fields. However, these peasants often argue that this attribution lacks contextualisation and fails to acknowledge the crucial role played by global-scale cattle agriculture in driving these dynamics.[57] In addition, a sizeable proportion of ethnic minorities reside in and around protected areas, particularly those adjoining neighbouring nations. This area is a resource of non-timber forest products (NTFPs), which refers to materials collected from natural or man-made forests and riverine habitats and used to support local livelihoods. It is essential for all rural communities to contribute significant proportions of food and nutrition, especially during seasons of rice shortage, and contribute to the household economy through the sale of high-value products.[58]

It has been noted that the growing dependency on agriculture among the Hmong community is under threat due to climate change. The linkage of climate change, agriculture, food production, and community forest has been seen from the data of the World Bank in 2012.[59] The

[52] Ibid.

[53] Locatelli et al. (n 51) 437.

[54] Warner and Afifi (n 18) 2.

[55] Chamura et al. [37].

[56] Adger et al. (n 20) 755.

[57] Charoenratana et al. [38].

[58] Carew-Reid and Taylor (n 25) 8.

[59] Inclusive Green Growth: The Pathway to Sustainable Development, (World Bank, 2012).

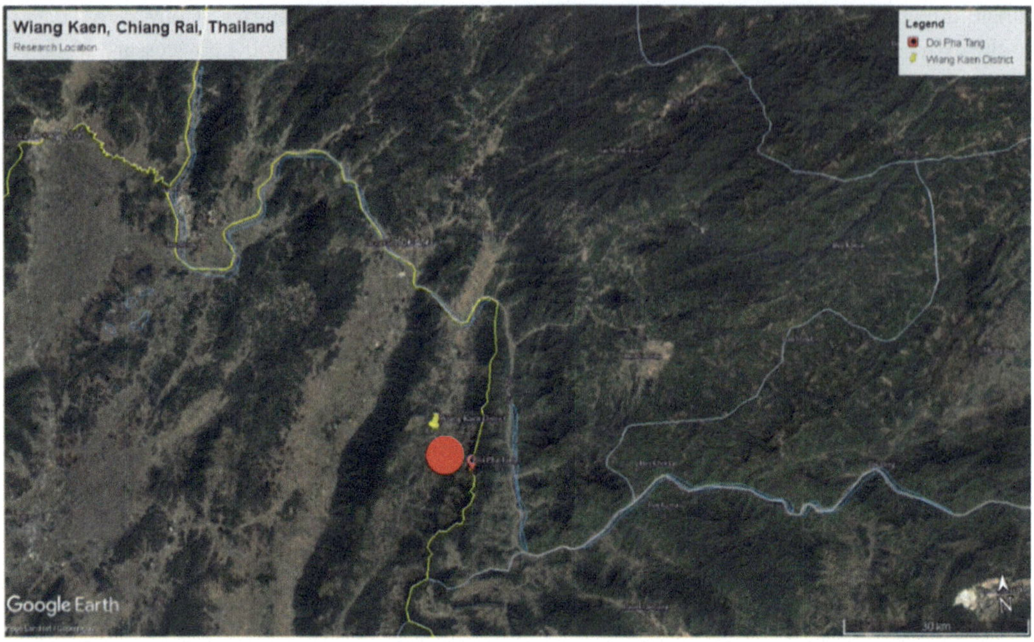

Fig. 1 Research site in Wiang Kaen, Chiang Rai Province. *Source* https://earth.google.com/web/search/Wiang+Kaen+District,+Chiang+Rai

data predicted that between 35 and 122 million people in 2030 would be in the poverty scenario because of climate change. Climate change can potentially exacerbate poverty and food insecurity among forest communities,[60] notably in Thailand, rendering them more susceptible to migration. Hence, it is imperative to recognise the gravity of the impacts of climate change on forests, agriculture, and consequent human migration, especially on ethnic minorities who rely on agriculture, and promptly investigate pathways for its resolution.[61] With this background, this study elucidates the Hmong ethnic minority forest community, climate change, and migration. This study has been conducted through empirical study; as a result, the next topic elaborates on research methodology, which will be expounded upon in the subsequent part.

4 Methods Employed in the Research

This study used a qualitative research design using an ethnographic methodology. The primary strategy for data collection was the utilisation of structured interviews. The selected villages for the survey are located inside the administrative boundaries of Wiang Kaen district in the province of Chiang Rai, in the northern region of Thailand. Wiang Kaen is a landlocked district bordering Thailand and the Lao People's Democratic Republic (Fig. 1). The entire area of Wiang Kaen is 526 km^2. More than 40% of the national reserve forest in the area has been encroached upon by human settlements and agricultural activities.[62] The region has nine distinct ethnic groups that have traditionally depended on shifting agriculture and slash-and-burn practices for sustenance. The

[60] Hallegatte et al. [39].

[61] Phungpracha et al. [40].

[62] Sang-Arun et al. [41].

mean income of the inhabitants in the district is below the national average.[63]

Ethnic minorities from the Hmong group inhabit the community forest in the region. According to Lee [43], the Hmong community resides in elevated regions with a moderate climate and abundant flora. These geographical areas provide the Hmong people opportunities for engaging in agricultural practices, animal rearing, hunting, gathering, and limited fishing.[64] The community in question exhibits exemplary forest management practices, although its agricultural methods continue to rely on slash-and-burn tactics. The agricultural productivity in this area has been adversely affected by a prolonged period of severe drought spanning many years, forcing people to migrate.[65] However, migration is often seen as a strategic response to address nutritional security and foster economic growth, a phenomenon observed throughout history, particularly in the case of Chiang Rai in 1905.[66] Based on the literature assessment, the hypothesis posits that migration in the region is influenced by climate change, specifically concerning its effect on agricultural productivity and food security.

The field study was done throughout the period spanning from May 2023 to August 2023. Two interviews were conducted for each selected home, with twenty interviewees in the designated region selected using purposive sampling. The participants were asked for permission to be interviewed with ethical approval from Mae Fah Luang University, Chiang Rai, Thailand. Each interview was conducted for 30–45 min. The sampling was selected according to preliminary research in 2019 when the author had conducted research on the 'Forest Act' movement in Thailand, and this community showed the author the example of the sustainable slash-and-burn practice. Due to this reason, continuous research on community forestry has been considered, and gradually, the topic has been taken with an overview of environmental sustainability in climate change. The interviewees were selected through the inclusion and exclusion criteria, such as (1) experiencing more than ten years in agriculture, (2) from the Hmong community, and (3) practicing traditional agricultural practices. The interview provided precise insights into viewpoints about the danger of climate change and migratory patterns associated with climate-related factors. Each successive interview provided an opportunity to the residents to engage in further introspection regarding their utilisation of local knowledge in adapting to climate change, thereby facilitating a more comprehensive understanding of the ramifications of drought induced by climate change within the framework of the migrant household's social, economic, political, and environmental necessities. The primary emphasis of these interviews was forest management and agricultural practices, with a secondary discussion on climate change.

This study analysed the interviews conducted with village respondents with content analysis to determine the prevailing patterns related to their perceptions of climate change danger, social relationships, and emigration patterns. Examining the topics pertaining to understanding climate change, adaptation, and migration strategy among migrants concerning social resilience yielded three primary themes: (1) the impact of drought on the community's forced migration patterns; (2) the disparities in perceptions of human mobility between older and younger generations in response to the drought; and (3) the factors influencing well-being, such as food security and adaptive capacity, in the context of climate-related drought. The objective of the thematic analysis was to ascertain the primary themes and sub-themes. The following section provides a comprehensive analysis of these three themes. To ensure the confidentiality of the participants, their names and identification details have been altered.

[63] Kunstadter et al. [42].

[64] Lee [43].

[65] Yamsiri [44].

[66] Lee (n 64) 12.

4.1 The Influence of Climate Change (Drought) on Hmong Community Forest Management

To fully grasp the impact of climate change on a community, it is crucial to first ascertain the perception of climate change within that group. This understanding is essential for comprehending the subsequent experiences and knowledge that shape their capacity to adapt.[67] The present study revealed the perception of the Hmong community concerning climate change, as shown in Table 1.

Table 1 Hmong's perceptions of climate change over the last 10 years (2012–2022)

Indicator	Frequency = 20	
	Yes	No
1. Noticing the climate change within ten years	15 (75%)	5 (25%)
2. Perceiving the longer slow-onset disaster occurrence	15 (75%)	5 (25%)
3. Perceiving the temperature increase	12 (60%)	8 (40%)
4. Predicting weather difficulty	11 (55%)	9 (45%)

Table 1 focuses on four critical aspects of the climate change perspective. Approximately 15 respondents of the Hmong population have seen the onset of climate change. The authors observed that climate change significantly impacts agricultural productivity and food security. Several respondents noted that there has been a decline of around 30% in agricultural productivity. The respondents noted that the circumstances necessitated an increased expenditure on acquiring paddies of rice. Consequently, the respondents indicated that buying rice in higher amounts increased their feelings of uneasiness, given that their food dependency relies on their agricultural output.

> Due to climate change, rice production has been consistently declining, which has had a significant impact on food security. As a result, we have had to purchase more rice or paddies from external sources, leading to increased expenditure on rice, said one of the respondents.

A small proportion, namely five respondents, have not seen the climate change phenomenon since they believe there has been no substantial alteration in air quality and environment. However, they acknowledged the reduction in the availability of food items. It is seen in their consistent emphasis on maintaining enough consumption within their households.

> In my household, I have an ample supply of rice from my field, although currently, I do not have a surplus that I can sell. However, I believe that the weather and surroundings remain unchanged. I feel the same, said one of the participants.

Similar to the community's observation of climate change, it has been shown that a significant majority, namely 15 respondents, perceived that the prolonged onset of a disastrous occurrence continues for a period that exceeds their memory from ten years ago. It has been observed that the duration of drought has increased, leading to significant implications for agricultural practices, such as the slash-and-burn technique. To engage in the practice of slash-and-burn agriculture for an extended period of around 30 days beyond the seasonal dry season, the resources must be plentiful. This time often occurs during the Songkran period, from March to April. As a result, the initiation of agricultural practices must be postponed, with concerns raised over the potential impact on livelihood security and income. Although a significant proportion of respondents perceive slow-onset disasters as being of longer duration, it is worth noting that a minority, including around 5%, have a different perspective on slow-onset disasters as being of longer duration. The respondents asserted that they continue to maintain adequate consumption levels and perceive the temperature in a manner consistent with their previous experiences. The continuation of agricultural practices enables farmers to sustain their agricultural output by cultivating alternative crops, such as vegetables, and traditional crops, like rice, corn, and maize.

---

[67] Manh and Ahmad [45].

In addition, the perception of temperature rise and unpredictability of weather has the most significant affirmative response rate of 60% and 55%, respectively. The respondents noted that they learn from climatic patterns, such as recurring rainfall in a cyclic pattern of 'one week of precipitation followed by one week of dryness.' The respondents noted that farmers have several challenges when changes in weather conditions impact their agricultural processes. Furthermore, the respondents involved in paddy cultivation have encountered weather changes and temperature fluctuations, which have considerably impacted the quality of rice produced and the quantity of paddy harvested. Additionally, respondents said they could foresee weather conditions by using their familiarity with local weather patterns.

Climate change has emerged as a significant determinant impacting the quality of life and overall well-being. The notable decline in food availability and security threatens their livelihood. Besides, Hmong communities residing inside or close to protected forests, including special-use forests such as national parks or nature reserves, are subject to stringent regulations restricting their legal entitlement to use non-timber forest products in these areas.[68] As a result of the limited economic prospects and choices available in mountainous regions, some of the Hmong people, particularly those who are economically disadvantaged, see forest exploitation, including the unauthorised cutting of wood, as a potentially feasible means of sustenance.

In contrast, most ethnic minority groups, especially the Hmong, who are local or traditional inhabitants, possess distinct perspectives and behaviours around climate change alongside their unique values and goals. These factors aid their potential to actively contribute to the mitigation and adaptation efforts associated with climate change.[69] Based on their familiarity with the local context and their position within the social hierarchy, these ethnic minorities might be categorised as a vulnerable group; however, their adaptation is reflected in the pattern of their agricultural activity.

It is to be noted that the preservation of a diverse ecosystem, landscape, and cultural and linguistic heritage is contingent upon traditional communities serving as custodians of resilient social-ecological systems. In the present scenario of forest management at the local level in Thailand, it is customary for officials at the village and commune levels to proactively safeguard their constituents against forest protection authorities. To ensure climate and political justice, it is imperative that the global community consistently prioritises the recognition of the rights of these communities.

4.2 The Age Difference in Migration Perspective

The historical account of Hmong migration to Chiang Rai reveals that the Hmong community has long included migration as an integral aspect of their experiences. The respondents in question have a good emotional disposition and are actively engaged in constructing a new community resembling their prior group via assimilating their cultures and traditions.[70] In the late nineteenth century, their migration to the Nan Province of Northern Thailand was induced potentially by livelihood needs, especially for the cultivation and sale of opium (*Papaver somniferum L.)* in the region.[71] Later, due to the destruction of the numerous opium fields by the Thai Army in 1967, a tremendous number of Hmong from the Nan province moved to the highlands of Laos. However in 1975, another Hmong migration influx of 116,000 people took place from Laos seeking refuge in Thailand, due to the Communist rule over Laos that continued until 1990.[72] In fact, the Hmong people reside in 13 northern provinces, with the most important

[68] Son and Kingsbury [46].

[69] Ishaya and Abaje [47].

[70] Jakae et al. [48].

[71] Culas and Michaud [49].

[72] Ibid.

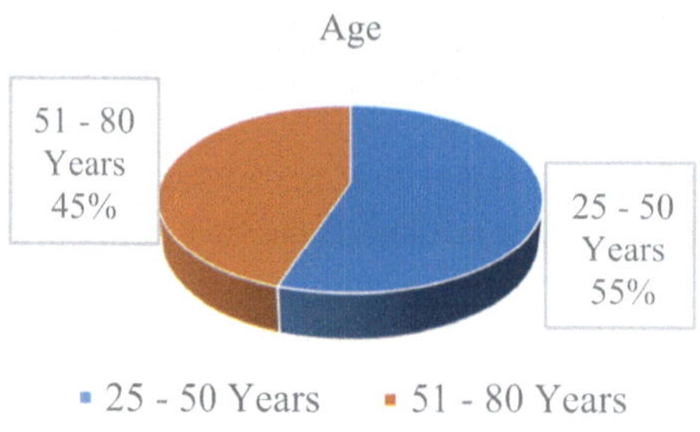

Fig. 2 Age differences of study

ones located in the mountainous area of Chiang Rai.[73]

In addition to discussing the perspective of climate change, this study also provides an in-depth analysis of the lived experiences of the respondents with respect to migration. The demographic characteristics of key respondents have been categorised into several age groups to enhance the perceptions about migration. Figure 2 presents the distribution of age groups, indicating that 11 respondents (55%) fall into the 25–50-year age range, while nine respondents (45%) belong to the 51–80-year age range.

Four primary points are emphasised on the causes influencing migration: quality of life, economic considerations, climate change, and adaptability. The majority of the respondents agreed with the notion that migration has positively impacted their standard of living and economic conditions. Throughout their migratory history, the respondents have engaged in the process of movement with the primary objective of cultivating crops for commercial endeavours. Indeed, the younger lot in the group agreed that climate change and consequent reduction in agricultural productivity contribute to migration. The researchers also discovered that a growing demand for resources and daily financial constraints prompted individuals to seek more secure employment opportunities actively.

> Climate change has a substantial influence on migration. I need relocation from my current community in order to establish a stable source of income. There are abundant job opportunities not only within the city but also in other provinces, said one of the respondents.

In contrast, some others in the younger generation refrain from engaging in migration. They continue to hold the belief that their agricultural endeavours provide enough security. The respondents, especially young women, stated that a significant portion of their existence is dedicated to this geographic region.

> As a woman, it is more advantageous for me to remain in the village despite the fact that climate change has had a greater impact on agricultural production. For my home, the current amount is still enough for consumption. In order to ensure a stable income, I engage in small-scale entrepreneurship, such as creating handmade crafts and selling them, said one of the participants.

Furthermore, the authors observed that older people do not consider nature as a danger to them but try to preserve the forests and advocate living in harmony with nature and its ecosystems.

> We hold the belief that nature is essential for our existence. We must comprehend the functioning of nature because nature bestows onto us the gift of life as well. We must save the environment since, without it, we cannot pass on a legacy to future generations. Protecting nature included safeguarding other organisms, ecosystems, and biodiversity, said the older groups.

Given that a majority of the younger demographic agrees with migration as a means to increase their quality of life, it is evident that they explicitly acknowledge the potential of migration to bolster their ability to adapt and demonstrate resilience. The individual's ability to generate revenue via a secure and consistent employment opportunity has transformed into an enduring and adaptable capability. For instance, individuals who relocate to other regions for employment may be able to engage in entrepreneurial endeavours. Indeed, the process of their movement is operationalised inside the internal region of Thailand. The respondents indicated their intention to relocate for employment during the dry season, with plans to return

[73] *The Hill Tribes of Thailand.* Chiang Mai: (Tribal Research Institute 1995).

to their residences during the planting season or in May. Thus, the migration patterns (outward and inward) operate on an annual basis.

The older Hmong population prefers maintaining their residence within the area. They assume the role of forest custodians and depend on the economic use of forests for agricultural and produce cultivation. After engaging in seasonal crop cultivation, the primary agricultural outputs consist of soybeans, red beans, green beans, maize, peanuts, and other similar crops. In addition, they use community tourism initiatives to promote their products, enhancing their revenue stream. It is posited that, despite the frequency of drought occurrences, many of the older respondents in question believe that migration will not be pursued, as they possess the ability to adapt to the effects of climate change. The Hmong community can use their traditional or local knowledge to adapt and thrive within the context of the Anthropocene period. This group exhibits a notably positive outlook towards their circumstances. Using local forest management experience is seen as a highly productive strategy for enhancing adaptability and resilience instead of relying only on migration.

This study aims to understand the contrasting nature of the migration strategy among the older and younger populations of the Hmong community, specifically in terms of its resilience or vulnerability. Furthermore, the contrasting views of older and younger cohorts within the Hmong community regarding migration give rise to the need to make choices about one's life trajectory. The older generation does not see migration as a viable response to climate change. In contrast, the younger demographic places a higher emphasis on migration to achieve economic benefits rather than seeing it primarily as a reaction to climate change.

Overall, the research has expounded upon the impact of drought as a gradual-onset catastrophe on the compelled migration inside the Hmong community, specifically regarding their internal migration patterns as opposed to relocating to alternative metropolitan regions in Thailand. Migration has emerged as a viable option for the younger generation of the Hmong community, as they want to attain financial stability. However, it should be noted that many do return to their homeland during the agricultural planting season. The individual's happiness is centred on food security and the ability to adjust to changes in their crop patterns and explore other crops.

5 Conclusion

This study aimed to explore the existing climatic knowledge of the Hmong community residing in Wiang Kaen, Chiang Rai Province, Thailand. The agricultural production in the area has been adversely affected by the current drought, resulting in food insecurity and a subsequent rise in economic activity. The use of migration or human mobility strategies are applied as a means to address existing issues. The Hmong population sees climate change as a significant issue in their agricultural productivity and patterns. There exists a considerable divergence in viewpoints on climate change between the older and younger cohorts. The elderly Hmong population is endeavouring to adapt their traditional forest management strategies in response to climate change despite the prevalence of drought, over migration. The prominence of migration as a viable choice for young Hmongs seeking secure jobs and income is driven mainly by economic concerns rather than climate change per se, which refers to pattern two of Koko Warner's human mobility emphasising weather dependency within the domestic rural–urban movement in temporary adaptation mentioned by Carl Middleton et al. and Louis Lebel. Consequently, the study's findings and subsequent discussion provide a detailed analysis of the contrasting situations between the Hmong community members who have migrated and those who have yet to ascertain the resilience or vulnerability of migration strategies. It was also seen that the migratory patterns present among the Hmong community show that many young people continue to return to their homes to engage in agricultural activities. Furthermore, this study suggests the need for further research on the cultural capital dimension of migration and its impact on life satisfaction. This will contribute to a more thorough understanding of the migratory patterns within this community, particularly with respect to the effects of climate change, which have not been well addressed.

Acknowledgements This study is funded by Mae Fah Luang University Research Fund 2023.

References

1. Mark D (2011) Climate change and Thailand: impact and response. Contemp Southeast Asia 33(2):229–258
2. Homdee T et al (2016) A comparative performance analysis of three standardized climatic drought indices in the Chi River Basin, Thailand. Agric Nat Res 50(3):211–219
3. Garbero A, Muttarak R (2013) Impacts of the 2010 droughts and floods on community welfare in rural Thailand: differential effects of village educational attainment. Ecol Soc 18(4):27
4. Seeboonruang U (2016) Impact assessment of climate change on groundwater and vulnerability to drought of areas In Eastern Thailand. Environ Earth Sci 75:1–13
5. Burrows K, Kinney PL (2016) Exploring the climate change, migration and conflict Nexus. Int J Environ Res Public Health 13(4):443
6. Lebel L et al (2016) Impacts, perceptions, and management of climate-related risks to cage aquaculture in the reservoirs of Northern Thailand. Environ Manag 58:931–945
7. de Guttry C et al (2016) Challenging the current climate change—migration Nexus: exploring migrants' perceptions of climate change in the hosting country. J Geograph Stud 147(2):109–118
8. UPI (2010) Thailand: drought disaster in 53 Provinces. In: World News. http://www.upi.com/Top_News/World-News/2010/03/26/Thailand-Drought-disaster-in-53-provinces/UPI-51861269634544/. Accessed on 25 Apr 2023
9. Calzadilla A et al (2013) Climate change impacts on global agriculture. Clim Change 120:357–374
10. Zander KK et al (2019) Human mobility intentions in response to heat in urban South East Asia. Glob Environ Change 56:18–28
11. Curran SR, Meijer-Irons J (2014) Climate variability, land ownership, and migration: evidence from Thailand about gender impact. Washington J Environ Law Policy 4(1):37–74
12. Warner K, Afifi T (2014) Where the rain falls: evidence from 8 countries on how vulnerable households use migration to manage the risk of rainfall variability and food insecurity. Clim Dev 6(1):1–17
13. Neil Adger W et al (2014) Human security. In: Barros VR, Field CB (eds) Climate change 2014 impacts adaptation and vulnerability part global and sectoral aspects, contribution of working group II to the fifth assessment report of the intergovernmental panel on climate change. Cambridge University Press, p 755
14. Lebel L (2013) Local knowledge and adaptation to climate change in natural resource-based societies of the Asia-Pacific. Mitigat Adapt Strategy Glob Change 18:1057–1076
15. Sirimongkonlertkul N et al (2013) Multi-temporal analysis of haze problem in Northern Thailand: case study in Chiang Rai Province. Kasertsart J (Nat Sci) 47:768–780
16. Carew-Reid J, Taylor L (2014) Natural systems and climate change resilience in the lower Mekong Basin: future directions for biodiversity, agriculture and livelihoods in a rapidly changing environment. ICEM, p 15
17. Forsyth T, Evans N (2013) What is autonomous adaptation? Resource scarcity and smallholder agency in Thailand. World Dev 43:56–66
18. Hu J (2010) The role of health insurance in improving health service use by thais and ethnic minority migrants. Asia-Pacific J Public Health 22(1):42–50
19. Tongdhamachart N, Alwi A (2023) The cultural identity of mien ethnic group in a digital era. Int J Prof Bus Rev 8(1):1–15
20. Jantakad S (2022) From state multiculturalism to multiculturalism from below: a case study of Chiang Rai flower festival and the cultural square. J Mekong Soc 18(2):165–190
21. Oupra S (2009) Language ecology and language planning in Chiang Rai Province, Thailand. University of Adelaide
22. Culas C, Michaud J (1997) A contribution to the study of Hmong (Miao) migrations and history. Bijdragen tot De Taal-, Land-en Volkenkunde 2:211–243
23. Tatman AW (2004) Hmong history, culture and acculturation: implications for counseling the Hmong. J Multicult Counsel Dev 32:222–233
24. Warner K (2018) Coordinated approaches to large-scale movements of people: contributions of the Paris agreement and the global compacts for migration and on refugees. Popul Environ 39:384–401
25. Sakdapolrak P, Naruchaikusol S, Ober K, Peth S, Porst L, Rockenbauch T, Tolo V (2016) Migration in a changing climate. Towards a translocal social resilience approach. DIE ERDE–J Geogr Soc Berl 147:(2):81–94
26. Solomon MK, Sheldon S (2018) The global compact for migration: from the sustainable development goals to a comprehensive agreement on safe, orderly and regular migration. Int J Refugee Law 30(4):584–590
27. Klepp S (2017) Climate change and migration. In: Oxford research encyclopedia of climate science
28. Deshingkar P (2012) Environmental risk, resilience and migration: implications for natural resource management and agriculture. Environ Res Lett 7(1):015603

29. Bryan G et al (2014) Under-investment in a profitable technology: the case of seasonal migration in Bangladesh. Econometrica 82 9(52):1671–1748
30. Middleton C et al (2017) Migration and floods in Southeast Asia: a mobile political ecology of vulnerability, resilience and social justice. In: Living with floods in a mobile Southeast Asia, pp 1–21
31. Lebel L (2013) Climate change and migration in South-East Asia: evidence, expectations and new questions. On the Move: Critic Mig Themes ASEAN 12(24):24–49
32. Middleton C (2022) The political ecology of large hydropower dams in the Mekong Basin: a comprehensive review. Water Alternat 15(2):251–289
33. Khoa VN et al (2012) Rainfall, food security, and human mobility: case study Viet Nam. UNU-EHS Rep 8:1–98
34. Tacoli C (2011) Not only climate change: mobility, vulnerability and socio-economic transformations in environmentally fragile areas of Bolivia, Senegal, and Tanzania. Human Settlements Work Pap Ser—Rural-Urban Interact Livelihood Strat 28:589
35. Klasen S et al (2016) Vulnerability to poverty: theory, measurement and determinants, with case studies from Thailand and Vietnam. Springer, p 4
36. Locatelli B et al (2011) Forests and climate change In Latin America: linking adaptation and mitigation. Forests 2(1):431–450
37. Chamura DJ et al (2011) Forest responses to climate change in the Northwestern United States: ecophysiological foundations for adaptive management. For Ecol Manag 261(7):1121–1142
38. Charoenratana S et al (2021) Food sovereignty and food security: livelihood strategies pursued by farmers during the maize monoculture boom in Northern Thailand. Sustainability 13(17):9821
39. Hallegatte S et al (2016) Shock waves: managing the impacts of climate change on poverty. World Bank Publications, p 12
40. Phungpracha E et al (2016) Traditional ecological knowledge in Thailand: mechanisms and contributions to food security. Kasetsart J Soc Sci 37(2):82–87
41. Sang-Arun J et al (2016) Promoting plant residue utilization for food security and climate change mitigation in Thailand. Sustain Food Water Asian Perspect 343–352
42. Kunstadter P et al (2019) Farmers in the forest: economic development and marginal agriculture in Northern Thailand. University of Hawaii Press
43. Lee GY (2005) The shaping of traditions: agriculture and Hmong society. Hmong Stud J 6(1):1–33
44. Yamsiri T (2014) Water management in Thailand: dams and the voice of the affected and displaced people. Stated Environ Migr 236–250
45. Manh NT, Ahmad MM (2021) 'Ethnic minority farmers' perceptions and use of local knowledge to adapt to climate change: some insights from Vietnam. Singap J Trop Geogr 42(3):397–414
46. Son H, Kingsbury A (2019) Community adaptation and climate change in the northern mountainous region of Vietnam: a case study of ethnic minority people in Bac Kan Province. Asian Geogr 37(1):33–51
47. Ishaya S, Abaje IB (2008) Indigenous people's perception on climate change and adaptation strategies in Jema'a local government area of Kaduna State, Nigeria. J Geogr Reg Plan 1(8):138–143
48. Jakae O et al (2020) Construction of social space in Thai Hill tribe ethnic groups in Kamphaeng Phet Province. Kasetsart J Soc Sci 41(2):329–335
49. Culas C, Michaud J (1997) A contribution to study of Hmong (Miao) migrations and history. KITLV 2:211–243

Reni Juwitasari is a researcher at Disaster Resilience and Environmental Sustainability (DRES), the Asian Research Center for International Development (ARCID), School of Social Innovation at Mae Fah Luang University. She holds an M.Ed. degree in Educational Administration from the Faculty of Education, Mahasarakham University (Thailand). She is a Ph.D. student majoring in Development Science, Faculty of Humanities and Social Sciences, Khon Kaen University, Thailand. She has published several book chapters, academic journals, and articles on Disaster, Education, and Traditional Knowledge related to Sustainable Development and Sustainability in Global South. She is now working on her research topic about community resilience, education for sustainable development (ESD) and sustainability (EFS), and local knowledge for disaster and climate change, including ECO-DRR under Sumitomo Foundation, JSPS, MFU Research Fund and UNESCO-Chair (Kobe). She hopes her work can contribute to reviving "Cosmolocal" to preserve local knowledge for sustainability achievement.

Sensitisation of Disaster Relief Operations Towards Persons with Disabilities

Ishari Gunarathna and Priyadarshani Premarathne

Abstract

Persons with disabilities (PWDs) often face societal exclusion, particularly during disasters when communities become more vulnerable and reliant on external assistance. PWDs, who are already at higher risk, experience heightened challenges. They usually rely on caregivers, such as family, neighbours, or government support, for daily assistance. However, even these support networks become vulnerable when disasters strike, increasing dependence on government and external relief. During times of disasters, the process of evacuating individuals to relief camps establishes reciprocal relationships between PWDs and relief providers. In the realm of disaster management, Persons with Disabilities (PWDs) go beyond being passive recipients; they actively contribute to rescue missions alongside their families. Particularly in Sri Lanka, where advanced tools for those with mobility impairments are scarce, PWDs take on a hands-on role. They guide support providers lacking specific rescue training, offering instructions on how to assist and meet their support expectations. This active participation exemplifies what we term 'reciprocal relationships between PWDs and relief providers.' This mutual support is of paramount importance since the entire population is deemed at risk, and collaborative efforts significantly boost the efficiency of evacuation endeavours and the provision of support services. However, it is essential to acknowledge the distinctive vulnerabilities and interdependencies present within the disability community. This recognition emphasises the critical necessity for inclusive disaster preparedness measures. While prior studies have addressed the vulnerabilities of PWDs during and after disasters, a critical knowledge gap exists in understanding these issues in Sri Lanka. This chapter investigates the challenges of PWDs during and after disasters, explores the difficulties of service providers in assisting PWDs, and emphasises the need for inclusive disaster preparedness, highlighting the significance and role of social workers in emergencies in Sri Lanka. The study utilised a qualitative research design, conducting twenty in-depth interviews in two disaster-prone districts in Sri Lanka, Kandy, and Kegalle. The findings reveal that PWDs face a range of challenges, such as limited access to information, inaccessible shelters, transportation barriers, loss of assistive devices, and insufficient

I. Gunarathna (✉) · P. Premarathne
Department of Sociology, University of Peradeniya, Peradeniya, Sri Lanka
e-mail: ishari2015@arts.pdn.ac.lk

P. Premarathne
e-mail: pri.hapuarachchi@arts.pdn.ac.lk

S. Jolly et al. (eds.), *Climate-Related Human Mobility in Asia and the Pacific*, Sustainable Development Goals Series, https://doi.org/10.1007/978-981-97-3234-0_4

healthcare access. The Framework for Integrating Rights and Equality (FIRE) offers crucial recommendations to comprehensively address these issues. Service providers, confronted with resource constraints, accessibility problems, communication barriers, inadequate training, low trust levels, and policy gaps, must integrate these principles for improved disaster preparedness and response. This involves prioritising resource allocation, enhancing accessibility, improving communication, investing in training, fostering trust, and addressing policy gaps, aiming to create a more inclusive and resilient system that better serves the needs of PWDs during emergencies and identifies the importance and role of social workers in emergencies in Sri Lanka.

1 Introduction

Sri Lanka, a small island nation in the Indian Ocean with an area of 65,000 km^2, boasts a wet and warm climate that was once blanketed by lush forests, rich in biodiversity. Despite significant deforestation, the country's abundant natural resources and diverse cultural heritage support a thriving tourism industry. With a population of approximately 22 million, Sri Lanka's multi-ethnic and multi-cultural society adds to its cultural vibrancy. Nevertheless, the nation's vulnerability to climate change is amplified by its high temperatures, complex hydrological system, and exposure to extreme climate events.[1] The more significant hazards experienced include floods, cyclones/high winds, droughts, landslides, and lightning.[2]

Sri Lanka has been increasingly affected by climate-induced disasters, resulting in significant economic and social losses. For instance, in 2017, devastating floods and landslides caused several fatalities and displaced half a million people.[3] In 2023, the United Nations Climate Impact Fact Sheet highlighted Sri Lanka's high vulnerability to climate change. Despite its relatively elevated Human Development Index compared to numerous other developing nations, the country consistently ranks among the top ten at risk of extreme weather events according to the Global Climate Risk Index.[4] Sri Lanka faces recurrent climate vulnerabilities, marked by excessive rainfall and frequent droughts. These droughts can be categorised into meteorological droughts, which are linked to reduced precipitation, and hydrological droughts, resulting from diminished surface and sub-surface water flows.[5] The impacts of these droughts ripple across various domains, affecting development, water resources, agriculture, and land management practices. As a result, these vulnerabilities exacerbate livelihood challenges, poverty levels, and overall public health within the country. The 2021 Climate Risk Country Profile from the World Bank underscores Sri Lanka's significant exposure to extreme heat, with particular consequences for urban outdoor labourers who work in demanding conditions.[6] Communities in vulnerable areas, especially in the northern regions, face heightened risks, which harm farmers, household food security, and the tourism sector.[7]

Moreover, projections indicate an amplified threat due to a lack of adaptive measures, leading to more frequent and intense extreme precipitation events. This escalation heightens vulnerability, impacting individuals, livelihoods, and infrastructure, and fostering waterborne

[1] Climate Change Overview (World Bank Country Summary) Climate Change Knowledge Portal. https://climateknowledgeportal.worldbank.org/country/sri-lanka. Accessed on 12 July 2023.

[2] National Disaster Management Plan (The Disaster Management Centre, State Ministry of National Security and Disaster Management 2022) 23–24.

[3] Sri Lanka Rapid Post Disaster Needs Assessment (Ministry of Disaster Management and United Nations Development Programme 2017) 182.

[4] Climate Impact Sri Lanka (United Nations, Sri Lanka) UN Sri Lanka Climate Change Fact Sheet. https://srilanka.un.org/en/254230-fact-sheet-climate-impact-sri-lanka. Accessed on 15 January 2024.

[5] Chandrasekara et al. [1].

[6] Climate Risk Country Profile: Sri Lanka (World Bank Group 2021) 11.

[7] Ibid., 6.

disease transmission, ultimately exacerbating poverty and inequality. The diverse array of disasters occurring throughout the year in Sri Lanka impacts different segments of the population in varying ways, dependent on their level of vulnerability. The consequences of climate change and disasters reverberate across multiple sectors, influencing agriculture production, water management, energy generation, food security, and public health.[8]

Some studies indicate that climate change tends to disproportionately affect the most economically disadvantaged groups.[9] For instance, heavy manual labourers are highly vulnerable to productivity losses due to heat stress. The extent of the impact and the coping strategies employed by individuals are strongly influenced by various factors, including socio-economic status, household composition, disability, socio-cultural norms, access to resources, poverty, and gender.

Research has also demonstrated that the effects of disasters are not gender-neutral, with women and children constituting some of the most at-risk groups. Gender-based differences in time use, access to assets and credit, treatment by formal institutions, limited participation in policy discussions and decision-making, and the absence of sex-disaggregated data for policy changes all contribute to this differential impact. Given the non-neutral nature of disaster impacts, it is imperative to assess social vulnerabilities within communities, especially among specific populations such as individuals with disabilities.[10]

A recent multi-dimensional vulnerability study revealed that households with Persons with Disabilities (PWDs) exhibit significantly higher vulnerability, with 60.4% classified as vulnerable compared to 54.4% among households without PWDs.[11] In other words, this Multi-dimensional Vulnerability Index (MVI) assigns numerical values to vulnerability, and in this context, households with Persons with Disabilities (PwDs) have a higher MVI (0.229) than those without PWDs (0.200). This indicates that households who had PWDs face elevated vulnerability across various dimensions. The notable contrast in vulnerability rates highlights the substantial impact of disability on overall vulnerability. This emphasises the necessity for comprehensive policies and support systems to effectively address the challenges encountered by households with PWDs. In fact these findings underscore the fact that inclusive policy measures and robust social protection systems will help combat this deep impact of disability on overall vulnerability. These measures should prioritise enhancing livelihood opportunities for PWDs and their families, fostering inclusivity and equity within society.[12]

These existing researches do not adequately address the crucial aspects of enhancing resilience among PWDs. Specifically, the collection of sex/age/disability-disaggregated data to inform better planning, resource allocation, and the targeting of service delivery for PWDs, ensuring their full social and economic participation, remains underexplored. Additionally, there is a lack of investigation into the development of tailored programmes that facilitate PWDs' access to employment opportunities in both the public and private sectors, as well as their engagement in policy formation.

Therefore, it is imperative to map the vulnerabilities of PWDs during and after disasters, identify the challenges encountered by service providers in disaster preparedness and relief programmes, and analyse the role of socio-cultural factors in promoting sensitisation and the development of an inclusive framework. This research encompasses all five phases of emergency management, including prevention, mitigation, preparedness, response, and recovery.

This chapter explores avenues for sensitising disaster relief operations towards Persons with Disabilities in Sri Lanka during and immediately after disasters. Focussing on the sensitivity of service personnel, the study maps various

[8] Ibid., 22.

[9] Kjellstrom et al. [2].

[10] Gender Equality, Poverty Reduction, and Inclusive Growth (World Bank Group 2016) 33.

[11] Jubota and Alkire [3].

[12] De Silva and Kawasaki [4].

vulnerabilities experienced by PWDs and service providers. The specific sub-objectives of this chapter include:

- Investigating the challenges faced by PWDs including issues of their mobility during and after disasters.
- Identifying the difficulties and issues encountered by service providers in their interactions with PWDs in the disaster management process.
- Identifying strategies for the inclusion of PWDs in disaster management operations and sensitisation among officials.
- Recognising the role of social workers in facilitating PWDs in disaster management.

This empirical study, conducted using a qualitative research design, was carried out in two critical and disaster-prone districts, Kandy and Kegalle, as identified by the National Building Research Organisation (NBRO)[13] in Sri Lanka. (Note: The NBRO's administrative, financial, and managerial functions are managed by the Cabinet-approved Interim Management Committee (IMC), led by the Secretary of the Ministry of Disaster Management. The Director General acts as the organisation's CEO, overseeing operations in seven key topic areas).

The study involved ten in-depth interviews with PWDs, focusing on the challenges and issues within current disaster relief operations, particularly in the context of evacuation and post-disaster displacement. Adaptations were made to accommodate children with different disabilities and service requirement levels, including the use of lip-reading and sign language professionals for data collection alongside observations. Additionally, ten semi-structured interviews were conducted with professionals actively involved in working with PWDs during disasters in 2022 and 2023. These professionals represented government and non-governmental organisations dedicated to safeguarding the rights of PWDs. Ensuring the rights of Persons with Disabilities (PWDs) holds great significance within the framework of Article 12(2) of the Constitution of Sri Lanka, which explicitly prohibits discrimination against any citizen based on factors such as race, religion, language, caste, sex, political opinion, place of birth, or similar grounds. Organisations are obligated to uphold and safeguard these rights, with a particular emphasis on protecting the rights of PWDs.[14]

Furthermore, the Sri Lankan Constitution, specifically Article 12(3), guarantees the rights of PWDs, emphasising the elimination of discrimination in their access to resources. Additionally, they uphold Article 12(3) of the Sri Lankan Constitution, focusing on non-discrimination with respect to access. Moreover, following Article 11 of the National Policy on Disaster Management in Sri Lanka (2013), these organisations advocate for maintaining equity in resource distribution during disasters, prioritising vulnerable groups such as children, differently-abled individuals, senior citizens, and women, among others, based on the severity and vulnerability of the situation. Furthermore, the professionals interviewed represented the Ministry of Disaster Management, the National Building Research Organisation, the Divisional Secretariats of Kegalle and Kandy, We for Rights Organisation, and The Employer's Federation of Ceylon. The study strictly adhered to research ethics, ensuring data privacy and confidentiality.

This chapter comprises four sections: an introduction, a brief analysis of disaster, disability, and disaster management in Sri Lanka, a presentation of the study results in line with the objectives and discussion, and a concluding section that addresses implications and recommendations.

[13] National Building Research Organisation (NBRO) Sri Lanka.

[14] The Constitution of the Democratic Socialist Republic of Sri Lanka as amended up to 31 October 2022, revised edition 2023.

2 Disaster, Disability, and Disaster Management

Examining the relationship between disasters, disability, disaster management, and Disability-Inclusive Disaster Risk Reduction is complex. This complexity is aggravated by the disproportionately severe impact of disasters on economically disadvantaged nations, primarily due to their heavy reliance on natural resources and limited resilience to climate variations and extremes. In this section, we briefly touch upon the terminologies discussed in this chapter.

Disaster is defined as a profound disturbance to a community or societal functioning due to hazardous events interacting with conditions of exposure, vulnerability and capacity, resulting in losses across various aspects (human, material, economic, and environmental).[15] Disasters can have immediate or widespread impacts, potentially overwhelming a community's self-reliance and requiring external assistance. As previously mentioned, Sri Lanka is ranked among the top three in the Global Climate Risk Index published by Germanwatch in 2019, with events like floods, landslides, cyclones, and droughts contributing to its status as a multi-hazards hotspot.[16] The UNDP Climate Fact Sheet highlights that climate change poses a significant threat, exacerbating challenges related to food and water availability while diminishing agricultural outputs. The data reveals that 64.4% is attributed to floods, 10.3% to droughts, 3.5% to landslides, and 21.9% to cyclones, all of which have adversely affected the lives and livelihoods of individuals.[17] This has resulted in a decline in well-being, necessitating recovery efforts for housing, land, and climate-induced mobility and settlement for the affected population.[18]

Persons with Disabilities are identified as individuals who have long-term physical, mental, intellectual, or sensory impairments that, in interaction with various barriers, may hinder their full and effective participation in society on an equal basis with others.[19,20] In this chapter, we utilise the International Classification of Functioning, Disability, and Health (ICF), a WHO-developed framework offering a universally accepted system to describe and classify health, emphasising the interplay between an individual's health and their environment. The ICF illustrates the bio-psychosocial model of disability, explaining the interaction between body functions and structure, activity, and participation components. Additionally, it establishes a standard language and conceptual basis for defining and measuring health and disability, guided by four principles: universality, parity, etiological neutrality, and environmental influence.

Disability encompasses not only reduced mobility but also a broader range of conditions, including hearing impairments, visual impairments, intellectual impairments, and various forms of physical disabilities, among others. Regardless of whether a disability is permanent or temporary, PWDs face several challenges, such as mobility problems, visual or auditory impairments, cognitive difficulties, communication barriers, medical issues, and the need for support systems to optimise functioning. They also encounter challenges related to environmental sensitivities, psychiatric disturbances, panic attacks, and infirmity associated with old age (See Footnote 20). PWDs are disproportionately vulnerable during disasters and often

[15] Sendai Framework Terminology on Disaster Risk Reduction (UNDRR). https://www.undrr.org/terminology#D. Accessed on 20 April 2023.

[16] David Eckstein, and others 'Global Climate Risk Index 2019' (Germanwatch 2019) AS A SINGLE WORD 6.

[17] United Nations Fact Sheet: Climate Impact in Sri Lanka (November 2023). https://srilanka.un.org/en/254230-fact-sheet-climate-impact-sri-lanka. Accessed on 30 November 2023.

[18] Ibid., 16.

[19] UN General Assembly, Convention on the Rights of the Persons with Disabilities. 2007, Article 01.

[20] Disability-Inclusive Disaster Risk Reduction and Emergency Situations (UN DESA, Disability). https://www.un.org/development/desa/disabilities/issues/disability-inclusive-disaster-risk-reduction-and-emergency-situations.html. Accessed on August 30 2023.

face additional challenges, as exemplified by the 2015 earthquake in Nepal, where they received less support and were at a higher risk of violence in shelters.[21]

Disaster management encompasses organisation, planning, and the application of measures for preparing, responding to, and recovering from disasters.[22] Disaster risk reduction aims to prevent, reduce, and manage disaster risk, enhancing resilience and sustainable development.[23] Disability-Inclusive Disaster Risk Reduction (DiDRR) is a human rights-driven approach with the overarching goal of ensuring equitable access for individuals with disabilities to community-level Disaster Risk Reduction (DRR) efforts.[24] By fostering collaboration between emergency management and community service sectors, along with the active involvement of people with disabilities and their representatives at the local level, DiDRR strives to enhance the resilience of individuals with disabilities in the face of disasters. It seeks to integrate their needs and rights into broader disaster management and risk reduction strategies, making these efforts more comprehensive and equitable. The DiDRR recognises that people with disabilities are often at heightened risk during disasters and aims to address these vulnerabilities, ultimately contributing to the development of more resilient and inclusive communities.[25]

In this scenario, before assessing the need for prioritisation among relief personnel, it is essential to examine the current legal and policy measures to determine the extent to which the protection of PWDs is mandated and prioritised within the existing disaster management frameworks.

3 Disaster Management: Governing Policies and Frameworks

Disaster management activities in Sri Lanka primarily revolve around the government sector, making it the key stakeholder. Despite this central role, the system is effectively decentralised, with interventions extending from the smallest unit, such as the Grama Niladari division, to provincial and national levels. The involvement of non-governmental organisations (NGOs) and community-based organisations is widespread, providing substantial support for various activities. Government officials play a pivotal role in overseeing and monitoring these activities. Therefore, in Sri Lanka, the government is considered the primary stakeholder in disaster management and resilience. Sri Lanka has established a comprehensive set of laws, policies, and frameworks to address issues related to disasters, disabilities, and climate change. These efforts align with international conventions and treaties aimed at creating a disaster risk-free environment for communities and building resilience.

Concerning PWDs, notably, Sri Lanka is a signatory to the United Nations Convention on the Rights of People with Disabilities (UNCRPD). Furthermore, Article 12 of the Constitution of the Democratic Socialist Republic of Sri Lanka enshrines the right to equality and protection against discrimination as fundamental rights, providing a basis for safeguarding the rights of individuals with disabilities.[26]

The Protection of the Rights of Persons with Disabilities Act, No. 28 of 1996 in Sri Lanka, which predates the CRPD by a decade, holds significance in addressing discrimination against individuals with disabilities.[27] However,

[21] Ilyas [5].

[22] Ibid.

[23] Ibid.

[24] Ibid.; Background Paper for Regional Consultation on Facilitating Innovative Action on Disability-inclusive and Gender-responsive DRR, 'Review of Disability-inclusive and Gender-responsive Disaster Risk Reduction in Asia and the Pacific' (United Nations 2022) 7.

[25] Villeneuve et al. [6].

[26] The Constitution of the Democratic Socialist Republic of Sri Lanka as amended up to 31 October 2022, revised edition 2023 (n 14) Article 12.

[27] The Protection of the Rights of Persons with Disabilities Act, No. 28 of 1996 in Sri Lanka.

it exhibits several limitations when it comes to safeguarding the rights and welfare of this vulnerable population within disaster contexts. The Act has not evolved to encompass contemporary international standards and best practices, particularly those reflected in the CRPD. While the Act rightly focuses on combating discrimination in areas such as employment and access to educational institutions, it inadequately addresses disaster preparedness, which is specifically addressed at Article 11 CRPD. The Act lacks comprehensive disaster-related provisions and does not explicitly mention disaster risk reduction (DRR) or measures aimed at reducing the vulnerabilities of Persons with Disabilities to disasters. There is no clear guidance on how to integrate Disability- Inclusive Disaster Risk Reduction (DiDRR) into the existing legal framework, leaving a gap in the protection of this group in the event of disasters. To rectify this, legal and policy reforms are imperative. These reforms should align with contemporary international standards, particularly the CRPD, to ensure the inclusivity, accessibility, and overall well-being of individuals with disabilities during disasters, in line with Sri Lanka's international commitments.

Moreover, Sri Lanka operates within a legal framework for disaster risk management that encompasses several key components, which are outlined below:

A. Sri Lanka Disaster Management Act, No. 13 of 2005

This Act aims to establish a legal framework for managing disaster risks in Sri Lanka, detailing the legal responsibilities, powers, and functions of the authorities involved in disaster management. The Act establishes the National Council for Disaster Management, District Disaster Management Committees, and Disaster Management Centres. However, the Sri Lanka Disaster Management Act does not specifically address the unique needs and concerns of PWDs.[28]

2 National Policy on Disaster Management 2013

This policy supplements the Sri Lanka Disaster Management Act and aligns the country's disaster management efforts with international best practices. It emphasises disaster risk reduction, community participation, gender mainstreaming, and the importance of technological advancements in disaster management. While the policy acknowledges the need for a multi-dimensional approach to disaster management for vulnerable groups (considered an umbrella term covering women, children, and disabled people, as well as the elderly), it doesn't explicitly elaborate on how the concerns of PWDs will be addressed.[29]

3 Institutional Disaster Management Plan 2022–2030

Crafted to address distinct disaster management requirements in educational institutions, this plan delineates preparedness, response, and recovery measures to guarantee the safety of students and staff during disasters. Additionally, it offers guidelines for preparation applicable to all ministries, government institutions, departments, and agencies. Yet, it does not provide specific provisions for PWDs in educational settings.[30]

4 National Disaster Management Plan 2019–2030

In Sri Lanka, various legal frameworks, including the 2005 Disaster Management Act, have played a crucial role in disaster risk management. The National Disaster Management Plan (NDMP) (2013–2017) aimed at minimising disaster impact but lacked specific measures for PWDs. Similarly, the National Emergency Operation Plan 2013–2017 prioritised coordination but again lacked PWD-specific provisions. The Sri Lanka Comprehensive Disaster Management Programme (SLCDMP)

[28] Disaster Management Act No 13 of 2005, Sri Lanka.

[29] National Policy on Disaster Management 2013, Sri Lanka.

[30] Institutional Disaster Management Plan 2022–2030 (Disaster Management Center), Sri Lanka.

(2014–2018) aimed at enhancing safety but did not fully address concerns of PWDs. The National Disaster Management Plan 2019–2030 now takes precedence, highlighting the need to update approaches in line with current priorities and challenges.

The National Disaster Management Plan strives to significantly increase the adoption of national and local DRR strategies following the Sendai Framework of Disaster Risk Reduction (SFDRR).[31] It encompasses all aspects of disaster risk reduction, placing a strong emphasis on localising DRR strategies by 2020 to align with the SFDRR and the 2030 Agenda for Sustainable Development.[32] The plan establishes precise objectives, goals, indicators, and timeframes aimed at preventing, reducing, and enhancing resilience against disaster risks. Furthermore, it plays a crucial role in monitoring the progress of these policies and plans, particularly in supporting local governments in their formulation and execution.[33] It is worth noting however that the text does not explicitly mention provisions for persons with disabilities (PWDs), even though Paragraph 19(d) of the SFDRR expressly calls for addressing their specific situation. Nevertheless, the plan likely contributes to Community-Based Disaster Risk Management (CBDRM) by localising strategies and bolstering resilience at the community level, aligning with the priorities outlined in the SFDRR.[34]

The National-level Disaster Management Plan in Sri Lanka operates across various levels, employing a multi-tiered approach to address diverse aspects of disaster preparedness, response, and recovery. This plan involves coordination and collaboration among national, provincial, and district-level institutions/agencies, showcasing a vertical integration of disaster management efforts.

Vertical Disaster Management Planning in Sri Lanka refers to hierarchical coordination and collaboration across different administrative levels, from the national government down to the provincial, district, divisional, Grama Niladari (Head of the Village), and community levels. This approach ensures that disaster management strategies and activities are aligned with the specific needs and conditions of each administrative unit, representing various sectors such as health, agriculture, water, and livelihood.

Horizontal Disaster Management Planning involves collaboration and coordination across various sectors, organisations, and stakeholders at the same administrative level. In Sri Lanka, this includes the involvement of multiple agencies and organisations working at the national, provincial, and district levels to collectively deliver various services during disasters.

This multi-level and collaborative approach in disaster management planning allows for a more comprehensive and responsive strategy to address the diverse challenges posed by disasters in Sri Lanka, taking into account both the administrative hierarchy and the need for cross-sectoral collaboration. Implementation of Community-Based Disaster Risk Management (CBDRM) in Sri Lanka is commendable, responding effectively to the escalating frequency of disasters. Acknowledged as a key strategy for bolstering community capacity and resilience, CBDRM is making positive strides. However, it is crucial to highlight the lack of specific details concerning the mechanisms of community involvement, underscoring the need for a thorough examination of the authenticity and depth of community participation in the decision-making processes. Additionally, the level of collaboration with other stakeholders remains unclear and raises questions about the effectiveness of a multi-stakeholder approach for comprehensive disaster risk management. While the expansion of CBDRM coverage to additional districts is viewed positively, there is a call for a rigorous assessment of the

[31] National Platform to Implement Sendai Framework for Disaster Risk Reduction (SFDRR) in Sri Lanka (November 2023). https://apad.lk/national-platform-to-implement-Sendai-framework-for-disaster-risk-reduction-sfdrr-in-Sri-Lanka/. Accessed on 20 November 2023.

[32] Aslam Saja et al. [7].

[33] National Disaster Management Plan 2022–2030 (n 30).

[34] Community Resilience Framework Sri Lanka (Disaster Management Centre 2015).

sustainability and scalability of these initiatives to gauge their long-term impact. In essence, despite recognising the positive trajectory, it is imperative to address specific aspects such as community involvement, stakeholder collaboration, and sustainability to ensure the continued effectiveness of CBDRM in Sri Lanka.

The discussion on disability in Sri Lanka involves various models such as the charity, medical, social, and rights-based models. A recent study on Human Rights Status of Persons with Disabilities concluded that persons with disabilities in Sri Lanka surface from a complex interface between widespread structural factors like health conditions, social factors, living situations, and the misapprehension in the thinking on the part of society.[35] Therefore, the dimensions included social justice, health care, and the human rights of those with impairments. The rights-based model, centred on fundamental human rights, strives for universal application, ensuring the protection of rights for persons with disabilities globally. This approach influences attitudes, environments, and organisational practices to promote equal rights for all individuals, including those with disabilities. In the context of increased health risks for persons with disabilities, a Human Rights-Based Approach (HRBA) becomes crucial in disaster relief efforts, aiming to safeguard the human rights of affected individuals.

The application of Disability-Inclusive Disaster Risk Reduction (DiDRR) in Sri Lanka faces several challenges.[36] While awareness is growing, practical implementation remains a challenge. Some existing policies and legal provisions support the rights of PWDs' but lack integration with broader disaster risk reduction strategies. Capacity building, collaboration, community engagement, data collection, and research are essential areas that require improvement. Emergency preparedness and response plans need to be made disability-inclusive, and adequate funding and resources must be allocated. The ultimate goal of DiDRR is to enhance the resilience of individuals with disabilities, both physically and socio-economically, to address the long-term challenges they face in the aftermath of disasters. Addressing these aspects is pivotal in nurturing a more inclusive and equitable society in Sri Lanka.

Despite progress in recognising the importance of disaster management in Sri Lanka, there remain areas requiring improvement, including policy integration, capacity building, collaboration, community engagement, data collection, and ensuring fully inclusive emergency preparedness and response measures. The study explores the experience of persons with disabilities during disasters from a human rights-based perspective and advocates for increased awareness and sensitivity among relief personnel to address these challenges. The prevalence of disabilities in Sri Lanka, attributed to prolonged conflict and an ageing demographic, underscores the need for protective measures, urging a shift towards a more inclusive and equitable society. Marginalisation of individuals with disabilities, especially in disaster situations, exacerbates their vulnerability, emphasising the importance of adopting a human rights-based approach. The existing support system, often reliant on families, needs enhancement through sensitisation and structural improvements across all phases of disaster management. An inclusive approach should transcend physical barriers, addressing attitudinal and structural obstacles that contribute to the vulnerability and exclusion of persons with disabilities. Integrating human rights principles is essential to construct a comprehensive and equitable framework for disaster response in Sri Lanka. In conclusion, the qualitative data analysis underscores the areas requiring attention for the effective implementation of Disability Inclusive Disaster Management and Risk Reduction in Sri Lanka.[37]

---

[35] Jeewaka Saman Kumara, 'Human Rights Status of Persons with Disabilities in Sri Lanka: Challenges and Opportunities'. http://pgihs.ac.lk/downloads/books/12chapter.pdf. Accessed on 15 January 2024.

[36] Disaster Risk Reduction Sri Lanka, Situational Report, 2019.

[37] Ibid., 24.

4 Study Results: Mapping Vulnerabilities of Persons with Disabilities and Challenges Faced by Service Providers in Disasters

The primary objective of this chapter is to enhance the comprehension of the significance of addressing the rights of PWDs in disaster contexts. The subsequent sections will present the findings of our study, followed by a discussion offering insights into different facets and outcomes of the research.

Employing an interpretivist ontological stance, this qualitative research was undertaken in two critically disaster-prone districts, namely Kandy and Kegalle, as identified by the National Building Research Organisation (NBRO) under the Ministry of Disaster Management. Ten in-depth interviews were conducted with persons with disabilities (PWDs), focussing on challenges and issues within current disaster relief operations, particularly in the context of evacuation and post-disaster displacement. An interview guide was utilised for conducting semi-structured interviews with professionals actively engaged in working with PWDs and individuals with various disabilities and their caregivers who experienced emergencies during disasters in 2022 and 2023. Thematic analysis was employed for data analysis, ensuring adherence to research ethics, as well as maintaining data privacy and confidentiality.

In this section, we delve into the multi-faceted challenges faced by PWDs in various stages of disaster management, specifically focusing on evacuation and protection during and after displacement. The insights drawn from in-depth interviews and observations of ten PWDs in the Kandy and Kegalle districts are thematically presented to provide a nuanced understanding of these challenges. The following narratives, each encapsulating a distinct seven sub-theme structure, are intertwined with their life experiences.

Exploring the theme of inadequate access to information and warning systems, a 45-year-old man with visual impairment, impacted by flooding, shared the challenges arising from insufficiently accessible warning messages:

> In the recent flood in Kegalle, my family received warning messages on their phones. But they were only text or pictures. So, I didn't know what was going on. I knew when someone told me, and there was very little time to get ready to leave.

This narrative vividly illustrates the impact of information disparities during disasters, negatively affecting equal access to timely and relevant information for all. Similarly, a 25-year-old woman with hearing impairment recounted her displacement experience, shedding light on inaccessible shelters and relief services and inadequate facilities:

> Due to the flood, we lost everything, and we only had a few clothes and a blanket. We were taken to a nearby temple, which served as an emergency shelter. Despite the presence of several officers and volunteers, there was no sign language interpreter available. This made it difficult for me to effectively communicate my needs and wants. The unsympathetic treatments I received and the isolation made me feel ignored most of the time.

This emphasises the urgent need for inclusive shelter facilities, invoking principles of non-discrimination and removing social exclusion. Furthermore, discussing limited mobility, transportation, and evacuation barriers, a 30-year-old woman with a physical impairment recounted the challenges faced during evacuation:

> During the flood evacuation, I couldn't get to the temple because there were no ramps. I had to depend on others to carry me up the stairs, which was scary. Temples are often used as temporary shelters for displaced people, but they're usually on hills with lots of stairs and no ramps. So, I had to be carried on people's shoulders, and it sometimes felt bad to inconvenience others because of the lack of facilities.

This narrative indicates the relationship between poverty and poor physical infrastructure inadequacies and aligns with accessibility concerns. Additionally, a 40-year-old woman with physical/mobility impairment discussed the repercussions of loss or damage of assistive devices:

> Due to the landslide, the whole house faded away. My wheelchair also got damaged. Now I have to wait until someone gives me a new one because the earlier one was also a gift from someone. We don't have enough income to buy a new one or even to repair the damaged one.

This indicates the economic vulnerability of PWDs, emphasising the need for more comprehensive support systems in Sri Lanka. Highlighting the theme of access to health care and medication, the mother of a 12-year-old child with a neurological disability expressed concerns about providing medication during disaster emergencies:

> Due to disaster emergencies, we were asked to go to a safer place. My son, who is disabled, requires daily medication for his epileptic condition. During the first two days of the flooding and in the community centre, I had great fear in my mind about the kid as I couldn't give him medicine properly for his medical conditions.

This not only points to the medical vulnerabilities of PWDs but also draws attention to disparities in healthcare accessibility during crises.

Moreover, a 23-year-old woman with a locomotor disability expressed the absence of adequate disaster preparedness training:

> We have never been to a training program for disaster preparedness or how to act when there is an emergency. I have heard that there were some meetings for the community in Aranayake in the community centre, educating them on how to get ready when there is a warning for the disaster.

This highlights the oversight in community training programmes and aligns with the necessity for inclusive preparedness initiatives. In the context of social isolation and stigmatisation, a woman aged 42 with locomotor disability in Aranayake emphasised the emotional impact of post-disaster treatment:

> Although the people who arrive to help us evacuate are really helpful, we are treated like objects and are not given any consideration after the disaster has passed. We will never accept the way we were treated.

This narrative elucidates how social isolation and stigma permeate the post-disaster environment, affecting the emotional well-being of PWDs.

Furthermore, a 52-year-old man in Kadugannawa, Kandy, shed light on the economic challenges faced by PWDs, highlighting economic vulnerability and increasing dependency:

> Not like normal people, we always have difficulties finding employment. When we were displaced due to the disaster, I lost my job and now find it difficult to find a new source of income.

This narrative intertwines economic vulnerabilities with employment discrimination, emphasising the need for inclusive economic policies. Lastly, a 32-year-old caregiver in Kegalle discussed the overwhelming responsibilities and emotional toll:

> I am looking after my disabled mother due to occupational hazards. Now, again, nature is beating us. Earlier, I did a part-time job and worked as a nanny. Now, due to this displacement, I lost my job and needed to take care of my mom also.

This narrative delves into the psychological and emotional implications of disasters on caregivers, emphasising the need for holistic support systems and mental health considerations. Overall, the challenges encountered by PWDs during and after disasters transcend practical impediments, delving into societal perceptions and human rights principles. An all-encompassing disaster management approach must not only address physical barriers/mobility issues but also confront attitudinal and structural obstacles that contribute to the vulnerability and exclusion of PWDs. This necessitates a profound understanding that merges sociological insights with human rights principles, forging a more equitable and inclusive disaster response framework. The narratives vividly illustrate the myriad challenges faced by PWDs and their families, post-disaster. To effectively address these challenges, a holistic and inclusive approach to disaster management and preparedness is indispensable. Moreover, this underscores the urgency of implementing inclusive policies, providing comprehensive training, and fostering heightened societal awareness to strengthen the well-being and resilience of persons with disabilities in the aftermath of disasters. Integrating these diverse perspectives has the potential to transform and improve the effectiveness of disaster response frameworks, promoting inclusivity and equity.

In addressing the second objective of identifying challenges faced by service providers in interactions with PWDs during disaster management, qualitative data from interviews with officials and service providers in Kandy and Kegalle reveal a spectrum of issues. Viewed through sociological, human rights-based, and disability lenses, these challenges highlight critical aspects of disaster response, encompassing gaps in training, resource limitations, accessibility issues, lack of inclusive facilities, communication barriers, lack of preparedness, regulatory constraints, community conflicts, mistrust, and a focus on other priorities during disasters.

According to a Disaster Relief Service Officer at the Divisional Secretariat in Kegalle, resource constraints impact the equal treatment of PWDs in IDP camps, emphasising compromised dignity and detrimental effects on their health.

> Temporary shelters/camps provide the bare minimum of amenities due to financial constraints. The government's distribution is limited to medicine and dry rations, making it challenging to cater to the specific needs of PWDs. We find it difficult to pay attention to their particular requirements. In these shelters, temporary washrooms are entirely inaccessible for PWDs, forcing them and their family members to seek alternative methods of support. Unfortunately, this compromises their dignity and has a detrimental impact on the personalities and overall health of young individuals.

The Assistant Director, NBRO, Kegalle, portrayed the difficulties of accessibility in evacuation centres, situated on historical sites with limited modifications possible.

> Imagine someone stuck on a mountain; removing them would be challenging and pose a significant physical risk. We typically use community centers or religious locations as evacuation centers for internally displaced people. In Sri Lanka, many temples are on hills with multiple staircases, presenting difficulties for wheelchair users and those with mobility impairments due to the lack of access, including ramps and elevators. Some of these sites are historical and recognized as UNESCO Heritage sites, making it impossible to modify the temporary shelters or temple spaces. PWDs must wait until the situation stabilizes, or we can relocate them to a more accessible place. However, it's challenging because only a few people are trained, and they must efficiently manage a large number of voluntary workers.

The narrative exemplifies the reduction of mobility rights for PWDs due to challenges like the lack of ramps and elevators. Another Disaster Relief Service Officer, this time from Kandy and Kegalle emphasised communication barriers during disasters, underscoring the vital role of sign language interpreters for effective support.

> As disaster relief officers, we strive to make a difference, but we often lack the training and resources to support PWDs effectively. In a recent disaster, we experienced the importance of sign language interpreters. Without them, communicating with individuals with hearing impairment was a challenge we want to avoid in the future. We value inclusivity but face hurdles when the system lacks essential tools and support. As a center manager, it's disheartening to witness their frustration due to communication and support gaps.

The manager expressed the disheartening aspect of witnessing frustration due to communication gaps. A volunteer trained by the Disaster Management Centre in Kegalle shed light on the inadequacy of training and experiences, narrating the challenging situation of caring for a visually impaired person in an IDP camp.

> In the Aranayake IDP camp, I had to care for a visually impaired person. Despite the stress, I had no prior training or updated knowledge on handling PWDs. It was a challenging situation, but I felt good that I could help in some way.

The Assistant Director, NBRO, Kandy, also discussed the issue of low trust and misunderstanding, pointing out the hesitation of PWDs to trust officials due to past negative experiences.

> Often, right after a disaster, the community and volunteers assist along with officials. However, I've noticed that sometimes PWDs hesitate to

> trust us or prefer assistance from others due to past negative experiences. Overcoming this lack of trust or resistance can be challenging. Mostly, military and religious leaders lead immediate evacuation and provide informal relief services, potentially leading to low trust among people from different cultures and ethnicities.

The Director further highlighted the influence of military and religious leaders in immediate evacuation, potentially leading to low trust among diverse communities. A Director of Planning at the Ministry of Disaster Management acknowledged the challenges of focusing on saving lives and preventing property damage during disasters, potentially overshadowing the needs of PWDs. The constraints of government rules and regulations were expressed, with officials making rapid decisions to address urgent needs, sometimes leading to subsequent questioning.

> As government officials, we are bound by rules and regulations in our work. At times, we make rapid decisions to assist people and address urgent needs, but this may lead to subsequent questioning, providing valuable lessons. Due to these factors, some officials avoid pushing their boundaries.

Another Assistant Director, this time from the NBRO emphasised the lack of preparedness, highlighting deficiencies in essential resources, accessible vehicles, and disaster warnings for the visually impaired.

> Despite recurring disasters each year, we still lack essential resources. We don't have wheelchair-accessible vehicles, only a Suvasaiya ambulance when needed. There are no braille or embossed system-supported disaster warnings for the visually impaired and identified disaster risk areas like Kegalle lack improvised community centers as temporary shelters or multi-functional facilities. In terms of training and resources, I believe we remain unprepared.

A representative from the We for Rights Organisation outlined societal pressures, acknowledging prevalent nepotism and the need to prioritise specific groups over persons with disabilities.

> In our society, nepotism is prevalent. We often face pressure to prioritize specific groups, as per the directives of local politicians and influential factions. Sometimes, we must give precedence to vulnerable groups over persons with disabilities.

Finally, the Director of Planning at the Ministry of Disaster Management emphasised the policy-implementation gap, noting that despite existing guidelines and norms, the system often neglects the voices of grassroot-level workers and those in need.

> While comprehensive guidelines and internationally-aligned norms exist in our policies, the real challenge lies in implementation. The system predominantly operates from the top-down, often neglecting the voices of those in need and ground-level workers.

This situation highlights a clear policy-implementation gap and a lack of inclusiveness. These challenges bring attention to the complex interplay of social structures affecting disaster response. Training gaps, resource constraints, and accessibility issues signal broader systemic flaws entwined with societal norms. Specifically, using a structural–functional approach helps understand issues like nepotism and prioritising certain groups, shedding light on the wider social context influencing decision-making in hierarchical cultures like Sri Lanka.

The challenges also point to potential violations of the rights of PWDs, especially in terms of dignity, equal treatment, and access to crucial services. The narrative of compromised dignity and health effects due to resource limitations in IDP camps stresses the urgent need for a rights-based perspective in disaster management.

Furthermore, these challenges uncover persistent barriers limiting the full inclusion of PWDs in disaster response. Issues like the lack of ramps, elevators, inaccessible washrooms, and communication gaps collectively infringe upon the mobility and communication rights of PWDs. This underscores the crucial necessity of integrating disability-inclusive strategies into disaster planning and response efforts.

5 Way Forward: Strategies for the Inclusion of PWDs in Disaster Management and Role of Social Workers

The qualitative data illustrates the escalating vulnerability of both the general public and PWDs during disasters, with PWDs facing heightened risks. This underscores the pressing need for inclusivity in disaster relief operations to uphold their rights, dignity, and overall well-being. Despite Sri Lanka's existing disaster management plan partially aligning with international standards, it operates hierarchically and lacks provisions for the inclusion of PWDs, neglecting the safeguarding of their human rights and well-being.

The study reveals numerous challenges faced by PWDs and service providers in disaster scenarios, prompting proposed strategies with a bottom-up approach as suggested by the researched community. Macro-level strategies involve policy revision through a bottom-up approach, annual policy reviews, thorough post-disaster source analysis, bi-annual evaluation and policy amendments, and adopting modern technology for PWD communication. Mezzo-level strategies include connecting village committees to district disaster officials and DMC coordinators, enhancing service provider coordination, and providing training for divisional and district officers. Micro-level strategies encompass empowering village groups to support PWDs, offering training on PWD evacuation (including swimming for some PWDs) and camp management, planning early, mapping PWDs, identifying shelters, providing PWD self-advocacy and disability training, and conducting sensitisation campaigns to raise awareness, such as emphasising that disability is not fate or doom but a matter of human rights for all.

Drawing from a comprehensive field study conducted within the country, in this sub-section, we delve into the multi-faceted role that social workers play in promoting inclusivity, and equity, and upholding the rights of PWDs. The subsequent section delineates the multi-faceted role of social workers in disaster management.

Social workers play a crucial role in promoting inclusivity for Persons with Disabilities (PWDs) in various ways. Their role begins with identifying PWDs within affected communities and delving deep into their specific needs, vulnerabilities, and preferences within the unique context of the island nation. They meticulously explore various aspects, including understanding the types of disabilities prevalent, mobility constraints faced, communication requirements essential for effective interaction, and the existing support networks that can be leveraged for assistance. They are instrumental in identifying and assessing PWDs, using tools like the International Classification of Functioning, Disability, and Health or the ICF model to understand their specific needs, vulnerabilities, and preferences. Additionally, social workers advocate for the rights of PWDs, raise awareness, and actively promote non-discrimination and equal access in disaster management contexts. For instance, they collaborate with local communities and relief service providers to ensure that PWDs have equal access to disaster-related services and resources, drawing from their expertise in navigating the diverse cultural landscape of Sri Lanka.

Currently, capacity-building efforts for PWDs are undertaken by individuals lacking proper training, such as officers in district secretariats, NGOs, and community workers. In this context, it is social workers who emphatically contribute by providing comprehensive training programmes for both PWDs and service providers, equipping them with essential knowledge and skills. They ensure that PWDs can access disaster-related information in accessible formats like sign language, braille, or audio.

Working closely with PWDs, social workers craft individualised support plans that cater to the unique requirements of each individual. They ensure access to evacuation strategies that are adapted to Sri Lanka's geographical and infrastructural considerations. This may involve providing mobility aids, arranging accessible transportation options, and addressing specific medical needs based on Sri Lanka's healthcare infrastructure. By translating messages into accessible formats, such as sign language, braille, or audio, social workers help in addressing the language diversity and accessibility requirements of different communities across the island and make it possible for PWDs to have easy access to important information regarding disasters.

Beyond just addressing practical needs, social workers also extend psychosocial support to help PWDs cope with the emotional and psychological challenges they may face during disasters. In Sri Lanka, which has a history of natural disasters and internal conflicts, these efforts are crucial in helping individuals overcome trauma and distress effectively. Social workers act as intermediaries between the PWDs and relief agencies, ensuring that service providers are well-informed about the specific needs of PWDs and effectively coordinate necessary services. They establish feedback mechanisms to capture inputs from PWDs and their caregivers. This feedback loop is invaluable in Sri Lanka's context for continuous service improvement and responsiveness to the evolving needs of the affected population. Furthermore, social workers engage in research to gain insights into the experiences of PWDs during disasters within the Sri Lankan context. These findings help shape evidence-based strategies and policies that enhance inclusivity and better safeguard the rights of PWDs in disaster management, aligning with Sri Lanka's unique socio-cultural fabric.

6 Conclusion and Recommendations

The findings of this study shed light on the complex challenges faced by PWDs during disaster management, emphasising the need for a comprehensive and inclusive approach. The qualitative data reveals a multi-faceted landscape, encompassing issues such as limited access to information, mobility constraints, loss of assistive devices, and inaccessible shelters for PWDs. Moreover, challenges faced by service providers, including inadequate training, resource constraints, and communication barriers, further complicate the already intricate dynamics of disaster response.

These challenges underscore the complex web of social structures and norms influencing disaster response. The hierarchical nature of the disaster management system in Sri Lanka, coupled with issues like nepotism and prioritisation of specific groups, reflects the broader social context impacting decision-making. Adopting a structural–functional approach helps unravel these complexities, emphasising the systemic shortcomings that intersect with societal norms and poor infrastructures and expertise. The concept of 'ableism' becomes highly relevant here in explaining why adequate measures were not adopted to address the situation of PWDs during the evacuation. Ableism, which involves discrimination and prejudice against people with disabilities, helps frame the discussion in terms of societal attitudes and biases that contribute to the marginalisation of PWDs in disaster contexts.

Moreover, the challenges highlight potential violations of the rights of PWDs, particularly in terms of dignity, equal treatment, and access to essential services. The narratives vividly illustrate how resource constraints in IDP camps compromise dignity and contribute to detrimental health effects, emphasising the urgent need for a rights-based perspective in disaster management. Overall, this study reveals persistent barriers limiting the full inclusion and participation of PWDs in disaster response. The lack of ramps, elevators, inaccessible washrooms, and communication gaps collectively infringe upon the mobility and communication rights of PWDs. These barriers underscore the importance of incorporating disability-inclusive strategies into disaster planning and response efforts.

In conclusion, the challenges encountered by persons with disabilities during disasters, such as limited access to information, mobility constraints, loss of assistive devices, and inaccessible shelters, demand urgent attention for the advancement of equity and inclusivity in disaster relief efforts. Simultaneously, officials and service providers engaged in disaster relief confront obstacles like inadequate training, resource constraints, and limited accessibility, impeding their effectiveness in assisting PWDs. Addressing these challenges necessitates a comprehensive, rights-based, and inclusive disaster management approach, focusing on the specific needs of PWDs to ensure their inclusion, dignity, and well-being. Urgently required is extensive training in disaster relief programmes, covering all planning phases with a specific emphasis on evacuation and displacement. The neglect of these specific needs perpetuates the vulnerability of PWDs, constituting a violation of their human rights and reinforcing societal stereotypes. As a signatory to international agreements recognising the rights of PWDs, Sri Lanka bears the responsibility to protect and promote these rights, particularly in the context of disaster management. The active participation of affected communities, including PWDs, in decision-making and preparedness is essential, emphasising greater inclusion and consideration of their unique needs in disaster mitigation projects. The study underscores the importance of fostering self-reliance and life skills among PWDs to enable them to protect themselves during disasters. Shifting away from attributing challenges solely to fate and destiny and recognising them as rights violations can lead to more sustainable outcomes for individuals and communities. In summary, a comprehensive and inclusive approach, emphasising empowerment and the recognition of rights, is crucial to addressing disaster-related challenges for PWDs and calls for a paradigm shift in disaster management towards a rights-based approach.

The proposed strategies and recommendations delineated below aim to formulate a more inclusive and resilient disaster response framework, accounting for the distinctive needs of PWDs and other marginalised communities across all phases of disaster management. Recommendations for the Sri Lankan context are rooted in various aspects, including the six dimensions of the Framework for Integrating Rights and Equality (FIRE), the Sendai Framework, and the human rights approach to disasters. These recommendations span three critical phases—pre-disaster, during the disaster, and post-disaster—ensuring effective coordination for a comprehensive disaster response.[38]

Recommendations

Addressing the specific context of displacement, the following recommendations focus on preparation, protection during evacuation, and durable solutions.

Pre-displacement

Comprehensive Vulnerability Mapping: Conduct thorough vulnerability mapping to identify areas prone to displacement and individuals, with a specific focus on PWDs at higher risk.

Accessible Evacuation Plans: Develop evacuation plans that are accessible and inclusive, considering the unique needs of PWDs, including transportation, communication, and shelter accessibility.

Community-Based Training: Implement community-based training programmes to educate both PWDs and the broader community on evacuation procedures, emphasising inclusive practices.

During Displacement

Specialised Evacuation Support Teams: Establish specialised evacuation support teams trained to assist PWDs, addressing their distinct needs during the evacuation process.

Accessible Shelter Facilities: Ensuring accessible shelter facilities guarantee that temporary shelters are fully accessible for PWDs, equipped with amenities such as ramps, elevators,

[38] See Scott et al. [8].

accessible washrooms, and designated spaces catering to various types of disabilities.

Nevertheless, challenges persist, notably in instances where some evacuation sites, such as temples, are located in elevated areas, making them inherently inaccessible due to various reasons. In such scenarios, it is essential to pre-identify specific locations suitable for PWDs within the community before a disaster occurs. However, these locations should not segregate PWDs from their families, close friends, or the broader community.

Emergency Communication Accessibility: Integrate accessible communication methods, such as sign language interpreters and alternative formats, to ensure crucial information reaches all individuals, irrespective of their abilities.

Post-displacement

Durable Housing Solutions: Prioritise the development of durable housing solutions that consider the long-term needs and accessibility requirements of PWDs, including necessary modifications and accommodations.

Inclusive Livelihood Programmes: Implement programmes that support the economic well-being of displaced PWDs, recognising and accommodating their skills and abilities.

Community Integration Initiatives: Promote initiatives fostering understanding, respect, and support for PWDs to reduce stigma and social isolation post-displacement.

Mental Health Support and Follow-Up: Establish mental health support services addressing the psychological impact of displacement on PWDs and their caregivers, emphasising holistic well-being.

These refined recommendations aim to enhance the preparedness, protection, and post-displacement support tailored to the needs of Persons with Disabilities, ensuring a more inclusive and effective response during and after the displacement process. Rooted in human rights principles, inclusivity, and effective governance, these recommendations strive to create a more equitable and resilient disaster response framework in the Sri Lankan context, with the overarching goal of 'building back better' and leaving no one, especially PWDs, behind in the face of adversity.

References

1. Chandrasekara SSK (2021) Drought in South Asia: a review of drought assessment and prediction in South Asian countries. Atmosphere 12:369
2. Kjellstrom T et al (2016) Heat, human performance, and occupational health: a key issue for the assessment of global climate change impacts. Ann Rev Public Health 37:97–112
3. Jubota A, Alkire S (2023) Understanding multidimensional vulnerabilities: impact on people of Sri Lanka. UNDP 56
4. De Silva MMGT, Kawasaki A (2018) Socioeconomic vulnerability to disaster risk: a case study of flood and drought impact in a rural Sri Lankan community. Ecol Econ 152:131
5. Ilyas M (2015) Women and girls with disabilities face unique risks in Nepal. https://disabilityrightsfund.org/women-and-girls-with-disabilities-face-unique-risks-in-nepal/. Accessed 10 July 2023
6. Villeneuve M et al (2019) Disability inclusive disaster risk reduction framework and toolkit: a report produced as part of the disability inclusive and disaster resilient Queensland project series. University of Sydney, p 5
7. Aslam Saja AM et al (2020) Implementing sendai framework priorities through risk-sensitive development planning: a case study from Sri Lanka. Prog Disaster Sci 5:3
8. Scott M et al (2023) FIRE: a framework for integrating human rights and gender equality in disaster risk reduction and climate change adaptation. Clim Develop 15:622

Ishari Gunarathna has been working as a Lecturer in the Department of Sociology, University of Peradeniya, since 2021. She completed her Bachelor's Degree in Social Work in 2010 and obtained MSc in Development Communication and Extension from the University of Peradeniya in 2015. Currently reading for her Master of Social Work at the University of Peradeniya and her areas of interest are Social Work and Direct Practice Methods, Social Work and Disability, and Community Development.

Priyadarshani Premarathne a Senior Lecturer at the University of Peradeniya, earned her Ph.D. in Sociology from the University of Bremen, Germany, along with a MPhil in Sociology from the University of Peradeniya and a Postgraduate Diploma in Counselling and Psychosocial Work from the University of Colombo. She instructs courses in Research Methodology, Sociology of Work, Medical Sociology, and Introduction to Sociology at various Sri Lankan Universities. Additionally, she serves as the postgraduate coordinator for the department. Her research focuses on job quality, well-being, work-life balance, decent work, and social issues impacting women, children, individuals with disabilities, and other marginalized communities.

Rising Waters, Stagnant Paths: Gendered Experiences of Flooding and Restricted Mobility in Can Tho City, Viet Nam

Danang Aditya Nizar and Ly Quoc Dang

Abstract

Viet Nam's response to its vulnerability to climate change impacts is marked by the integration of climate change concerns into sectoral laws. The Government of Viet Nam has also acknowledged the interlinkages of climate change, gender equality, and mobility within its climate regulatory framework. However, the focus on climate-related mobility has predominantly centred on macro-level mobility, which refers to the broad categorisation of migration, displacement, and planned relocation, with limited attention given to micro-level mobility. This chapter understands micro-level mobility as the smaller-scale and day-to-day movements which involve individual decisions regarding daily activities, destinations, and modes of travel. Furthermore, challenges remain in translating national commitments into local action, as evidenced by urban flooding in Can Tho City. Complex interactions between political, economic, social, and environmental elements contribute to recurring floods, with a disconnection between central Government commitments and local implementation, resulting in gendered impacts. Women in Can Tho City bear disproportionate impacts during recurring flood events, restricting their micro-level mobility in the household and public sphere. Women's decision to remain during floods results from a complex interplay between risk perception, socio-economic elements which contribute to vulnerability, and limited macro-level mobility options. However, at the same time, these women also actively employ strategies to cope with their restricted mobility, which signifies their agency in negotiating the associated risks and adapting to recurring floods. These insights into women's mobility behaviour during flooding offer a valuable starting point for policymakers to integrate gender and mobility, both at the macro and micro-level dynamics, into concrete climate actions. The findings also underscore the need to pay more attention to micro-level mobility within the climate-related mobility discourse.

1 Introduction

Over the past decade, there has been a growing scholarly interest in exploring the intricate interrelationship between climate change,

D. A. Nizar (✉)
Raoul Wallenberg Institute of Human Rights and Humanitarian Law, Jakarta, Indonesia
e-mail: danang.nizar@rwi.lu.se

L. Q. Dang
Mekong Delta Development Research Institute, Can Tho University, Can Tho, Viet Nam
e-mail: lqdang@ctu.edu.vn

S. Jolly et al. (eds.), *Climate-Related Human Mobility in Asia and the Pacific*, Sustainable Development Goals Series, https://doi.org/10.1007/978-981-97-3234-0_5

gender, and mobility.[1] This heightened attention reflects the recognition that these factors are not isolated but are deeply intertwined, influencing one another in multifaceted ways. Scholars have focused on understanding how climate change impacts mobility patterns and how these impacts are experienced differently based on gender dynamics.[2] However, the literatures on this nexus are still predominantly focused on the macro-level mobility discourse and less on micro-level mobility framing.[3]

This thriving body of knowledge has started making its mark on different countries' climate policy documents, including Viet Nam. As one of the most vulnerable countries to the adverse impacts of climate change, Viet Nam has taken significant measures to improve its climate change governance and regulatory framework.[4] In parallel, Viet Nam has also strengthened its commitment to climate change mitigation as reflected in the Nationally Determined Contribution (NDC) document. In the updated 2022 NDC, Viet Nam has increased its unconditional contribution from 9 to 15.8% and its conditional contribution from 27 to 43.5%.[5] Furthermore, Viet Nam to a certain extent has also incorporated gender equality and mobility issues into the climate regulatory framework.

This chapter examines the systematic integration of gender principles and mobility considerations into Viet Nam's climate governance and regulatory framework. It considers the extent to which these principles have been incorporated and how they impact the gender-specific consequences of climate-related micro-level mobility at the local level, using the case study of recurrent urban flooding in Can Tho City, Viet Nam. This is achieved by analysing Viet Nam's climate and gender regulatory frameworks, focusing on how the government has taken steps to promote the development of a gender-responsive climate regulatory framework. The analysis also delves into how mobility issues are being framed and positioned within these frameworks and to what extent they have been implemented at the local level. The next section provides an explanation of the methodology used to obtain these findings, and the subsequent section presents and further elaborates the results of this analysis.

This result is exemplified in the subsequent section through a case study concerning the restricted mobility of women in the context of urban flooding in Can Tho City, Viet Nam. The urban flooding phenomenon in Can Tho City is assessed through the lens of human–environment interactions, synthesised into six dimensions.[6] Moreover, various categorisations of how micro-level mobility is disrupted during extreme weather events are employed to gain insights into how women adapt to restricted mobility during floods.[7] Finally, the last section highlights the insights drawn from the case study analysis, which reinforce the call for integrating gender equality and mobility components into the climate regulatory framework in an integrated manner, from national to local levels. Moreover, the findings also emphasise the need for broader discussions on micro-level mobility within the gender and climate-related mobility discourse.

2 Methodology

This research, spanning a duration of 10 months from May 2019 to February 2020, provides a comprehensive exploration of the experiences of 24 women in Can Tho City before, during, and after flood events. Can Tho City was selected as the research site due to its prominence as the largest city in the Mekong Delta, with a

[1] See Chindarkar [1]; Evertsen and van der Geest [2]; Lama et al. [3]; Boas et al. [4].

[2] See Lama et al. [3]; Middleton et al. [5].

[3] See Akyelken [6]; Amankwaa and Gough [7].

[4] Binh [8].

[5] Government of Viet Nam, 'Nationally Determined Contribution (NDC)' (2022).

[6] Dang [9].

[7] Ebenezer Amankwaa and Katherine Gough (n 3) 235–524.

population of around 1.2 million, and plays a pivotal role as the economic hub of the region.[8] The study employs a qualitative approach, relying on semi-structured, in-depth individual interviews, and participant observation to capture the nuanced impact of floods on the lives of women residents. Participants were purposively selected to ensure representation across diverse social backgrounds, varied family structures, and different parts of the city, considering their multifaceted experiences with floods.

In addition to primary data collected directly from the women respondents and other relevant actors, the study incorporates secondary data from academic journals and grey literatures. The temporal focus of the research aligns with the rainy season from September to November when floods and high tide events are prevalent in Can Tho City. The investigation concentrates specifically on 48 floods and high tide events occurring within a three-month timeframe, shedding light on the intricacies of these phenomena. The research involves visits to 14 households during flooding, providing real-time insights into the challenges faced by women residents. For the remaining homes, visits occurred during non-flood times, prompting retrospective reflections from respondents on their previous flood experiences. This methodological approach ensures a holistic understanding of the multifaceted implications of floods on the city, communities, and, notably, the lives of women residents.

3 Governance and Regulatory Framework of Climate Change and Gender Equality in Viet Nam

3.1 Climate Change

Viet Nam has been recognised by different climate risk indices as one of the countries with the highest susceptibility to extreme changing climate conditions. For instance, Viet Nam was ranked as the third most vulnerable country in Southeast Asia and as the tenth in Asia by the 2021 University of Notre Dame Global Adaptation Initiative (ND-GAIN) [9]Index. The ND-GAIN Index calculates the country's vulnerability based on six essential sectors, including food, water, health, ecosystem services, human habitat, and infrastructure, and three cross-cutting components: (i) exposure to climate-related hazards; (ii) sensitivity to the impact of climate-related hazards; and (iii) the adaptive capacity to cope with climate-related hazards.[10]

In the 2019 INFORM Risk Index, Viet Nam is also ranked first as the country with the highest exposure to river and coastal floods.[11] Viet Nam scored 9.9 and 9.6 for river and coastal floods, respectively, with 10 representing the maximum exposure to specific hazard.[12] Similarly, the 2020 Global Climate Risk (GCR) Index developed by Germanwatch also identified Viet Nam as the sixth most affected country by extreme weather events between 1999 and 2018 among 181 assessed countries.[13] The GCR Index focuses on the quantification of loss and damage caused by extreme weather events, based on the analysis of four indicators: (i) number of deaths; (ii) number of deaths per 100,000 inhabitants; (iii) sum of losses relative to Purchasing Power Parity (PPP); and (iv) losses per unit of Gross Domestic Product (GDP).

Viet Nam's 2022 NDC document also demonstrates a similar analysis. For 2020 alone, when considering the overall cost of

[8]Global Covenant of Mayors for Climate and Energy, 'Member Profile: Can Tho, Vietnam' (2022).

[9]University of Notre Dame, 'Country Index Rankings' (*Notre Dame Global Adaptation Initiative*). https://gain.nd.edu/our-work/country-index/rankings/. Accessed 12 October 2023.

[10]Chen et al. [10].

[11]Thow et al. [11].

[12]The World Bank Group and the Asian Development Bank, 'Climate Risk Country Profile: Vietnam' (2021).

[13]David Eckstein et al., 'Global Climate Risk Index 2020: Who Suffers Most from Extreme Weather Events? Weather-Related Loss Events in 2018 and 1999 to 2018' (Germanwatch e.V. 2019).

environmental degradation, the government estimated that the country experienced a climate-related economic loss of around USD 10 billion, equivalent to 3.2% of the country's GDP.[14] The government also identified climate change as one of the factors that could hinder the country's development.[15] It is projected that, without effective adaptation and mitigation measures, Viet Nam could experience a potential cumulative economic loss of USD 400–523 billion or 12–14.5% of the country's GDP by 2050, due to climate change.[16] The Government of Viet Nam has adopted various regulatory measures responding to this phenomenon, among them, notably, the decision to mainstream climate change considerations into different sectoral laws, instead of enacting an individual law to address climate change issues.

For example, the Law on Environmental Protection (No. 72/2020/QH14), which supersedes the Law on Environmental Protection (No. 55/2014/QH13), has stipulated that environmental protection planning should incorporate the climate change trend prediction and its impacts, and climate change analysis and responses should also be integrated into the Social-Economic Development Strategy (SEDS) and the General Social-Economic Development Plan (GSEDP).[17] This stipulation is a significant measure to address climate change issues because both the SEDS and GSEDP are the overarching documents that guide the state's political, economic, and social direction, including in terms of regulatory and legislative development. Similarly, the Law on Natural Disaster Prevention and Control (No. 33/2013/QH13), Law on Land (No. 45/2013/QH13), and Law on Meteorology and Hydrology (No. 90/2015/QH13) have also specified the requirement to mainstream climate-related considerations into strategies and plans in their respective sectors.[18] The government has enacted 19 regulations and policies related to GHG emissions reduction and climate change adaptation, which then translated into 20 resolutions, master plans, plans, and programmes.[19]

Even though there is no specific climate change law in Viet Nam, the government has mandated the Ministry of Natural Resources and Environment (MONRE) as the leading state agency responsible for addressing all climate-related issues, including regulation and policy formulation. The MONRE Minister and the Deputy Prime Minister act as the vice-chairmen of the National Climate Change Committee (NCCC). The NCCC, established in 2012, is the primary institutional body that proposes strategic advice to the Prime Minister related to the country's direction in addressing climate change issues. The MONRE also functions as the permanent acting agency of the NCCC and the host agency for the Committee Office, as it is the NCCC's operational assisting body.[20] Furthermore, the MONRE is also mandated as the National Focal Point to lead the implementation of the United Nations Framework Convention on Climate Change (UNFCCC), Kyoto Protocol, and Paris Agreement, as well as other relevant climate-related international treaties, in Viet Nam.[21]

The Government of Viet Nam has taken steps to incorporate mobility within its climate regulatory framework, particularly concerning climate-related migration. One of the earliest strategic documents that mention this issue is the 2018 Third National Communication to the UNFCCC, which recognised that climate change has a detrimental impact on the economic well-being of families, and thus serves as a primary

[14] Government of Viet Nam (n 5) 17.

[15] Ibid.

[16] The World Bank Group and the Asian Development Bank (n 12) 18.

[17] Government of Viet Nam, 'Law on Environmental Protection' (2020).

[18] Du et al. [12].

[19] Government of Viet Nam (n 5) 4 and 18.

[20] Government of Viet Nam, Ministry of Natural Resources and Environment (MONRE), 'Report National Adaptation Plan for the Period 2021–2030, with a Vision to 2050' (2022).

[21] Ibid, 54.

driver of increased migration.[22] Furthermore, the country's 2022 NDC document highlights the increasing trend of climate-related migration. It recognises how climate change contributes to the rise in migration, compelling tens of thousands of households to experience permanent relocations, thereby putting at risk the preservation of their cultural identity and local knowledge, constituting a form of non-economic loss and damage.[23]

Planned relocation, as a distinct form of mobility, has also been mentioned in Viet Nam's National Adaptation Plan (NAP) 2021–2030 in the context of the government's response to disaster events. The government illustrated the importance of planned relocation as part of its climate strategies through the implementation of the Programme on Population Resettlement for Areas Vulnerable to Natural Disasters, Extremely Challenging Areas, Border Regions, Islands, Free Migration, and Special-Use Forests for the Period 2013–2015 and the Orientation to 2020 (Decision No. 1776/QD-TTg dated November 21, 2012) during the period of 2011–2015. In this programme, more than 70,000 out of 85,900 households (over 60%) residing in disaster-prone regions were relocated to establish stable settlements.[24] This effort encompassed the planning of climate-resilient residential settlements, the development of flood-resistant residential areas in the Mekong Delta, and the population resettlement programme for areas vulnerable and exposed to disasters.[25] Furthermore, Viet Nam has introduced resettlement as part of its disaster management strategies, particularly in the Mekong Delta region, since the early 2000s and has subsequently enacted around ten policies on resettlement in alignment with the country's climate actions.[26] With this understanding, the following section evaluates the legal framework regulating the gender issues in Viet Nam.

3.2 Gender Equality

The promotion of gender equality and women's rights in Viet Nam can be traced back to the pre-independence period. Ho Chi Minh, the founder of Viet Nam's communist movement and the country's first president, repeatedly positioned women at the centre of his liberation movement; he noted,

> In building socialism, one of the important issues is the struggle for equal rights for women. Equality is not only about politics, but also from the real life of social and family life. Therefore, we have to respect women. We do revolution to gain equality. Boys and girls are equal. It is necessary to truly liberate women and respect women's rights, only when women are freed from the remnants of customs, feudal thought, then they are confident enough to rise up to contribute their efforts to build a socialist country.[27]

In this statement, communism and socialism are depicted as the sole paths to achieving women's emancipation, contrasting with the portrayal of the Western and capitalist societies as morally compromised systems that perpetuate patriarchal family structures and the subjugation of women.[28]

This concept of gender equality and women's rights within socialism and the nation's liberation discourse was then institutionalised in the country's regulatory framework after its independence in 1945. For example, the 1946 Constitution in Article 24 explicitly stipulates that 'women of the Democratic Republic of Viet Nam have equal rights with men in terms of political, economic, cultural, and

[22] Government of Viet Nam, Ministry of Natural Resources and Environment (MONRE), 'The Third National Communication of Vietnam to the United Nations Framework Convention on Climate Change' (2018).

[23] Government of Viet Nam (n 5) 17.

[24] Government of Viet Nam, MONRE (n 20) 41.

[25] Ibid.

[26] See United Nations in Viet Nam, 'Migration, Resettlement and Climate Change in Viet Nam' (2014); Danh and Mushtaq [13].

[27] Dung [14].

[28] Hoang [15].

social activities and families.'[29] Furthermore, this commitment is also present in the 2013 Constitution, where Article 26 mandated that 'male and female citizens have equal rights in all fields. The state shall adopt policies to guarantee the right to and opportunities for gender equality.'[30] This assertion on the importance of equal rights between men and women becomes the basis for further promotion of gender equality principles in the country's regulations and policies. Furthermore, the Government of Viet Nam also ratified and adopted the Convention on Elimination of All Forms of Discrimination against Women (CEDAW) in 1982, which is among the earliest in Southeast Asia.[31] This demonstrates the government's strong commitment to promoting gender equality and women's rights because the CEDAW is one of the main human rights conventions adopted by the United Nations, which 'obligates Governments to promote, protect and fulfil the equal rights of men and women in their own jurisdictions.'[32]

Another landmark and key regulation to promote gender equality and women's rights are the Law on Gender Equality (No. 73/2006/QH11), enacted by the government in 2006. This law stipulates gender equality to become a cross-cutting issue and requires the mainstreaming of gender issues in multi-sectorial legal documents.[33] This approach is different from how the Government of Viet Nam addresses the climate change issue, where no individual law specifically addresses climate concerns. More operational gender equality actions are governed under the National Strategy on Gender Equality (NSGE) 2021–2030. The NSGE 2021–2030 aims to address gender equality issues in the following six areas: (i) politics; (ii) economics and labour; (iii) family and gender-based violence; (iv) health; (v) education; and (vi) training.[34]

The Ministry of Labour, Invalids, and Social Affairs (MOLISA) is the designated state agency to lead the country's efforts related to gender equality, including the formulation of legal frameworks.[35] Thus, the MOLISA is also responsible for overseeing and monitoring the implementation of the Law on Gender Equality (No. 73/2006/QH11). Furthermore, four other key institutions at the national level have specific mandates related to gender equality, which are: (i) Ministry of Culture, Sports, and Tourism (MOCST); (ii) Parliamentary Committee for Social Affairs (PCSA), under the National Assembly; (iii) Viet Nam Women's Union (VWU), and (iv) National Committee for Advancement of Women (NCFAW).

The MOCST is the focal point for implementing and monitoring the Law on Prevention and Combat against Domestic Violence (No. 13/2022/QH15), which supersedes the previous Law on Domestic Violence Prevention and Control (No. 02/2007/QH12).[36] Moreover, the responsibility to ensure the mainstreaming of gender principles into the state's legislations lies under the jurisdiction of the PCSA.[37] Another important institution in the state's efforts to promote and ensure gender equality and women's rights is the VWU, which is instrumental in shaping policies and carrying out Viet Nam's progressive stance on gender equality.[38] The VWU was established in 1930 and played a central role in mobilising women's support during the liberation movement and stands as one of the most influential institutions dedicated to gender issues, with around 15 million members across all administrative levels.[39] The President of the VWU, together with the Deputy Minister and

[29] Dung (n 27) 162.

[30] Binh (n 4) 47.

[31] Sharifah Syahirah [16].

[32] Hintjens [17].

[33] Binh (n 4) 41.

[34] Government of Vietnam, 'National Strategy on Gender Equality 2021–2030' (2021).

[35] Linh et al. [18].

[36] Ibid, 41.

[37] Ibid.

[38] Hoang (n 28) 299.

[39] Ibid.

Minister of MOLISA, holds the critical role as the chair and vice-chair of the NCFAW, respectively. The NCFAW is an inter-sectoral committee that coordinates the implementation of Viet Nam's strategic direction on gender equality.[40]

The government's continuous commitment to the advancement of gender equality has played a significant role in enhancing Viet Nam's position in the gender-related global indices, such as the Gender Development Index (GDI), the Gender Inequality Index (GII), and the Global Gender Gap Index (GGGI). Developed by the Human Development Report Office in 2014, the GDI assesses gender disparities in health, education, and economic resource control using sex-disaggregated Human Development Index (HDI) data.[41] Viet Nam's GDI stands at 0.997, categorising Viet Nam within Group 1, denoting a high level of gender equality between males and females. Furthermore, the GII assesses gender disparities in three aspects: (i) reproductive health; (ii) empowerment, and (iii) economic participation.[42] In 2019, Viet Nam scored 0.314 on this index, placing it at 65th among 162 countries.[43]

Lastly, the GGGI measures the extent of gender-based disparities across critical sectors, including the economy, education, health, and political representation.[44] Viet Nam has demonstrated a positive trajectory in its ranking from 2020 to 2023. Within East Asia and the Pacific region comprising 20 countries, Viet Nam has risen from the ninth rank to the sixth.[45] Moreover, globally, its ranking has improved significantly, moving it from the 87th position to the 72nd out of 153 countries.[46] This upward trend signifies that Viet Nam has successfully bridged 71% of its gender gap during this period.

[40] Linh et al. (n 35) 41.

[41] Conceição et al. [19].

[42] Ibid, 361.

[43] Ibid, 362.

[44] World Economic Forum, 'Global Gender Gap Report 2023' (2023).

[45] Ibid, 25.

[46] Ibid.

Nevertheless, despite the abovementioned achievements, it is essential to acknowledge that a disparity persists between the principles outlined in gender equality legal frameworks and the realities on the ground. This gap primarily stems from the prevalence of traditional customs over legal mandates, perpetuating male dominance and a patriarchal paradigm in Viet Nam.[47] Gender norms influence various facets of life, encompassing the political, social, economic, and cultural, including the division of labour along gender lines. Consequently, this dynamic results in gendered livelihood options, distinct perceptions of risk, and interpretations of climate change and its environmental consequences.[48] Therefore, it is crucial to assess how males and females in Viet Nam are affected by the adverse impacts of climate change differently and to what extent the government has incorporated gender-responsive principles into its climate actions and initiatives.

4 Gender-Responsive Climate Regulatory Framework in Viet Nam

Gender equality and empowerment have been widely recognised as inseparable from climate mitigation and adaptation actions. During the twentieth session of the Conference of the Parties (COP 20) in 2014, the parties adopted the Lima work programme on gender (Decision 18/CP.201). This programme is designed to facilitate the advancement of gender-responsive climate policies and directives throughout all aspects of negotiations.[49] This commitment was further strengthened in the 2015 United Nations Climate Change Conference (COP21 or CMP 11) in Paris, France, where the parties pledged their commitment to uphold, promote,

[47] Le [20].

[48] Phan et al. [21].

[49] United Nations Women, Implementation of Gender-responsive Climate Action in the Context of Sustainable Development (2016).

and incorporate their respective responsibilities regarding human rights, along with gender equality, the empowerment of women, and intergenerational equity in their efforts to combat climate change.[50] Furthermore, the Paris Agreement underscores the importance of a gender-responsive approach to addressing climate change, as articulated in Article 7 and Article 11.[51]

This global commitment is essential because research has shown that women are disproportionately affected by poverty when compared to men, exhibit lower capacity to adapt to current and impending climate change effects, and tend to have less participation and contribution in advancing the understanding of gender-specific climate adaptation or mitigation strategies.[52] Gender-based disparities in climate vulnerability not only mirror pre-existing gender inequalities, but also perpetuate them. For example, the gendered divisions of labour within households often pose greater obstacles for women when it comes to adjusting to shifts in environmental conditions.[53] Consequently, this reduces their prospects for sustainable livelihoods and exacerbates resource shortages.

Numerous studies have also demonstrated similar findings in the context of flooding in Viet Nam. For example, in the case of flooding in Hue City, women who are typically more connected to their families as compared to men felt a heightened responsibility to secure additional income to support their families in the context of flooding.[54] Despite these significant contributions, women received less information about Disaster Risk Reduction (DRR), and their limited participation in flood-related planning stemmed from a lack of confidence and the substantial time they dedicated to household responsibilities.[55] This phenomenon is also found in the case of urban flooding in Da Nang City, where the heavy household responsibilities of women have resulted in lower participation in flood mitigation planning.[56] However, despite their limited involvement in formal planning processes, women still played a crucial role in flood management at the household level due to their responsibilities in the home.[57]

Considering this empirical evidence illustrating the gendered impacts of climate change in Viet Nam, the government has formally recognised this phenomenon within its climate strategies. For instance, the NAP 2021–2030 highlights disproportionate climate change impacts on women as follows:

> Women are more vulnerable to climate change than men given their limited opportunity, condition in accessing education, literacy, income-generating opportunities, land ownership, access to credit, and mental and sexual violence. Women are more dependent on land and ecosystems that are highly exposed to climate change.[58]

This direct correlation between climate change and gender-specific issues was also mentioned in the country's 2022 NDC,

> Rising temperatures, prolonged heatwaves, and droughts increase stress, pressure and anxiety in households, which can lead to violence against women. Climate and temperature related illnesses are increasing in frequency, which poses significant challenges to maternal health.[59]

Nevertheless, at the same time, the government acknowledged that the formulation of the country's climate change and disaster risk management policies and strategies has lacked a foundation in gender analyses.[60] This has hindered the ability to discern the distinct climate change impacts on men, women, and groups experiencing differential vulnerability, thereby

50 Binh (n 4) 47.

51 Ibid.

52 Van Aelst and Holvoet [22].

53 Eastin [23].

54 Pham and Lam [24].

55 Ibid, 40.

56 Anh et al. [25].

57 Ibid, 36.

58 Government of Viet Nam, MONRE (n 20) 32.

59 Government of Viet Nam (n 5) 16.

60 Government of Viet Nam, 'Viet Nam's Voluntary National Review on the Implementation of the Sustainable Development Goals' (2018).

impeding the development of tailored solutions for each demographic. Furthermore, the government also emphasised the lack of synergy among the country's regulatory framework related to GHG emission reduction, climate change adaptation, and socio-economic development, including advancing gender equality and women's rights.[61]

The Government of Viet Nam has implemented specific measures to address this issue by incorporating gender-specific language and integrating principles of gender equality. The Law on Natural Disaster Prevention and Control (No. 33/2013/QH13), for example, explicitly articulates the need for humanity, fairness, transparency, and gender equity to be upheld in natural disaster prevention and control.[62] Moreover, the law also clearly outlines the importance of the active involvement of women's organisations and individuals in activities such as integrating communication and educational information on disaster prevention and control, as well as participating in developing Disaster Risk Reduction (DRR) and control measures.[63]

Another example is the Law on Environmental Protection (No. 72/2020/QH14), which has integrated more explicit directives aimed at mainstreaming gender considerations and ensuring the active involvement of women and women's organisations.[64] The law also mandates Civil Society Organisations (CSOs), including women's organisations, to adhere to environmental protection laws and actively participate in environmental preservation activities. Furthermore, women's organisations are granted the right to offer input, consultations, advice, criticism, participation, and proposals concerning local environmental protection matters.[65] Viet Nam's NAP 2021–2030 also has mainstreamed gender equality principles in all its key tasks and programmes, such as through the development of gender mainstreaming guidelines to enhance women's engagement in climate-related policy formulation and implementation, as well as to increase women's participation in initiatives related to DRR and climate change adaptation.[66]

Specific mention of climate-related mobility to women can be found in Viet Nam's 2018 Third National Communication to the UNFCCC, which underscores the role of climate change as one of the primary catalysts for the surge in migration, with a notable increase in the number of female migrants from rural areas to central cities and industrial parks for better opportunities.[67] Moreover, the document highlights how female migrants often face worse living conditions than their male counterparts, particularly with respect to housing and environmental sanitation.[68]

Despite this incorporation of gender equality principles to create a gender-responsive climate regulatory framework, there remain gaps in the national commitment and its implementation at the local level.[69] In the next section, a case study of urban flooding in Can Tho City, Viet Nam, will assess and illustrate this gap.

5 Case Study: Urban Flooding in Can Tho City, Viet Nam

5.1 Urban Flooding: Interactions of Six Dimensions

There has been a consensus in the DRR and climate change discourse that 'disaster' needs to be approached within the framework of 'vulnerability,' resulting from human and environmental interactions.[70] In the DRR school of

[61] Government of Viet Nam (n 5) 25.

[62] Government of Viet Nam, 'Law on Natural Disaster Prevention and Control' (2013).

[63] Ibid. ch 2.

[64] Government of Viet Nam (n 17) ch 11.

[65] Ibid, art 158.

[66] Government of Viet Nam, MONRE (n 20) 124.

[67] Government of Viet Nam, MONRE (n 22) 34.

[68] Ibid.

[69] Binh (n 4) 42.

[70] Birkmann [26].

research, disaster risk is understood as the equation of 'hazard' interacting with 'exposure' and 'vulnerability,' which further developed into the 'resilience' concept that introduces 'capacity' within the equation.[71] Furthermore, vulnerability should not be perceived only as a societal precondition prior to disaster events, but can also be shaped and potentially exacerbated by various coping strategies adopted in the aftermath of a disaster at the individual or institutional level.[72] Similarly, the climate change schools of thought view vulnerability as the potential damage caused to a system by particular climate-related hazards.[73] This, for example, is apparent in the Intergovernmental Panel on Climate Change (IPCC) Sixth Assessment Report (AR6), which defines vulnerability as 'the propensity or predisposition to be adversely affected.'[74] This reflects that the framing and depiction of vulnerability in AR6 are largely influenced by an impact-oriented viewpoint, which focuses on the immediate adverse impacts of climate change on different facets of human lives. In this chapter, the urban flooding phenomenon is understood as the result of human–environment interactions, which is synthesised into six dimensions: (i) geographical location and topography; (ii) urbanisation and local planning; (iii) climate change; (iv) tides; (v) upstream and downstream planning; and (vi) land subsidence.[75]

In terms of geographical location and topography (dimension i), Can Tho City is the largest city in the Mekong Delta and is located in a low-lying area on the south of the Hau River, which is the bigger branch of the Mekong River, about 80 km from the sea, and thus faces both natural flooding from upstream of the Mekong River and tide flows from the sea downstream (Fig. 1).[76] The city also has an extensive network of rivers, channels, and canals with remarkably high density. The total length of this network exceeds 1.5 km, covering a water surface area of over 1 km^2, or equivalent to around 6% of the city's total area.[77] In the scenario where a sea-level rise of 100 cm occurs, almost one-fourth of the total area of Can Tho City will be inundated. Furthermore, Can Tho City has experienced a massive transformation of the built-environment landscape, with more buildings and industrial zones replacing green spaces and agricultural lands, which was caused by rapid urbanisation from surrounding areas (dimension ii).[78] For instance, there has been an increase of 7863 people in the area over 15 years (from 2003 to 2019), which has led to the conversion of agricultural land for alternative purposes, reducing it from 51.72 ha in 2003 to 15.17 ha in 2019.[79] Consequently, green spaces have also declined, shrinking from 187,668 m^2 in 2004 to 20,808 m^2 in 2014.[80] This phenomenon has also exerted additional pressure on the city's groundwater extraction practices, leading to the exacerbation of land subsidence issues in Can Tho City (dimension vi).[81]

This rapid urbanisation is also closely correlated with climate change (dimension iii) and tide dimensions (dimension iv), where climate-related hazards such as sea-level rise, saline intrusion, and tidal inundation pose significant threats to rural livelihoods, which contribute to influence people's decision-making to permanently migrate from rural to urban areas, such as

[71] See Hewitt [27]; Wisner et al. [28].

[72] Birkmann (n 70) 15.

[73] Brooks [29].

[74] Intergovernmental Panel on Climate Change, 'Synthesis Report. Contribution of Working Groups I, II and III to the Sixth Assessment Report of the Intergovernmental Panel on Climate Change' (2023) 130.

[75] Dang [30].

[76] Department of Statistics of Can Tho City, 'Can Tho City Statistical Book 2019' (2019).

[77] Do [31].

[78] Department of Statistics of Can Tho City (n 77) 3.

[79] Ibid.

[80] Ibid.

[81] Dang (n 75) 69.

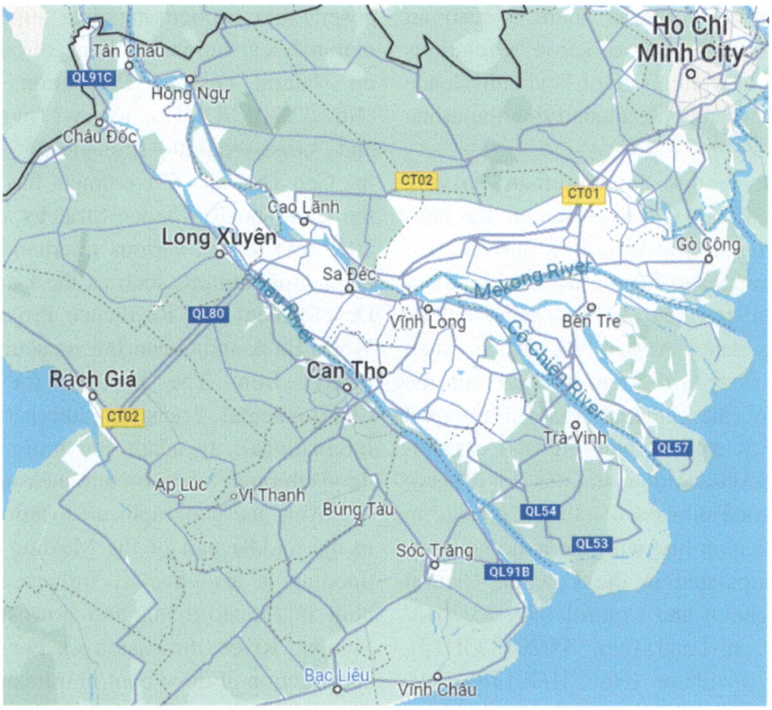

Fig. 1 Location of Can Tho City, Viet Nam[82]

Can Tho City.[83] The annual rainfall in Can Tho City has only slightly fluctuated between 1980 and 2010, with no identifiable trend of increased annual rainfall.[84] Authors like Nguyen Ngoc Huy et al. even argue that most of the flooding events in Can Tho City between 1978 and 2013 were mostly influenced by the natural phenomenon of high tidal flow (dimension iv), instead of the city's rainfall intensity.[85] However, the MONRE has projected that climate change (dimension iii) will increase the annual rainfall rate in Can Tho City to 10% by 2035, as depicted in Table 1.

This escalation will intensify the pressure on Can Tho City, because the city is not designed or equipped with adequate landscape, infrastructure, and other urban planning measures (dimension ii) to absorb the heightened precipitation

Table 1 Percentage of annual rainfall changes when compared with the period of 1986–2005[86]

City	RCP4.5[87] scenario			RCP8.5 scenario		
	2016–2035	2046–2065	2080–2099	2016–2035	2046–2065	2080–2099
Can Tho	10.5% (6.6–14.4)	13.7% (4.5–23.6)	15.1% (2.8–26.6)	10.7% (4.0–18.0)	18.3% (13.5–23.6)	21.2% (12.3–30.7)

82 Google Maps, 'Map of Can Tho City, Viet Nam' accessed 8 November 2023.

83 Linh et al. (n 35) 33.

84 Nguyen Ngoc Huy er al., 'Planning for peri-urban development and flooding issues: The story of new urban areas in Can Tho City' (2018).

85 Ibid, 10.

86 Government of Viet Nam, Ministry of Natural Resources and Environment (MONRE), 'Climate Change and Sea Level Rise Scenarios for Viet Nam' (2016) 57.

87 Representative Concentration Pathways (RCP) refers to the scenarios that include a time series of emissions and concentrations of the full suite of greenhouse gases and aerosols and chemically active gases, as well as land use/land cover. The word 'representative' signifies that each RCP provides only one of many possible scenarios that would lead to the specific radiative forcing characteristics.

rate effectively.[88] Moreover, there is also an absence of groundwater extraction regulations to mitigate the alarming rate of land subsidence in the city (dimension vi), intensifying the urban flooding risk.[89]

In response to this limited urban carrying capacity, the Can Tho City Government has integrated some climate considerations into its 2020 Master Plan on Socio-Economic Development with a vision towards 2030 (Decision No. 1533/QD-TTg dated August 30, 2013), such as through the prioritisation of studying the construction of the regional and sub-regional dyke system, as well as building and expanding automatic observation stations to observe the environmental impacts of climate change in the city.[90] This is in line with the national commitment as stipulated in the Law on Natural Disaster Prevention and Control (No. 33/2013/QH13), Law on Land (No. 45/2013/QH13), and Law on Planning (No. 21/2017/QH14). Nevertheless, the effective implementation of this master plan is yet to be seen because there are still identifiable gaps in the coordination among city departments and agencies and between relevant agencies at the national and local levels.[91] For instance, despite the intricate and cross-sectoral nature of urban flooding, the responsibility for addressing this issue still falls under the siloed approach of different sectoral departments. As a result, relevant departments and agencies, such as the Department of Construction, Department for Planning and Investment, Can Tho City Climate Change Coordination Office, and Committee for Flood and Storm Control, have yet to formulate a comprehensive strategic response to the existing flood risk.[92] This governance issue remains despite the explicit mandate stipulated in the national legislation, such as through the Law on Natural Disaster Prevention and Control (No. 33/2013/QH13), to have more integrated and cross-sectoral strategies.[93] Additionally, the Can Tho City Government has also enacted the Can Tho Resilience Strategy until 2030 in 2019 and hosted various international development programmes, such as the Can Tho Urban Development and Resilience Project from The World Bank and Green Infrastructure Feasibility Project from The Institute for Social and Environmental Transition–International (ISET-International), to adapt and mitigate the urban floods through infrastructure measures.[94]

Given its geographical location, which is in the middle part of the Mekong Delta, urban flooding is also heavily influenced by events both in the upstream and downstream of the Mekong River (dimension v). For instance, the construction of flood control infrastructure in the upper delta of the river has disrupted the river's flowing rate and pattern to the downstream.[95] Another example is the rice intensification programme in the Mekong Delta region as part of the government's agriculture development initiative. Infrastructures that were developed to support the programme have also altered the natural water flows of the river, which has contributed to the urban flooding occurrence in Can Tho City, as expressed by one of the experts from Can Tho University,

> Local governors and planners in the upstream have constructed massive dykes and embankments to support the rice intensification program. However, this has increased the water flow to the downstream part of the river and put more pressure on the regions in the area, including Can Tho City.[96]

This reflects the recurring disconnection between the climate commitments at the national and local levels. The Law on Irrigation

[88] Huong and Pathirana [32].

[89] Ibid.

[90] Government of Viet Nam, 'Decision No. 1533/QD-TTg of August 30, 2013, approving the master plan on socio-economic development of Can Tho city through 2020, with a vision toward 2030' (2013).

[91] Do (n 78) 19.

[92] Garschagen [33].

[93] Government of Viet Nam (n 60) ch 2.

[94] 100 Resilient Cities and others 'Resilience Accelerator Can Tho, Vietnam: Designing for Nature-based Infrastructure' (2019).

[95] Dang (n 75) 71.

[96] Ibid 80.

(No. 08/2017/QH14) expressly stipulates that climate change concerns must be integrated into the design, construction, and operation of the irrigation works and infrastructure.[97] Furthermore, Chap. 4 of the Law also requires the incorporation of scenarios of water resources when impacted by climate change conditions within the irrigation plan.[98] Unfortunately, these considerations have been overlooked when planning and constructing the infrastructure to support the rice intensification programme, which heightened the urban flooding risk in Can Tho City.

In summary, the urban flooding phenomenon in Can Tho City is a complex outcome of intricate interactions across six different dimensions. These dimensions encompass environmental, economic, social, and political aspects, forming a comprehensive systemic framework that highlights the multifaceted nature of this challenge. This framework provides the foundation for comprehending the gendered impacts of urban flooding on the female population of Can Tho City, particularly in terms of their restricted mobility.

5.2 Restricted Mobility: Between Vulnerability and Coping Strategy

Within the climate change and mobility nexus, women are frequently portrayed as one of the most vulnerable demographic groups with less capacity and opportunities to be mobile than their male counterparts, due to unequal power relations and gendered norms.[99] In the context of urban poor communities, research has shown that their mobility practices result from a complex interplay between personal aspirations, societal expectations, and financial uncertainties, which is often highly gendered.[100] However, there has been much debate on this framing since the dichotomy between mobility and immobility in the climate change context tends to perpetuate gender dualism, where men are often linked to mobility and women to fixed spaces and immobility.[101] Furthermore, this binary categorisation may obscure the complexity of the decision-making processes behind mobility (or immobility) practices and disregard the agency of women who choose to remain in a place.[102] The recognition of someone's agency to determine their actions is central to ensuring the incorporation of a human rights-based approach into climate actions.[103] Therefore, immobility should not be oversimplified as a result of vulnerability, but instead be considered as part of adaptive capacity and coping strategy, depending on the context.

Understanding the mobility patterns and behavior in the face of climate change is essential to inform decision-makers on how to address this issue systematically at the policy level. To fully comprehend this phenomenon, it is imperative to consider micro-level mobility alongside macro-level mobility rather than exclusively concentrating on the latter. The Climate Change Adaptation Framework categorises climate-related mobility into three distinct forms: (i) displacement, generally understood as forced movement of persons; (ii) migration, characterised by its voluntary nature; and (iii) planned relocation, referring to the planned process of moving persons or group of persons from one location to another.[104] This chapter refers to these forms of mobility as climate-related macro-level mobility, encompassing the temporary or permanent movement of individuals or groups of individuals outside of their places of habitual residence, influenced by climate-related phenomenon. On the other hand, climate-related

[97] Du et al. (n 18) 1289.

[98] Ibid.

[99] Boas et al. [4].

[100] Akyelken [34].

[101] Boyer et al. [35].

[102] Evertsen and van der Geest [36].

[103] Scott et al. [37].

[104] The Nansen Initiative, 'Agenda for the Protection of Cross-border Displaced Persons in the Context of Disasters and Climate Change' (2015).

micro-level mobility is understood as the smaller-scale and everyday movements and decisions made in response to climate-related events, which often relate to individual decisions regarding daily activities, destinations, and modes of travel.[105] This concept of smaller-scale and everyday mobility is often used to analyse the interlinkages between gender and mobility in the urban context, particularly related to the gendered patterns of trip purpose, trip distances, and access to private and public transport.[106] This chapter argues that this form of mobility is valuable for assessing the lived experience of individuals in responding to climate-related events, including floods, particularly for those who do not leave their residences during these occurrences (i.e. macro-level mobility). Limited information on lived experiences at the individual level will not only constrain the understanding of the nuanced realities of living with climate-related risks, but also obscure the multifaceted challenges involved in coping with these risks. In the context of extreme weather events, Ebenezer Amankwaa and Katherine Gough provide insights into how micro-level mobility could be disrupted in three different forms: (i) postponed when people need to restrict their movement; (ii) improvised, referring to the strategies that people apply to navigate the disruption; and (iii) assisted, when people require specific supports to overcome the disruption.[107]

The Government of Viet Nam has recognised the interlinkages between climate change and mobility in its climate actions. For example, the 2022 NDC mentioned that 'climate change is also one of the reasons for increasing migration and forcing tens of thousands of households to permanently relocate.'[108] Furthermore, the NAP 2021–2030 also introduced a dedicated section on the impacts of climate change on migration, which underscores how climate change conditions could influence the decision-making of economic migrants, particularly in the context of agricultural settings.[109] The Law on Natural Disaster Prevention and Control (No. 33/2013/QH13) also mentions displacement under search and rescue activities in response to disaster events. Nevertheless, this recognition does not manifest in the practical aspects of climate-related mobility governance and maintains a limited focus on macro-level mobility. Moreover, using the term 'migration' as a catch-all phrase when referring to mobility carries the risk of masking the multi-causality dimension of the interlinkages between climate change and mobility, and it may neglect the significance of micro-level mobility, especially among populations who choose to stay and do not engage in the macro-level mobility.[110]

In the context of Can Tho City, urban flooding has been a recurring phenomenon ingrained in the daily experience of its residents. Consequently, the normalisation of these events has influenced some individuals, including women, to remain in their homes during flood events.[111] The decision of women to stay results from complex negotiations, taking multiple factors into account. This includes the societal expectations of women as the primary caregivers responsible for household duties during flood events and their proactive response to the absence of structural adaptive intervention at the government level.[112] Furthermore, risk perception also plays an important role in influencing the decision to remain or to move, particularly related to the water depth, flood duration, frequency, and floor space of the place of residence.[113] Given that frequent floods in Can Tho City characterised by generally shallow water levels (approximately up to 50 cm on average) and a relatively brief duration (around 5–8 h within a span of 3–10 days) have become part of the urban experience of its residents, the

105 Amankwaa and Gough (n 7), 237.

106 See Hanson [38]; Uteng [39].

107 Amankwaa and Gough (n 7), 238.

108 Government of Viet Nam (n 5) 17.

109 Government of Viet Nam, MONRE (n 20) 34.

110 Boas, Pater and Furlong (n 100) 2.

111 Akyelken (n 101) 1590.

112 Dang (n 6) 35.

113 Thinh et al. [40].

associated risk may be perceived as negotiable, and thus, adaptable.[114] Nevertheless, for some individuals the decision to remain is also influenced by their vulnerability and limited capacity to move, as articulated by one of the respondents below,

> Yes we will do that [moving out]. But we still don't know, because we need to find enough money in order to build our own house. ... We have to find a better place to live because it is flooded here. Sometimes, we cannot do anything.[115]

Therefore, the decision to remain or to move during floods is the result of intricate interactions of various social, economic, and environmental aspects, as well as continuous negotiation of risk and capacity, rather than being driven by a single causal factor. Furthermore, individuals who decide not to engage in macro-level mobility, whether due to vulnerability, conscious choice, or a combination of both, must actively navigate and adapt to the challenges presented by the disrupted micro-level mobility.

Both women and men experience restricted mobility in the household and public sphere. Within the household sphere, both genders need to navigate carefully around the house to prevent injuries from slippery surfaces.[116] However, women are often exposed to floodwater inside the house more frequently, as they tend to stay at home while their husbands are outside for work. This makes them more prone to skin diseases and other health problems, resulting in additional expenses for health-related costs.[117] In response, some women have taken measures to improvise their restricted mobility by using protective equipment, such as gloves and plastic shoes, to reduce the risk of associated injuries and diseases.[118] On the other hand, women with higher economic status have experienced this restricted mobility within the household sphere differently because they have the financial means to hire additional assistance for managing the flooded house or to acquire water pumps to accelerate the drying process.[119] This emphasises the importance of not simplifying the gendered impacts of climate change into the gender dualism between men and women since both genders are not homogenous groups. Therefore, 'it is important not to see men and women as isolated and impervious categories ... because intersectionality with class, caste, religion, age, etc. affects the resources, rights and responsibilities that any woman has.'[120]

Women who need to leave their homes regularly, especially those involved in employment and community work, must continually navigate their choices to address their restricted mobility in the public sphere. For example, they need to adjust their mode of transportation, as moving around on a bicycle during flood times can be challenging and, if not done cautiously, may result in injuries, as expressed by a Can Tho resident,

> Sometimes I have to hold my bicycle and walk when I pass the deep flooded areas. I am so afraid riding a bicycle during floods as I have fallen twice before.[121]

Women who are engaged in paid employment also often experience the added pressure of working overtime to compensate for the hours they lose when they arrive late to work due to the floods, as a consequence of postponed mobility.[122] Often, these women are also expected to take on the additional responsibility of managing the flooded house, resulting in both physical and emotional fatigues. This similar additional burden of work is also experienced by women who are involved in informal employment sectors, such as highlighted by a Can Tho City resident who works as a lottery ticket seller,

---

[114] Thinh et al. [41].

[115] Dang (n 75) 157.

[116] Dang (n 75) ch 4.

[117] Danh [42].

[118] Dang (n 75) 122.

[119] Ibid.

[120] Sultana [43].

[121] Dang (n 75) 131.

[122] Dang (n 6) 32.

> My husband could not solve the flood water at our house. I had to deal with it, even though I was already exhausted from selling lottery tickets.[123]

Moreover, negative emotive words, causing pain, hurt, worry, fear, tiredness, anxiety, and hopelessness, are frequently associated with women's experiences during the floods.[124] To a certain extent, this also contributes to the higher prevalence of women experiencing non-economic loss related to mental health issues in the face of urban flooding.[125]

Nevertheless, this restricted mobility does not prevent women from utilising their social networks to cope with their restricted mobility, particularly regarding financial access. Women are assuming the role of finding and accessing financial credits, both from formal and informal sources.[126] In the context of informal sources, women rely on their social networks within their public sphere, such as through their families, neighbours, and friends. The following statement illustrates this perceived role when a Can Tho City resident was asked who would be responsible for finding and borrowing money,

> Myself of course! My husband would not dare to do that. As a man, he is too proud *(sỹdiện)* to ask for money from others.[127]

This expected gendered role regarding financial loans could be explained by the gender distinction in social capital, which is closely related to social norms and interpersonal relationships. Social capital is generally categorised into (i) bonding; (ii) bridging, and (iii) linking ties.[128] Bonding ties are established within a group of individuals who share common interests, experiences, or belief systems, often associated with survival and recovery in post-disaster and conflict settings.[129] On the other hand, bridging ties are more related to horizontal networks among groups of individuals who come from different social identities but have shared interests, while linking ties refer to traverse group boundaries in a vertical direction.[130] Bridging and linking ties are valuable for goods and information exchange and usually occur in more formal group settings.[131]

Women are often associated with bonding ties because they engage more in informal gatherings and collective communal activities, which provide more opportunities to strengthen and maintain their bonding ties with relatives and neighbours.[132] These strong bonding ties will yield enhanced access to socio-economic support between group members, leading to women receiving more support from their social networks in the aftermath of disasters.[133] The illustrative case of Can Tho City residents above reflects how strong bonding ties established through informal networks can harness collective support in difficult times. Furthermore, this strategy employed by women to access additional financial means from their social networks reiterates women's strategies to cope and adapt with the risk from remaining in their places of residence in the time of floods. Even though the decision to abstain from temporary evacuation or permanent relocation may be influenced by their vulnerability and restricted options, women in this study demonstrate a proactive stance in utilising their agency and refusing to passively accept the risks that arise from living with floods.

Unfortunately, this restricted micro-mobility and the limited opportunities to engage in macro-level mobility experienced by the women residents have not been recognised as citizen responses that need to be facilitated by the government. In the Can Tho Resilience Strategy until 2030, there is no reference to any forms of

[123] Dang (n 75) 96.

[124] Ibid, ch 5.

[125] Danh (n 118) 19.

[126] Dang (n 75) ch 4.

[127] Ibid, 162.

[128] Mark Pelling, *The Vulnerability of Cities: Natural Disaster and Social Resilience* (Earthscan 2003).

[129] Harrison [44].

[130] Ibid, 527.

[131] Ibid.

[132] Phan, Jou and Lin (n 48) 10.

[133] Ibid.

mobility as components of the strategies aimed to enhance the city's resilience.[134] Two potential actions that could address the restricted micro-level mobility faced by the residents during floods include development and retrofitting of resilient housing and the improvement of the quality and accessibility of the public transportation system.[135] Moreover, the document does not include any initiatives to facilitate potential macro-level mobility options for residents at the time of floods. Similarly, the international development programmes hosted by the Government of Can Tho City to mitigate the flood risk focus solely on physical and infrastructure interventions, overlooking considerations of macro and micro-level mobilities.[136]

Insights from this study are expected to offer evidence on the interplay between gender and mobility within the context of flood occurrences in Can Tho City, underscoring the importance of integrating these factors into the city's flood adaptation and mitigation strategies. This could be achieved, for example, by offering additional health incentives and implementing flood-proofing systems to ensure the continued operation and accessibility of the public transportation system during flood times, thereby supporting the residents' restricted micro-level mobility. While the Government of Can Tho City may not consider evacuation or permanent relocation as necessary measures, given the typical characteristics of shallow and short-duration floods in the city, it remains crucial to explore the facilitation of macro-level mobility options. This could involve the provision of financing mechanisms to assist residents from flood-prone areas in relocating to other parts of the city with lower flood risks. In this manner, the decision of the residents regarding mobility, whether to move or remain, could be respected, facilitated, and safeguarded by the government.

[134] People's Committee of Can Tho City, 'Can Tho Resilience Strategy until 2030' (2019).

[135] Ibid.

[136] 100 Resilient Cities and others (n 94) 5.

6 Conclusion

The chapter has highlighted the response of Viet Nam to its high vulnerability to the adverse impacts of climate change. It is noted that the government has mainstreamed climate change issues into different sectoral laws, such as the Law on Environmental Protection (No. 72/2020/QH14), Law on Natural Disaster Prevention and Control (No. 33/2013/QH13), Law on Land (No. 45/2013/QH13), and Law on Meteorology and Hydrology (No. 90/2015/QH13).[137] Furthermore, the government has enacted 19 regulations and policies concerning GHG emissions reduction and climate change adaptation, resulting in the creation of 20 resolutions, master plans, and programmes.[138]

The government has also recognised the interlinkages between climate change issues with gender equality and mobility issues in the climate regulatory framework. For instance, the country's NAP 2021–2030 and 2022 NDC specifically mention climate change impacts on women's rights.[139] Other examples are the Law on Natural Disaster Prevention and Control (No. 33/2013/QH13) and Law on Environmental Protection (No. 72/2020/QH14), which explicitly stipulate the mainstreaming of gender principles into their respective sectors, as well as promoting the active participation of women's organisations in the realms of DRR and environmental protection, respectively.[140] In terms of climate-related mobility, the focus in the NAP 2021–2030 and 2022 NDC has been predominantly on macro-level mobility through the recognition of climate change as a contributing factor to migration.[141] Nevertheless, the gap still persists in the implementation of this

[137] Du et al. (n 18) 1289.

[138] Government of Viet Nam (n 5) 4 and 18.

[139] Government of Viet Nam, MONRE (n 20) 32; Government of Viet Nam (n 5) 16.

[140] Government of Viet Nam (n 60) ch 2; Government of Viet Nam (n 17) art 4.

[141] Government of Viet Nam, MONRE (n 20) 34; Government of Viet Nam (n 5) 17.

national level commitment at the local level, as evidenced in the case of urban flooding in Can Tho City.

The complex interactions between six dimensions, which is a combination of political, economic, social, and environmental elements, have resulted in the recurring floods in Can Tho City.[142] The disconnection between climate commitments made by the Central Government and their implementation by local authorities has also played a role in these instances. For example, the absence of climate considerations in the local planning (dimension ii) and fragmentation in planning between the upstream and downstream areas along the Mekong Delta (dimension v) have contributed to this disparity.[143] Furthermore, Can Tho City's local planning is also predominantly focusing on physical and infrastructure measures to address the recurring floods, with limited consideration to the intersection between gender and mobility in the context of urban flooding.[144]

Women residents of Can Tho City have been disproportionately affected by the urban flooding, including those who choose to stay at their homes during the floods instead of evacuating or relocating to other areas. Their decision to remain is influenced by various intertwined factors, such as risk perception, socio-economic elements which contributed to vulnerability, and limited macro-level mobility options, instead of due to a singular causal factor. Staying behind during flood times also means that women must contend with restricted micro-level mobility in the household and public sphere that is highly gendered. For instance, the gendered role that expects women to be the primary caretakers in the house imposes an additional burden on them and exposes them to greater health risks associated with floodwater.[145] On the other hand, women have employed various measures to improvise their restricted mobility and have assumed a key role in seeking and securing additional financial resources to support the well-being of their households.[146] Thus, the decision of women to stay back, even though it may influenced by their limited capacity to engage in macro-level mobility and result in micro-level restricted mobility including its associated risks, should not be reduced to mere passivity. Instead, it reflects the result of active negotiations of their agency aimed at coping with and adapting to their own vulnerability in the time of floods. However, it is still imperative to recognise that those vulnerability aspects should be addressed to enhance women's resilience in facing recurring flood incidents.

These insights on women's mobility patterns and behaviour could provide an entry point for policy makers to incorporate the intricate interactions of gender and mobility in the context of urban floods into climate actions in more concrete terms. They highlight the significance of establishing an integrated climate governance system and regulatory framework from national to local levels, addressing the implementation gap between these tiers of governance. Furthermore, results of this study emphasise the significance of the government developing and implementing policies that respect, facilitate, and safeguard the macro- and micro-level mobility decision undertaken by individuals in response to urban floods. Additionally, these findings highlight the importance of including the micro-level mobility dimension in climate-related mobility discourse, especially at the local level.

References

1. Chindarkar N (2012) Gender and climate change-induced migration: proposing a framework for analysis. Environ Res Lett 7(2):1–7
2. Evertsen K, van der Geest K (2020) Gender, environment and migration in Bangladesh. Clim Develop 12(1):12–22
3. Lama P, Hamza M, Wester M (2021) Gendered dimensions of migration in relation to climate change. Clim Develop 13(4):326–336

[142] Dang (n 75) 70–72.

[143] Huong and Pathinara (n 88) 389.

[144] 100 Resilient Cities and others (n 93) 5.

[145] Dang (n 75) 35.

[146] Ibid, ch 4.

4. Boas I, de Pater N, Furlong BT (2023) Moving beyond stereotypes: the role of gender in the environmental change and human mobility nexus. Clim Develop 15(1):1–9
5. Middleton C, Elmhirst R, Chantavanich S (eds) (2018) Living with floods in a mobile Southeast Asia: a political ecology of vulnerability, migration and environmental change. Routledge, New York
6. Akyelken N (2020) Living with urban floods in Metro Manila: a gender approach to mobilities, work and climatic events. Gender Place Cult 27(11):1580–1601
7. Amankwaa E, Gough K (2023) We are at the mercy of the floods! Extreme weather events, disrupted mobilities, and everyday navigation in Urban Ghana. Singap J Trop Geogr 44:235–524
8. Binh VN (2020) Gender equality and women's empowerment in climate change adaptation and disaster risk reduction: laws, policies and practices in Vietnam. Vietnam J Family Gend Stud 14(1):40–57
9. Dang LQ (2022) Patterns of vulnerability among women in urban flooding in Can Tho City, Vietnam. Asian Soc Sci 18(3):1911–2025
10. Chen C et al (2023) University of Notre dame global adaptation initiative: country index technical report. University of Notre Dame
11. Thow A et al (2022) INFORM climate change: quantifying the impacts of climate and socio-economic trends on the risk of future humanitarian crises and disasters. Office of the European Union
12. Du H et al (2023) A framework for reviewing laws and policies for climate resilience: the case of the Vietnamese Mekong delta. J Environ Plan Manag 66(6):1280–1304
13. Danh VT, Mushtaq S (2011) Living with floods: an evaluation of the resettlement program of the Mekong delta of Vietnam. In: Stewart MA, Coclanis PA (eds) Environmental change and agricultural sustainability in the Mekong delta. Springer, New York
14. Dung VV (2019) Ho Chi Minh's thoughts on women's equal rights. Eur J Polit Sci Stud 3(1):159–168
15. Hoang LA (2020) The Vietnam women's union and the contradictions of a socialist gender regime. Asian Stud Rev 44(2):297–314
16. Sharifah Syahirah SS (2015) Regional-global governance network on women's rights: CEDAW and its implementation in ASEAN countries. Proced Soc Behav Sci 172:519–524
17. Hintjens H (2008) UNIFEM, CEDAW and the human rights-based approach. Develop Change 39(6):1181–1192
18. Linh NS et al (2021) The state of gender equality and climate change in Viet Nam. United Nations Environment Programme and United Nations Women
19. Conceição P et al (2020) Human development report 2020: the next frontier—human development and the Anthropocene. United Nations Development Programme
20. Le TQ (2013) Vietnam gender equality in the present. Vietnam J Soc Sci 2:76–81
21. Phan LT, Jou SC, Lin JH (2019) Gender inequality and adaptive capacity: the role of social capital on the impacts of climate change in Vietnam. Sustainability 11:1257
22. van Aelst K, Holvoet N (2016) Intersections of gender and marital status in accessing climate change adaptation: evidence from rural Tanzania. World Develop 79:40–50
23. Eastin J (2018) Climate change and gender equality in developing states. World Develop 107:289–305
24. Pham TDM, Lam TTS (2016) Gender needs and roles in building climate resilience in Hue City, Vietnam. Asian Cities Climate Resilience Working Paper Series, p 33
25. Anh TT et al (2016) Gender analysis in building climate resilience in Da Nang: challenges and solutions. Asian Cities Climate Resilience Working Paper Series, p 35
26. Birkmann J (2013) Measuring vulnerability to promote disaster-resilient societies and to enhance adaptation: conceptual frameworks and definitions. In: Birkmann J (ed) Measuring vulnerability to natural hazards: towards disaster resilient societies. United Nations University Press
27. Hewitt K (ed) (1983) Interpretations of calamity: from the viewpoint of human ecology. Routledge, New York
28. Wisner B et al (2004) At risk: natural hazards, people's vulnerability and disasters. Routledge, New York
29. Brooks N (2003) Vulnerability, risk and adaptation: a conceptual framework. Working Paper 38
30. Dang LQ (2021) Vulnerable lives of women and empowerment process in urban flooding in Can Tho City, Viet Nam. DPhil thesis, Chiang Mai University
31. Do H (2018) Taking advantage of water spaces in the organization of architectural landscape and climate change adaptation. Institute for Social and Environmental Transition-International
32. Huong HTL, Pathirana A (2013) Urbanisation and climate change impacts on future urban flooding in Can Tho City, Vietnam. Hydrol Earth Syst Sci 17(1):379–394
33. Garschagen M (2015) Risky change? Vietnam's urban flood risk governance between climate dynamics and transformation. Pac Aff 88(3):599–621
34. Akyelken N (2020) Living with urban floods in Metro Manila: a gender approach to mobilities, work and climatic events. Gender Place Cult 27(11):1580–1601
35. Boyer K, Mayes R, Pini B (2016) Narrations and practices of mobility and immobility in the maintenance of gender dualisms. Mobilities 12(6):847–860
36. Evertsen K, van der Geest K (2020) Gender, environment and migration in Bangladesh. Clim Develop 12(1):12–22
37. Scott M et al (2022) FIRE: a framework for integrating human rights and gender equality in disaster risk reduction and climate change adaptation. Clim Develop 15(7):622–627

38. Hanson S (2010) Gender and mobility: new approaches for informing sustainability. Gender Place Cult 17(1):5–23
39. Uteng TP (2011) Gender and mobility in the developing world. The World Bank
40. Thinh DC et al (2017) Flood loss models and risk analysis for private households in Can Tho City, Vietnam. Water 9(313):1–22
41. Thinh DC et al (2016) Multi-variate analyses of flood loss in Can Tho City, Mekong delta. Water 8(6):1–21
42. Danh VT (2014) Household economic losses of urban flooding. In: Asian cities climate resilience working paper series, p 12
43. Sultana F (2010) Living in hazardous waterscapes: gendered vulnerabilities and experiences of floods and disasters. Environ Hazards 9(1):43–53
44. Harrison J, Montgomery C, Bliss J (2016) Beyond the monolith: the role of bonding, bridging, and linking social capital in the cycle of adaptive capacity. Soc Nat Resour 29(5):525–539

Danang Aditya Nizar is a development practitioner with extensive experience working in various thematic areas, including education, disaster risk reduction, displacement, gender, and sustainable agriculture, with several NGOs and UN agencies. Currently, he holds the position of Programme Officer for Climate Change and Human Mobility at Raoul Wallenberg Institute of Human Rights and Humanitarian Law (RWI), Regional Asia Pacific Office. He initiated the reactivation of the Asia Pacific Academic Network on Disaster Displacement, which is a networking platform for Asia Pacific scholars who focus on the intersection of climate change, human mobility, and human rights nexus. Additionally, he has led the development and publication of notable research products at RWI, including 'A Bibliometric Analysis of Research at the Nexus of Climate Change, Human Mobility, and Human Rights' and 'Climate Change, Mobility, and Human Rights: Slow-onset Environmental Change and Displacement in the Mekong Region'. Danang holds a Master's degree in Anthropology of Development and Social Transformation from the University of Sussex, UK, with a specialisation in refugees, displacement, and humanitarian response.

Ly Quoc Dang is currently working at Mekong Delta Development Research Institute, Can Tho University. His work focuses on gender perspectives in climate risks and water management, sustainability. Moreover, he is founder of Gender Action Research Network which focuses on gender lenses in the Mekong Sub-Region

Establishing a Human Rights-Based Approach to Climate Change-Induced Internal Displacement in the Regime of Bangladesh: Challenges and Way Forward

Md. Abdul Awal Khan

Abstract

The rise of global temperature and climate change creates multiple adverse impacts on the life and livelihood of human beings, as well as on the environment and non-human biodiversity. One of the significant effects of climate change is that the sea level will rise and submerge substantial parts of low-lying countries, which is one of the significant reasons for being displaced internally. Bangladesh's marginalised and displaced population is perhaps one of the worst victims of climate change. Despite the given fact, establishing a legal relationship between climate change and human rights law remains one of the daunting tasks of policymakers. There is numerous research on Disaster Risk Reduction and disaster management adaptation in Bangladesh to identify normative gaps for protecting internal climate-displaced people in Bangladesh. Despite significant developments in law and policy measures, the normative gaps and lack of protection mechanisms and durable solutions to address climate change-induced internal displacement still exist significantly in Bangladesh. This paper primarily revisits the current policy framework to understand the challenges of applying a human rights-based approach in national law and policy framework in Bangladesh. Secondly, the chapter identifies the shortcomings and barriers of prevention and preparedness mechanisms for displacement and recommends durable solutions that are most needed to protect climate change-induced displaced populations.

1 Introduction

Human rights implications due to climate change have been an emerging issue in national and international jurisdictions.[1] Among the various impacts of climate change, displacement creates multidimensional challenges in society, especially in a developing country like Bangladesh.[2] Usually, the increased number of internal displacements is observed mostly from rural areas to urban areas due to multiple reasons, including employment opportunities. For example, recent research identified that farmers of Sirajgonj district of Bangladesh were forced to seek alternative livelihoods in urban areas and that Dhaka city has already received more

M. A. A. Khan (✉)
Independent University, Dhaka, Bangladesh
e-mail: awalkhanlaw@iub.edu.bd

[1] Preston [1].

[2] Khan [2].

S. Jolly et al. (eds.), *Climate-Related Human Mobility in Asia and the Pacific*, Sustainable Development Goals Series, https://doi.org/10.1007/978-981-97-3234-0_6

than ten million people displaced due to climate change.[3] Whenever these people arrive in urban areas, they face economic and social disparity, with limited access to resources, leading to further vulnerability.[4] Their mass movement to a new location, if they are not systematically relocated, is likely to create other social disturbances, such as not getting access to drinking water, sanitation, etc. There is evidence that the increase in the number of vector-borne diseases is also the result of the adverse impact of climate change, and displaced people suffer the most from vector-borne diseases.[5] When people are displaced from their usual place of living, critical human rights violations happen, and the right to life substantially declines as displaced people lose their homes and livelihoods.[6]

Therefore, a human rights-based approach or an HRBA is crucial in both legal and policy measures. Though an HRBA is not a prerequisite to addressing displacement in the context of disasters and climate change, the approach does help to identify the legal and policy gaps in protecting the human rights of the victims.[7] An HRBA is also essential to assess the status of law, policy, and practice in line with international standards.[8] It can determine the shortcomings between the responsibility and international obligation of states, especially under the international conventions on the respective subject matter. A human rights-based approach and its practice are not clearly and systematically adopted in local legislation in Bangladesh. This chapter highlights a number of human rights issues to understand the realities of law and practice of displacement in the context of disasters and climate change. This chapter then examines the existing legal and policy framework from a human rights-based approach. Secondly, the chapter identifies the shortcomings and barriers of prevention and preparedness mechanisms for displacement and recommends durable solutions which are most needed to protect climate change-induced displaced populations.

The chapter explicitly attempts to address how far an HRBA has been ensured in the climate legislation of Bangladesh and what are the major challenges to establishing it in Bangladesh's legal system. The chapter is divided into five sections including a background discussion on the topic in the first section and a concluding discussion in the last section. Section 2 describes some forms of climate-related human mobility in Bangladesh and Sect. 3 describes the conceptual framework of HRBA along with particular focuses on the legal and policy framework. Section 4 describes various challenges to mainstreaming an HRBA in the legal system of Bangladesh.

2 Climate-Related Human Mobility in Bangladesh

Bangladesh is ranked the seventh most at risk nation in the Global Climate Risk Index 2021,[9] although its contribution to global warming is meagre (around 0.48% of global emissions).[10] Bangladesh's geographical location and high population density make it one of the world's most disaster-prone countries.[11] A recent satellite altimetry data from the Department of Environment reveals that the average sea-level rise in the coastal zone of Bangladesh is

[3] Rozario and Chowdhury [3].

[4] Ibid.

[5] National Adaptation Plan of Bangladesh (2023–2050) [4].

[6] Awal Khan (n 2)7.

[7] Scott [5].

[8] Universal Values, 'Principle One: Human Rights-Based Approach', https://unsdg.un.org/2030-agenda/universal-values/human-rights-based-approach, accessed on 30 September 2023.

[9] Global Climate Risk Index, https://germanwatch.org/sites/default/files/Global%20Climate%20Risk%20Index%202021_1.pdf, accessed on 30 Sept 2023.

[10] UNDP, Global Climate Promise, https://climatepromise.undp.org/what-we-do/where-we-work/bangladesh, accessed on 30 Sept 2023.

[11] IDMC, https://www.internal-displacement.org/countries/bangladesh, accessed on 1 Sept 2023.

3.8–5.8 mm/year over the last 30 years, and about 12.34–17.95% of the coastal area will be submerged during this century.[12] Significant hazards[13] that impact the life and livelihood of the people of Bangladesh are 'floods, cyclones and surges, tornados, earthquakes, riverbank erosion, landslides, salinity intrusion, drought, tsunamis, lightning, arsenic contamination, human-induced hazards, and health hazards.'[14] For instance, in 2022 the Northeastern region (Sylhet) of Bangladesh experienced floods twice, where more than 9 million people were displaced and faced enormous human rights sufferings.[15]

Disasters are the main trigger of displacement in the country, particularly during the June–September monsoon season, when floods displace an average of a million people yearly.[16] Moreover, increased salinity causes crop damage, low production, cropping and irrigation pattern change, limited land for agricultural production, and loss of livelihoods and biodiversity, which ultimately causes a high risk of internal displacement.[17] Sea level rise and coastal vulnerability are considered the most significant threat for Bangladesh among all the disaster events.

Among the various disaster events, cyclones displace internally an average of 110,000 a year.[18] The Internal Displacement Monitoring Centre (IDMC) also reported that 427,000 people were displaced as of the end of 2022.[19] The flash flood of 2022 severely affected nine Northeastern districts, damaging 1133 km^2 of croplands, 44,254 water ponds, and 49,885 sanitation facilities.[20]

It should be noted that most of the disaster scenarios described above result in internal displacement. Internally displaced persons (IDPs) have been defined by the National Strategy on Internal Displacement Management (NSIDM) 2021 based on the definition drafted in the Peninsula Principles.[21] The NSIDM defined IDP as:

> Persons, group of persons, households, or an entire community who have been forced or obliged to flee or to leave their homes or places of habitual residence temporarily or permanently or who have been evacuated as a result of disasters caused by sudden and slow-onset climatic events and processes, and who have not crossed an internationally recognised State border.[22]

A recent research study mentioned that the human rights implication of climate change in Bangladesh is still severe and all sorts of human rights, such as the right to life and security, right to health, right to water, housing, food, sanitation, and education, were violated due to climate change.[23] Women in the southwest coastal area

---

[12] Climate Change Initiatives of Bangladesh, Achieving Climate Resilience [6].

[13] 'Hazard means any unnatural incident which is created by natural law, due to technical faults or by humans and as a result bring down the normal lifestyle of peoples into danger and risk through occurring devastation and create sorrows and sufferings including devastating and irreparable damages to necessary items to maintain livelihood' Disaster Management Act, 2012, Act no 34 of 2012, sec 2 (2).

[14] National Plan for Disaster Management (2021–2025) (Ministry of Disaster Management and Relief, 2020) iv.

[15] Sylhet Flash Floods: Situation & Support, https://reliefweb.int/report/bangladesh/sylhet-flash-floods-situation-support, accessed on 26 September 2023.

[16] National Plan for Disaster Management (2021–2025), (n 14) iv.

[17] National Adaptation Plan of Bangladesh (2023–2050) (Ministry of Environment, Forest and Climate Change, Government of the People's Republic of Bangladesh, 2022) 45.

[18] IDMC, (n 11).

[19] Ibid.

[20] National Adaptation Plan of Bangladesh, (n 17) 25.

[21] The Peninsula Principles on Climate Displacement Within States, 2013. https://reliefweb.int/report/world/peninsula-principles-climate-displacement-within-states-2013, accessed on 1 September 2023.

[22] National Strategy on Internal Displacement Management (Ministry of Disaster Management and Relief.
Government of the People's Republic of Bangladesh 2021), 5.

[23] Climate Injustice vs Gender Justice: Why should this Matter? https://cprdbd.org/study-sharing-seminar-climate-injustice-vs-gender-justice-why-should-this-matter/ accessed on 27 September 2023.

of Bangladesh have been facing several violations of their gender rights.[24] Climate change-related displacement also creates a severe impact on elderly people due to their generally weaker physical and mental condition and dependency on young adult family members.[25] Therefore, the human rights of climate-affected people not only require proper addressing in the normative discourse but also in practice.

3 Conceptual Framework of an HRBA

Since the adoption of the Charter of the United Nations (UN) in 1945, human rights issues have been the key foundation of the activities of the UN.[26] To respect, protect, and promote human rights have been the mandate of all the member states of the UN based on the Charter of the UN.[27] Therefore, an HRBA is argued for, encouraged, and promoted at the policy level of the government as a fundamental principle of state policy, for a country such as Bangladesh.[28] An HRBA could also be understood as a:

> conceptual framework for the process of human development that is normatively based on international human rights standards and operationally directed to promoting and protecting human rights. It seeks to analyse inequalities which lie at the heart of development problems and redress discriminatory practices and unjust distributions of power that impede development progress and often result in groups of people being left behind.[29]

An HRBA is 'expressly grounded in international human rights law that regulates the conduct of states ('duty bearers') in relation to the people living within the state's jurisdiction ('rights holders'). Principles, standards, and guidelines developed through the international system for the protection of human rights are integral to this approach.'[30] This approach can be understood from the report of the Office of the UN High Commissioner for Human Rights which says:

> Mere charity is not enough from a human rights perspective. Under a human rights-based approach, the plans, policies and processes of development are anchored in a system of rights and corresponding obligations established by international law. This helps to promote the sustainability of development work, empowering people themselves—especially the most marginalized—to participate in policy formulation and hold accountable those who have a duty to act … A human rights-based approach identifies *rights holders* and their entitlements and corresponding *duty-bearers* and their obligations and works towards strengthening the capacities of rights-holders to make their claims and of duty-bearers to meet their obligations.[31]

Major elements of an HRBA include under governance: transparency and accountability; under procedural functions: participation, consultation, and access to information; under substantive policies: express focus on Fundamental Rights, non-discrimination, and equality; and a strong focus on traditionally marginalised groups.[32] These HRBA elements are important not only for the smooth realisation of human rights for the socially marginalised and vulnerable groups, but also to hold duty bearers responsible for human rights violations and guarantee remedies to the victims. So, an HRBA focuses on

[24] Ibid.

[25] Malak et al. [7].

[26] Universal Values, Principle One: Human Rights-Based Approach (n 9).

[27] See articles 1, 13, 55, 62, 68, 76 of the Charter of the UN.

[28] Human rights matters are stated in Chapter II of the Constitution of Bangladesh, The Constitution of the People's Republic of Bangladesh, Act of 1972, Chapter II.

[29] Universal Values Principle One: Human Rights-Based Approach, https://unsdg.un.org/2030-agenda/universal-values/human-rights-based-approach, accessed on 18 November 2023.

[30] Scott et al. [8].

[31] Office of the United Nations High Commissioner for Human Rights (n 18).

[32] Md Abdul Awal Khan and Matthew Scott, Bangladesh National Law and Policy Report, Displacement in the Context of Disasters and Climate Change (Raoul Wallenberg Institute of Human Rights and Humanitarian Law, Sweden and Independent University, Bangladesh, 2020) 2.

developing the capacities of both 'duty-bearers' to meet their obligations and 'rights-holders' to claim their rights.[33] It also empowers people to claim their rights primarily under a state mechanism.[34] The most crucial aspect of the approach might be in incorporating international human rights principles into national laws, policies, and institutional frameworks.

3.1 The National Legal and Policy Framework

Bangladesh has been playing an active role in international negotiations. It is a party to major climate change instruments including UNFCCC, the Kyoto Protocol, and the Paris Agreement. Bangladesh has made many legal and policy frameworks to streamline action to address climate change-related impacts in line with international mandates and commitments.

Bangladesh signed and ratified the Paris Agreement in 2016.[35] The Paris Agreement requires all state parties to report regularly on their emissions and implementation and to put forward and strengthen their best efforts to mitigate climate change through the Nationally Determined Contributions (NDCs).[36] Bangladesh submitted its only NDC report in 2021. However, the NDC is presumed to be a narration of the achievements, initiatives, and actions taken so far by the Government of Bangladesh against climate change and disaster without a particular focus on disaster and displacement from a human rights point of view.

It is also important to mention that for a long time, the climate change regime did not account for human rights concerns, nor did it account for the linkage between climate change and disaster. There was an explicit reference to the linkage between climate change and disaster management with the adoption of the Cancun Adaptation Framework and the Outcome Statement of the Rio +20 conference in 2012.[37] Regarding a human rights dimension, human rights issues mentioned in the Paris Agreement contain their very first mention in a climate change treaty.[38] The Preamble to the Paris Agreement contains an acknowledgement 'that climate change is a common concern of humankind' and that 'Parties should when taking action to address climate change, respect, promote and consider their respective obligations on human rights.'[39] Based on these international commitments, a policy and legal framework on climate displacement, with a human rights-based approach, has been developed in Bangladesh.[40]

(i) *The Constitution of Bangladesh*:

The Constitution of Bangladesh has specific provisions to protect and promote the human

[33] UNSDG, Human Rights Based Approach, https://unsdg.un.org/2030-agenda/universal-values/human-rights-based-approach, accessed on 30 September 2023.

[34] Empowerment, Inclusion, Equality: Accelerating sustainable development with human rights, https://www.ohchr.org/sites/default/files/Documents/Issues/MDGs/Post2015/EIEPamphlet.pdf, accessed on 30 September 2023.

[35] Bangladesh, https://unfccc.int/node/61013, accessed on 30 April 2023. Paris Agreement adopted by 196 Parties at the UN Climate Change Conference (COP21) in Paris, France, on 12 December 2015, entered into force on 4 November 2016.

[36] 'Each Party shall prepare, communicate and maintain successive Nationally Determined Contributions that it intends to achieve. Parties shall pursue domestic mitigation measures, with the aim of achieving the objectives of such contributions.' Article 4 (2), Paris Agreement.

[37] See Cancun Agreement, Future We Want—Outcome document, https://sustainabledevelopment.un.org/futurewewant.html, accessed on 1 October 2023.

[38] 'Acknowledging that climate change is a common concern of humankind, Parties should, when taking action to address climate change, respect, promote and consider their respective obligations on human rights, the right to health, the rights of indigenous peoples, local communities, migrants, children, persons with disabilities and people in vulnerable situations and the right to development, as well as gender equality, empowerment of women and intergenerational equity.' Preamble. Paris Agreement. See also Duyck et al. [9].

[39] Paris Agreement (n 35) preamble.

[40] Subsequent sections discuss the policy and legal framework in detail.

rights of its citizens particularly their judicially enforceable Fundamental Rights[41] as well as the Fundamental Principles of State Policy,[42] which are not judicially enforceable. The Preamble of the Constitution expressly guarantees fundamental human rights and freedom, equality, and political, economic, and social justice for all citizens and the government is responsible for ensuring these rights.[43] Article 11, although contained in the non-binding part of the Constitution, also guarantees these fundamental human rights and freedoms and respect for the dignity and worth of a human being.[44] The Constitution, though it does not explicitly mention climate change, has provisions pertaining to the protection and care of the environment. Protection and improvement of the environment are mentioned as Fundamental Principles of State Policy.[45] Moreover, provisions for ensuring necessities, such as food, clothing, shelter, education, and medical care, are also mentioned as fundamental principles.[46] As climate change directly impacts those rights, therefore, the government bears a constitutional responsibility to protect them. Again, rural development and agricultural revolution,[47] public health and morality,[48] equality of opportunity, including the participation of women in all spheres of national life, and the right to work[49] are also identified as Fundamental Principles of State Policy of the Government of Bangladesh, as are equality of opportunity in public employment,[50] right to protection of law,[51] and right to property.[52] The right to be free from discrimination on the basis of disability is also protected as a fundamental right under the Constitution of Bangladesh.[53] Therefore, the Government of Bangladesh is constitutionally responsible if the above-specified rights are directly or indirectly affected in the context of disasters, including in situations of disaster displacement.[54]

(ii) *Bangladesh Climate Change Strategy and Action Plan (BCCSAP), 2009.*

The BCCSAP refers to all forms of human mobility, including by migration, displacement, and planned relocation (or 'resettlement'). It highlights the need to strengthen coastal embankments to prevent coastal outmigration, riverbank erosion, and saline intrusion.[55] It mentions that increased riverbank erosion and saline water intrusion in coastal areas will likely displace hundreds of thousands of people.[56] It states that if sea-level rise is higher than currently expected and coastal polders are not strengthened or new ones built, six to eight million people could be displaced by 2050 and will have to be resettled.[57] Specifically, the BCCSAP states that people in many parts of the country will be so adversely affected by climate change that they will be compelled to relocate, both inside and outside the country, and that close monitoring is necessary, along with adequate institutional support, for their proper resettlement.[58] The BCCSAP therefore emphasises building multi-purpose cyclone shelters for coastal Bangladesh as one way to address

[41] The Constitution of the People's Republic of Bangladesh (n 28) Art 26-47A.

[42] Ibid, Articles 8–25.

[43] Ibid, Preamble.

[44] Ibid, Article 11.

[45] Ibid, Article 18A: 'The State shall endeavour to protect and improve the environment and to preserve and safeguard the natural resources, bio-diversity, wetlands, forests and wild life for the present and future citizens.'.

[46] Ibid, Article 15.

[47] Ibid, Article16.

[48] Ibid, Article 18.

[49] Ibid, Article 20.

[50] Ibid, Article 29.

[51] Ibid, Article 31.

[52] Ibid, Article 44.

[53] Ibid, Article 28.

[54] Md Abdul Awal Khan and Matthew Scott (n 32) 8.

[55] Government of the People's Republic of Bangladesh, Bangladesh Climate Change Strategy and Action Plan (BCCSAP) 2009 (Ministry of Environment and Forests, 2009), 1.

[56] Ibid.

[57] Ibid, 1.

[58] Ibid, 59.

displacement risk in sudden onset contexts.[59] Notably, the BCCSAP not only identifies climate change threats but also calls for enhanced monitoring of the internal and external migration of adversely affected people and requests support through capacity building to enable their rehabilitation in a new environment.[60] The BCCSAP addresses specific programmes on migration via six pillars that include developing a monitoring mechanism for internal and external migrations, developing a protocol to provide adequate support for Climate Displaced People's (CDP) resettlement and rehabilitation, and building capacity through education and training to facilitate CDP's resettlement in a new environment.[61] The BCCSAP acknowledges the contribution of civil society, NGOs, research organisations, and development partners for its formation. However, implementation remains the government's responsibility via internal and external financing.[62] The document does not describe the details of the monitoring mechanism of the BCCSAP.[63] The BCCSAP targets coastal regions as the most vulnerable area for CDP and states that the people's livelihoods in these areas will be lost and agricultural production is significantly reduced.[64] There are 44 programmes[65] listed in the BCCSAP based on the six pillars[66] of action.

Despite the various plans and programmes of the BCCSAP, there remain gaps between national and sectoral policies, plans, and practices, such as weak implementation processes, and coordination gaps regarding the implementation of the project activities during and after the project.[67] Moreover, it was identified that only infrastructural support was provided among the victims under the project of the BCCSAP, with no other durable solutions provided.[68] Less priority was given to 'Food Security, Social Protection and Health' with respect to project allocation under the BCCSAP's thematic area.[69]

(iii) *National Adaptation Plan (NAP), 2023–2050*

The National Adaptation Plan (NAP) has been prepared to fulfil the country's obligation at the international level. The NAP has been adopted as one of the core planning and investment documents for Bangladesh to adapt to climate change up to 2050. The NAP primarily covers eight specific areas, including, 'water resources; disaster, social safety, and security; agriculture; fisheries, aquaculture, and livestock; urban areas; ecosystems, wetlands, and biodiversity; policies and institutions; and capacity development, research, and innovation.'[70] The NAP also aims to achieve six specific goals: '(i) ensuring protection against climate change and disasters; (ii) developing climate-resilient agriculture; (iii) building climate-smart cities; (iv) protecting nature for adaptation; (v) integrating adaptation into planning; and (vi) ensuring capacity-building and innovation in adaptation.'[71] Moreover, other cross-cutting issues, such as infrastructure, health, gender, elderly persons, persons with disabilities, youth among ethnic communities, and other socially disadvantaged groups, and the private sector are also covered.[72] The NAP also

---

[59] Ibid, 12.

[60] Walsham [10].

[61] BCCSAP (n 55), 2–3.

[62] Ibid. The MoEF is tasked with developing and overseeing the implementation of the BCCSAP. The involvement of other stakeholders to implement the BCCSAP is not officially included. For further discussion, see Sect. 4.2.3.4.

[63] Ibid, xv.

[64] Ibid, 3.

[65] Ibid, 32.

[66] The six pillars of the BCCSAP are food security, social protection, and health; comprehensive disaster management; infrastructure; research and knowledge management; mitigation and low carbon development; and capacity building and institutional strengthening.

[67] Shakila Yasmin, Implementation of Bangladesh Climate Change Strategy and Action Plan (BCCSAP, 2009): Gaps between Policy and Practices, (2018) 3 (4), *European Journal of Social Sciences Studies*, 25.

[68] Ibid.

[69] Ibid, 22.

[70] National Adaptation Plan of Bangladesh (2023–2050), (n 17) iii.

[71] Ibid.

[72] Ibid.

aims to build Bangladesh as a climate-resilient country with six national adaptation goals[73] targeted to be fulfilled through '23 broad-scale strategies and 28 outcomes encompassing diverse aspects of safeguarding against climate-induced disasters.'[74] The NAP gives importance to planned resettlement options to reduce and manage climate-induced displacement or migration.[75] A few initiatives have already been taken, such as housing and shelter for landless, homeless, and displaced people under the Ashrayan (shelter) Project at Langolerhat, Rangpur, and Cox's Bazar.[76] Moreover, the Government of Bangladesh has built a few crop storage and shelters for livestock in northeastern Haor areas. The NAP also mentions the development of city climate action plans for major urban and peri-urban areas, highlighting the need to create strong resilience for urban-poor communities, particularly for the climate-displaced people.[77]

The NAP acknowledges that substantial financial support is required from national and international authorities to implement these strategies.[78] Locally Led Adaptation (LLA) in line with the community preferences' approach is one of the interesting approaches of the NAP, incorporating social inclusion in different adaptation interventions.[79] A monitoring and evaluation (M&E) framework is suggested in the NAP to identify and assess outcomes of the implementation process, but there are no detailed provisions in this regard.[80] However, a human rights focus on internal displacement remains inadequate under the NAP, as there is no clarity regarding how the LLA approach could be applied to internal displacement.

Besides the above-specified climate policies, the Government of Bangladesh has adopted a detailed legal framework governing disaster management, which is also relevant in the context of climate change displacement.

(iv) *Disaster Management Act, 2012 (DMA)*

The sole legally binding instrument relating to disaster management in Bangladesh was enacted to 'make the activities about disaster management coordinated, object-oriented and strengthened and to formulate rules to build up the infrastructure of effective disaster management to fight all types of disaster.'[81] By adopting the DMA, climate change-induced vulnerabilities have been covered within the jurisdiction of the definition of disaster.[82] The DMA defines

[73] Ibid, Goal 1: Ensure protection against climate change variability and induced natural disasters; Goal 2: Develop climate-resilient agriculture for food, nutrition, and livelihood security; Goal 3: Develop climate-smart cities for improved urban environment and well-being; Goal 4: Promote nature-based solutions for conservation of forestry, biodiversity, and well-being of communities; Goal 5: Impart good governance through integration of adaptation into the planning process; and Goal 6: Ensure transformative capacity building and innovation for climate change adaptation (CCA).

[74] Ibid.

[75] Ibid, 58.

[76] Ibid, 5. Under the scheme totalling 3,700,160 inhabitants have been rehabilitated. 'Specially designed Machang & houses for hilly and char areas are considered in this scheme along with housing for other vulnerable areas... "Khurushkul Special Ashrayan Project" launched in 2014–15, which is one of the world's biggest housing projects for climate refugees. Under this project, 139 five storied buildings with modern facilities have been set to be constructed in Cox's Bazar, with a capacity to rehabilitate 4409 climate refugee families. Till date, 20 buildings have been constructed and a total of 640 climate victim families have been allotted a 406 sq. ft. flat each in its first phase. Training programmes and loan disbursements are going on to make the rehabilitated families financially self-reliant. The project also focuses on mitigation through plantation of 1.5 million trees, rainwater harvesting, the solar panel based alternate power sources, improved cook stoves etc. Besides this, 50,104 families have been rehabilitated under the "Guchchhogram" project.'.

[77] National Adaptation Plan of Bangladesh (2023–2050), (n 17) 65.

[78] Ibid.

[79] Ibid.

[80] Ibid.

[81] Disaster Management Act, 2012, Act no 34 of 2012, Preamble.

[82] Abdullah Al Faruque, Environmental Law: Global and Bangladesh Context, (New Warshi Book Corporation, 2017) 280. Disaster Management Act, 2012, Act no 34 of 2012, sec 2 (f). 'Disaster means any such incidents mentioned below created by nature or human or created due to climate change and its massiveness and devastation cause such damage....Any unnatural incident or a misfortune causing massive life loss and damage;'.

disaster as an incident caused by nature, human-made, and climate change.[83] The Act aims:

> to mitigate overall disaster, conduct post-disaster rescue and rehabilitation program with more skill, provide emergency humanitarian aid to vulnerable communities by bringing the harmful effect of disaster to a tolerable level through adopting disaster risk reduction programs, and enact rules to create effective disaster management infrastructure to fight disaster to make the activities of concerned public and private organizations more coordinated, object-oriented and strengthened to face the disasters.[84]

At the institutional level, the National Disaster Management Council was established by the DMA to provide guidelines to respective departments and authorities about formulating relevant plans and policies, such as the Standing Orders on Disaster (SOD).[85] The enforcement of the SOD, including effectively incorporating the displacement-related provisions into the domestic legal system, is also mentioned in the DMA.[86] In the DMA, disaster-induced displacement issues are identified in section 16, where providing resources, services, emergency shelter, transport, and other facilities is mentioned as one of the responsibilities of the National Disaster Response Coordination Group.[87] 'Disaster zone declaration and participation of different forces'[88] to manage disaster are also relevant provisions by which displacement and other related vulnerabilities have been addressed under the DMA.

Enacting a legally binding instrument like the DMA has made Bangladesh's governmental responsibility to manage disasters mandatory. Based on the DMA, various policies and frameworks have been developed[89] which is a significant development in the domestic legal structure, as it might help reduce disaster-related vulnerabilities, including displacement. In addition, the elements of an HRBA can be detected in various provisions of the DMA, such as, 'protection and risk reduction for ultra-poor and underprivileged community especially the older persons, women, children and handicapped persons while providing assistance.'[90] Compensation for disaster caused due to negligence or intention of persons or organisations to be granted through the civil courts has also been mentioned in the DMA.[91]

Nevertheless, the provisions of the DMA seem to have overwhelmed duty bearers (government) with all the responsibilities and have led to the formation of too many committees.[92] On the other hand, there are no specific provisions for the rights holders (disaster-affected people) with respect to their rights and the methods of how to ensure them through appropriate mechanisms. It is relevant to note that beyond the DMA, another statutory law, namely the Climate Change Trust Act, which was adopted in 2010 attempts to 'redress the adverse impact of climate change on Bangladesh and to take measures on other matters relating thereto'[93] and one of the objectives of the Act is to 'initiate and implement suitable action plan for implementation of special programme regarding climate

[83] Disaster Management Act (n 81) section 2(11).

[84] Ibid, Preamble.

[85] Ibid, section 4.

[86] Md Abdul Awal Khan and Matthew Scott, (n 32) 18.

[87] Disaster Management Act (n 81) section 16.

[88] Ibid, Chapter 3.

[89] Such as The Standing Orders on Disaster 2019, National Adaptation Plan, etc.

[90] Disaster Management Act, 2012, Act no 34 of 2012, section 27. 'The government, through specified method by rule, may take program to provide necessary assistance for proper rehabilitation or to reduce the risk of affected or vulnerable community due to disaster: Provided that preference would be given on protection and risk reduction for ultra poor and underprivileged community specially the older persons, women, children and handicapped persons while providing assistance.'.

[91] Ibid, section 49.

[92] National Disaster Management Council, comprising 41 members headed by the Prime Minister, National Disaster Response Coordination Group, comprising 13 members, seven various National level disaster management committees and six Local Level Disaster Management Committees and Groups.

[93] Climate Change Trust Fund Act, 2010 Act no LVII of 2010, Preamble.

change.'[94] The DMA has been further enhanced by the Standing Orders on Disaster 2019.

(v) *The Standing Orders on Disaster 2019 (SOD)*

The Standing Orders on Disaster (SOD) have been prepared to determine roles and responsibilities and detailed work plans of each ministry, division, department, and agency to manage disaster-related risks at every stage of a disaster.[95] The objective of the SOD is also to take necessary measures to implement emergency management in various projects and Disaster Risk Reduction-related activity. The provisions of the SOD were drafted after considering the Sustainable Development Goals (SDGs), Sendai Framework for Disaster Risk Reduction, and other international agreements.

The SOD is divided into seven chapters. 'Disaster Risk Management Policy Structure' is discussed in Chap. 2. Disaster risk management coordination at the national and local levels is discussed, respectively, in Chaps. 3 and 4.[96] The responsibilities and functions of the ministries, divisions, departments, and agencies for Disaster Risk Management are discussed in Chap. 5. The responsibilities of field-level officials, representatives of local government, and humanitarian agencies are discussed in Chapter 6. More importantly, the role of the National Disaster Emergency Coordination and Operation Centre and humanitarian response activities are discussed in the last chapter. Although the SOD does not address any concrete measures to reduce the risk of displacement at the initial stage, it provides enhanced protection measures during evacuation and throughout displacement, which is highly appreciable.[97] The SOD determines the tasks of the various committees, where performance and support of the local committees are found weaker to response and policy implementation.[98] There is also evidence that a lack of interaction and coordination between the project implementation and the district, union, and *upazila* (sub-district) level officer and the district committees is a common scenario.[99] Therefore, non-compliance of the SOD is apparent at the ground level. Disaster management has also been given a fillip with the adoption of the National Strategy on Internal Displacement Management.

(vi) *National Strategy on Internal Displacement Management (NSIDM) 2021*

The National Strategy on Internal Displacement Management (NSIDM) is the sole policy document addressing displacement as a central issue in the context of disasters and climate change in Bangladesh. The specific focus of the NSIDM includes 'disaster displacement and human mobility challenges in local, national and regional DRR strategies; to integrate strategies with regional climate change actions; and to initiate systematic data collection on disaster displacement.'[100] Therefore, the approaches of the NSIDM cover cross-border displacement issues as well, although a viable action plan on regional DRR might be too ambitious. Other focus areas of the NSIDM include 'prevention of displacement and protection during displacement, durable solutions, institutional arrangements and funding monitoring and evaluation.'[101] The National Action Plan (2022–2042) to implement the NSIDM was also adopted in 2022.[102] Voluntary mobility is internationally

[94] Ibid, section 6.

[95] Government of the People's Republic of Bangladesh, 'Standing Orders on Disaster' (Ministry of Disaster Management and Relief, 2019) 1.

[96] Ibid.

[97] Md Abdul Awal Khan and Matthew Scott, (n 32) 18.

[98] Ahmed et al. [11].

[99] Alam and Ray-Bennett [12].

[100] National Strategy on Internal Displacement Management (Ministry of Disaster Management and Relief, Government of the People's Republic of Bangladesh, 2021) 3.

[101] Ibid, 2.

[102] Ibid, 4.

considered as a climate change adaptation strategy, which is reflected in the NSIDM.[103]

The NSIDM is also framed as setting out an inclusive and realistic rights-based policy framework that ensures the protection of the rights of disaster and climate-induced internally displaced persons in pre-displacement, during displacement, and post-displacement phases.[104] The objectives of the NSIDM also include 'preventive and adaptive measures to minimise the internal displacement caused by climate-related disasters.'[105] One of the significant provisions of the NSIDM is outlining strategic responses as durable solutions. Suggested durable solutions of the NSIDM are:

- Return to the place of origin after the disaster is over.
- When return to the place of origin is not possible, local integration in the place of self-relocation.
- Where none of the above-mentioned solutions are feasible, in those cases, planned relocation to a third location under a government initiative.[106]

The NSIDM has formed departmental and sub-departmental committees for monitoring and evaluation purposes and an oversight committee with relevant national and international stakeholders.[107] However, the NSIDM's monitoring and evaluation strategy require involving more civil society members along with government officials to ensure transparency and accountability. The participation of civil society stakeholders has been mentioned, but their comprehensive role and responsibilities are not detailed in the NSIDM. The National Plan for Disaster Management 2021–2025 promulgated also covers broader aspects of disaster management.

[103] Ibid.

[104] Ibid, 4.

[105] Ibid, 6.

[106] Ibid, 29.

[107] Ibid, 40.

(vii) *National Plan for Disaster Management 2021–2025 (NPDM21–25)*

The National Plan for Disaster Management (NPDM21–25) is an ambitious policy document of the Government of Bangladesh with 34 core targets to build a resilient nation by 2025. The NPDM21–25 is the upgraded version of the previous policy documents, namely the National Plan for Disaster Management (NPDM) 2016–2020, 2010–2015. The NPDM 2021–25 is essentially based on the SOD 2019 and is in line with the priorities of the Sendai Framework for Disaster Risk Reduction (SFDRR) and also connected to the Asian Regional Plan for Disaster Risk Reduction (ARPDRR).[108]

Among the various goals the NPDM21–25 targets various strategies to reduce economic losses in every disaster, such as Disaster Risk Reduction (DRR), Humanitarian Response, and Emergency Recovery.[109] The NPDM21-25 is the latest policy framework based on previous experiences in Disaster Risk Management and international DRR frameworks.[110] The NPDM21-25 is also drafted to implement the Disaster Management Act 2012 and Standing Orders on Disaster 2019 and prepare the Annual Work Plans for the respective Ministries/ Departments of the government and other relevant stakeholders.[111] Managing displacement issues is prioritised in Priority 4 of the document which reads as follows:

> Priority 4: Enhancing disaster preparedness for an effective response and to "Build Back Better' in recovery, rehabilitation, and reconstruction – Concerned action plans include strengthening forecasting and early warning systems, emergency response capacity building, sector-wise preparedness, inclusive recovery and rehabilitation, business continuity, and multi-hazard response and recovery measures.[112]

[108] National Plan for Disaster Management (2021–2025) (n 14) ii.

[109] Ibid, 36.

[110] Ibid.

[111] See National Plan for Disaster Management (2021–2025) (n 14)) ii.

[112] Ibid.

Under Priority 4 of the NPDM21-25 for 'Enhancing disaster preparedness for effective response, recovery, and reconstruction,'[113] a 'NSIDM' has been developed, and 'effective response to disasters and related displacement, including access to safe shelter, essential food, and non-food relief supplies, as appropriate to local needs' matters are included. However, the NPDM21-25 does not clearly focus on the measures that can be undertaken to prevent displacement when disaster strikes.[114] Although Disaster Risk Reduction issues are highly prioritised in the NPDM21-25, considering the complex phenomenon of displacement, a clear focus on essential aspects of an HRBA[115] is still missing. The target of the NPDM21-25 is building a resilient nation, particularly by acting to save lives and reduce economic losses in every disaster, humanitarian, and emergency response[116] within the given time frame, but practically, the current trend of the NPDM21-25 indicates that it would not be able to achieve the target completely by 2025 since less than 2 years are left to achieve the same.

Other relevant policy documents of the Government of Bangladesh on climate change and disaster include the Mujib Climate Prosperity Plan (MCPP) 2022–2041, Bangladesh Delta Plan 2100 (prepared in 2018), Bangladesh Climate Change Trust Act 2010, Bangladesh Climate Fiscal Framework 2020, National Action Plan for Clean Cooking 2020–2030, Plan of Action to Implement Sendai Framework for Disaster Risk Reduction 2015–2030, Bangladesh Energy Efficiency and Conservation Master Plan up to 2030, Renewable Energy Policy of Bangladesh 2008, Bangladesh National Action Plan for Reducing SLCPs 2012, updated in 2018, Cyclone Shelter Construction, Maintenance and Management Policy 2011 (CSCMMP).

[113] Ibid, 40.

[114] Md Abdul Awal Khan and Matthew Scott (n 32) 19.

[115] See section II for elements of HRBA.

[116] National Plan for Disaster Management (2021–2025) (n 14) 6.

3.2 Judicial Response to Climate Change and Displacement and Reliance on a Human Rights-Based Approach

Climate change litigation issues have received global attention in the last decade through huge academic development in legal and social sciences literature.[117] Due to the substantial growth of climate litigation literature focusing on case law, new emerging legal dimensions and avenues, types of actors, litigation objectives, and jurisdictions have created opportunities for climate litigation. However, there are no specific instances of climate change litigation in Bangladesh and there are no specific provisions regarding the role of the judiciary in addressing climate change-related impacts and governance.

Despite the given fact, there are options in the legal system of Bangladesh particularly through Public Interest Litigations (PILs) by which the High Court Division of the Supreme Court of Bangladesh can issue orders for protecting and promoting human rights, which are acknowledged as Fundamental Rights.[118] A PIL is considered one of the most useful tools for protecting the rights of the poor and marginalised people. Therefore, judicial responses to climate change and displacement matters may be invited through a PIL, by any bonafide citizen or organisation, if those matters are considered as a violation of Fundamental Rights. This constitutional remedy through the judiciary dealing with climate change and displacement could only be initiated if the court is satisfied that no other equally efficacious remedy is provided by law.[119] The higher judiciary can also take *suo*

[117] Peel and Osofsky [13].

[118] Article 102 of the Constitution of Bangladesh. 'The High Court Division on the application of any person aggrieved, may give such directions or orders to any person or authority, including any person performing any function in connection with the affairs of the Republic, as may be appropriate for the enforcement of any of the Fundamental Rights conferred by Part III of this Constitution….'.

[119] Ibid.

moto actions if any matter is drawn to its attention. Climate litigation is also possible through the mandate of Article 32 of the Constitution of Bangladesh where 'protection of right to life and personal liberty' has been ensured as one of the Fundamental Rights of the citizens. Moreover, if there are gaps in the domestic legislation in addressing any issue, the Supreme Court of Bangladesh may consider international conventions and protocols on that issue for appropriate guidelines and directives to be followed by all concerned until a national legislation is enacted on that issue.[120] Therefore, international law could be used as a guide for interpreting domestic law, if any climate change victims file litigation against the Government of Bangladesh for failure to carry out their climate change-related obligation. The *Leghari* case in Pakistan,[121] brought by a farmer, about failure to implement climate adaptation measures as well as the *Daniel Billy* case[122] at the Human Rights Committee—brought by Torres Straits Islanders also about failure to take adequate climate adaptation measures, is good examples of how the rights of climate affected people were addressed under domestic jurisdiction.

4 Major Challenges to Mainstreaming a Human Rights-Based Approach in the Legal System of Bangladesh

This section examines the significant challenges of managing the issue of internal displacement within Bangladesh with a human rights focus, in the present time of globalisation. Mainstreaming an HRBA into the legal system is not an easy task for developing countries like Bangladesh, where the human rights protection mechanism remains weak. Moreover, managing climate-displaced people within national jurisdiction presents many common challenges in Bangladesh, including resources, capacity, and finance. As disaster strikes suddenly, therefore, systematic coordination among the actors to respond to the extensive and immediate action remains a daunting task. Generally, cyclones, tropical storms, and floods take a heavy toll on the coastal areas of Bangladesh. There is evidence that the disaster situation gets worse in the remote islands of the central coastline of Bangladesh because access becomes difficult due to insufficient transportation facilities.[123] Therefore, people living in remote and coastal areas face more human rights suffering compared to other disaster-affected areas. Hence, an HRBA is lacking at the level of implementation, particularly for the people living in the vulnerable areas.

In addition, as discussed earlier, the primary focus of the SOD, NAP, DMA, and other relevant policy documents is to provide immediate humanitarian support for the disaster victims. Once the disaster is over, people usually go back to their original place of living from where they were displaced, because no long-term opportunities are available for them in the shelter area, such as education, health care, employment, and living conditions. Here, an HRBA becomes a pivot for mainstreaming the rights of the disaster victims to enable sustainable displacement management.[124] However, the Constitution of Bangladesh only guarantees the protection of fundamental civil and political rights,[125] while economic, social, and cultural rights, such as, the right to water, right to food, and right to shelter, health, and education, are not guaranteed

[120] Supreme Court of Bangladesh, BNWLA v. Government of Bangladesh, [2009], Petition No. 5916 [2008].

[121] Asghar Leghari v. Federation of Pakistan [2015] W.P. No. 25501/201.

[122] Daniel Billy and Others v Australia [2019] CCPR/C/135/D/3624/2019.

[123] The three 'R' challenges to disaster risk resilience in the coast, https://www.dhakatribune.com/epaper/292936/the-three-%E2%80%98r%E2%80%99-challenges-to-disaster-risk, accessed on 28 September 2023.

[124] Alam and Endacott [14].

[125] The Constitution of the People's Republic of Bangladesh, (n 28) Part III.

in the Constitution of Bangladesh because those rights are Fundamental Principles of State Policy which are judicially non-binding.[126] Due to this constitutional nature, the mainstreaming of the economic, social, and cultural rights of the climate-displaced people at the policy level remains challenging. On the other hand, the deep and complex relationship between climate change displacement and human rights has not been recognised so far in the legal system of Bangladesh, notwithstanding the endorsement of the approach in the NPIDM. Therefore, managing internal displacement and related issues will require long-term government plans by mainstreaming a human rights-based approach at a national policy level.

Generally, unexpected and sudden displacement results in internal disturbance, which might destabilise socio-political and economic conditions. Moreover, ensuring the core elements of a human rights-based approach, such as good governance, transparency, and accountability, in the legal system of Bangladesh, is also a challenge because of lack of political will. A recent scenario shows that only 15% of homeless people had received government support in the Sylhet district when the devastating floods affected many people.[127] According to media and government sources, 40,091 houses were damaged in the Sylhet district, but it has been pointed out by civil society that the actual number was higher than the reported number.[128] The local authority said that only 6000 of the 40,000 houses could be reconstructed with the government's allocation of money.[129] Therefore, the response strategy, early warning, and preparedness issues are identified as inadequate and should be strengthened.[130] Another report also found that lack of transparency, accountability, and discrimination observed on the government's part in dealing with recent disasters, particularly in respect of relief distribution and the construction of shelters and dams.[131]

The Government of Bangladesh has put serious effort into formulating various national plans, strategies, policies, and laws for improved and effective disaster management, yet the reality on the ground is quite different. Procedural gaps, lack of participation of the disaster-affected people in the decision-making process, inadequate consultation with relevant stakeholders, and limited access to information, transparency, and accountability remain significant shortcomings to effective disaster management.[132] Moreover, integrating international human rights standards and their application in national legislation always remains one of the bigger political challenges of developing countries like Bangladesh[133] because human rights violations are a frequent incident.[134] Again, legal and institutional capacity buildings are also challenges to implementing an HRBA in the local legislation of Bangladesh.

[126] Ibid.

[127] Sylhet Flood: Only 15% Homeless People to get Government Assistance, https://www.dhakatribune.com/nation/2022/07/16/sylhet-flood-only-15-homeless-people-to-get-govt-assistance, accessed on 22 September 2023.

[128] Ibid.

[129] Ibid.

[130] See also Govt. Support for Flood Victims Still Inadequate https://www.newagebd.net/article/174833/govt-support-for-flood-victims-still-inadequate, accessed 22 September 2023.

[131] Disaster management: TIB Finds Lack of Govt. Transparency, Accountability https://www.thedailystar.net/country/news/disaster-management-tib-finds-lack-govt-transparency-accountability-2016741, accessed 28 September 2023.

[132] National Human Rights Commission Bangladesh, introduction, http://www.nhrc.org.bd/site/page/1c65dfa1-f9c2-48e9-a66b-eab8de75d9b1/, accessed 28 September 2023.

[133] Human Rights in Bangladesh: A Midterm Assessment of Implementation During (The UPR 3rd Cycle, 2020).

[134] Bangladesh ranked 124th out of 139 countries in Global Rule of Law Index. WJP (World Justice Project) (2021) World Justice Project Rule of Law Index 2021. https://worldjusticeproject.org/rule-of-law-index/country/2021/Bangladesh/, accessed on 28 September 2023.

5 Conclusion

An HRBA in domestic jurisdictions, like Bangladesh, still presents academic and policy challenges due to its complex and unique nature. Moreover, the legal system of Bangladesh is not fully aligned with international human rights standards. The paper mostly catalogues some of the important legal and policy documents and then provides some insight into implementation challenges. This study has also explored judicial responses to address the rights of climate change victims and various challenges to implementing a human rights-based approach at the policy level in Bangladesh. Bangladesh has made significant progress in terms of law, policy, and practice embracing various aspects of climate change and human rights. It is also one of the only countries to expressly adopt a policy on disaster displacement with an expressly human rights-based approach. Despite this development, implementation, particularly an emergency response strategy in more vulnerable areas, remains a challenge. Planned relocation for displaced people in particular is still a daunting task in Bangladesh. Therefore, integrating systematic and planned relocation with an HRBA is required in relevant climate policy action. Therefore, the implementation mechanism should be stronger, particularly at the ground level of the administration. Otherwise, the reflection of a human rights-based approach at the academic and policy levels would not bring any positive result to the victims of climate change unless and until enforcement gaps exist. Substantial Loss and Damage financing is also crucial to cope with anticipated increases in disasters and displacement, but transparent and fair management of those funds should be ensured. This study attempts to cover those issues and argues for a stronger national law, policy, and institutional framework addressing the gaps identified in this research.

References

1. Preston BJ (2018) The evolving role of environmental rights in climate change litigation. Chin J Environ Law 2:131
2. Khan MA (2015) Impact of climate change on human rights of the displaced people: Bangladesh perspective. Ph.D. thesis (2015). https://researchdirect.westernsydney.edu.au/islandora/object/uws:32643. Accessed on 15 Dec 2023
3. Rozario SR, Chowdhury M (2023) Climate induced displacement in Bangladesh through the lens of loss and damage, 2 May 2023. https://researchinginternaldisplacement.org/short_pieces/climate-induced-displacement-in-bangladesh-through-the-lens-of-loss-and-damage/. Accessed on 30 Sept 2023
4. National Adaptation Plan of Bangladesh (2023–2050) Ministry of environment, forest and climate change, Government of the People's Republic of Bangladesh, 2022, p 45
5. Scott M (2022) Internal displacement in Asia and the Pacific: a human rights-based approach to law, policy and practice. Forced Migr Rev 69:64–67
6. Climate Change Initiatives of Bangladesh, Achieving Climate Resilience (2022) Ministry of Environment, Forest and Climate Change, p 2
7. Malak MA et al (2020) We are feeling older than our age: vulnerability and adaptive strategies of ageing people to cyclones in coastal Bangladesh. Int J Dis Risk Red, abstract 48
8. Scott M et al (2023) FIRE: a framework for integrating human rights and gender equality in disaster risk reduction and climate change adaptation. Clim Dev 15(7):622
9. Duyck S, Lennon E, Obergassel W, Savaresi A (2018) Human rights and the Paris agreement's implementation guidelines. Carbon Clim Law Rev 2(3):191
10. Walsham M (2010) Assessing the evidence: environment, climate change and migration in Bangladesh. IOM, pp 32–34
11. Ahmed et al (2016) Exploring implementation gaps between policy and practice for disaster management, Bangladesh. J Integr Dis Risk Manage 6(2):79, 86
12. Alam E, Ray-Bennett NS (2021) Disaster risk governance for district-level landslide risk management in Bangladesh. Int J Dis Risk Reduct 59(1):4
13. Peel J, Osofsky HM (2020) Climate change litigation. Annu Rev Law Soc Sci 16:21
14. Alam S, Endacott J (2022) Mainstreaming internal migration in law and policy frameworks in Bangladesh: analysis of a rights-based approach to a wicked policy problem. J Environ Dev 31(3):300

Dr. Md. Abdul Awal Khan has contributed to legal education, research, and training since 2005. Currently, he has been serving as a Professor of Law at the Independent University, Bangladesh (IUB). Dr. Khan has served as an Additional Director of the Institutional Quality Assurance Cell (IQAC), IUB, and his teaching position. He obtained his Ph.D. at the School of Law, Western Sydney University, Australia 2015. He was a visiting researcher at the School of Global Studies, University of Gothenburg, Sweden, and the Department of Social Sciences, University of Roehampton, London, UK, as an Erasmus Mundus Visiting Scholar in 2017. Dr. Khan completed several projects funded by Raoul Wallenberg Institute (RWI). He published numerous research articles in national and international journals. His area of expertise includes environmental law, climate change law, policy, and human rights.

Climate Change, Agriculture, and Internal Human Mobility in the Bhutan Himalayas

Om N. Katel, Anooja Nair, Ugyen Yangchen, and Chogyel Wangmo

Abstract

When it comes to the impact of climate change, Bhutan is one of the world's most vulnerable countries. Although Bhutan has not contributed anything to cause climate change, it is facing the brunt of it impact. Bhutan's climate is strongly influenced by the rugged mountainous terrain surrounding it as well as the Indian summer monsoon and changes in climate, which are increasing the vulnerability of vital economic sectors such as agriculture, forestry, and water resources. Climate change also impacts the country's rich biodiversity and fragile community livelihoods through the increased risks of hazardous events such as glacier lake outburst floods, flash floods, droughts, and forest fires. This chapter analyses the impacts of climate change on agricultural livelihoods and associated internal human mobility. Despite being a small country, Bhutan has been doing its best to adapt to the impacts of climate change, but its topography and inadequate resources due to the small economy have been constraining Bhutan's developmental philosophy of Gross National Happiness and its achievement of its global sustainable development goals.

O. N. Katel (✉) · C. Wangmo
Department of Environment & Climate Studies, College of Natural Resources, Royal University of Bhutan, Punakha, Bhutan
e-mail: ombhutan2016@gmail.com

A. Nair
Department of Food Science & Technology, College of Natural Resources, Royal University of Bhutan, Punakha, Bhutan

U. Yangchen
Department of Sustainable Development, College of Natural Resources, Royal University of Bhutan, Punakha, Bhutan

1 Introduction

Owing to its diverse mountain topography, rugged terrain, and large variations of altitude over short distances, Bhutan is one of the most vulnerable and exposed countries in the world to the peril of climate change. Hydropower, agriculture, and tourism are the major sources of revenue; however, these sectors happen to be highly sensitive to the impact of climate change. With changing temperatures and precipitation patterns, more than 90% of the river system in Bhutan is expected to be affected due to snow-melts and glacier retreats.[1] Although the carbon sequestration in Bhutan is about three times greater than its national greenhouse gas emissions, yet, the country faces the brunt of climate change impact.[2] One

[1] Molden et al. [1].

[2] National Environment Commission, Royal Government of Bhutan, 'Third National Communication' (2021) 71–118. http://www.nec.gov.bt/publications/climate-change, accessed 20 November 2023.

S. Jolly et al. (eds.), *Climate-Related Human Mobility in Asia and the Pacific*, Sustainable Development Goals Series, https://doi.org/10.1007/978-981-97-3234-0_7

of the impacts often dominating the discourses includes the forced human mobility of communities from their homes pushing them to extreme poverty.[3] Such an impact severely undermines the efforts of Bhutan in attaining critical development pathways to achieve the 2030 agenda for sustainable development. On the other hand, inaccessible and sparse hydrometeorological data makes it difficult to understand the interactions of natural ecosystems and the social dynamisms of climate impact. Considering such information gaps in the complexity of climatic change and its impact, this chapter covers the inter-linkages of climate change and agriculture-based livelihoods and internal human mobility including a review of the policies and strategies put in place by the Government of Bhutan.

2 Climate Change in the Bhutan Himalayas

According to Rao,[4] disaster results from a combination of endogenous and exogenous shocks to physical, economic, and social systems, whereby the adaptive capacity of communities weakens with the loss of economic and social support. This usually happens during the occurrence of climate extreme events. When extreme events occur with greater frequency, the adaptation capacity of most communities is lost or significantly reduced. In Bhutan, wind-storms, cyclones, floods, and landslides are the major climatic events that have caused significant loss and damage to lives, livelihoods, and property. There have been more than 12 major disaster incidences in Bhutan since 1985.[5] In these contexts, the need to maximise efforts by all entities, individually and collectively to minimise the effects on all fronts, is crucial.

The Hindu Kush Himalaya assessment report reveals that warming in the Himalayas will be higher than in other regions.[6] Bhutan is specifically located in the eastern Himalayas covering the foothills in the south to the inner Himalayas in the mid zone and the high Himalayas in the north within a range of 150 kms. In the context of Bhutan, natural hazards and extreme events have significantly increased in the recent past[7] and such extreme events appear to be associated with temperature rise. Topographically, the annual average temperature in Bhutan ranges from 25 °C in the sub-tropical south to minus 15 °C at the highest elevation in the north.[8] Such wide differences in temperature mean that the impact of global environmental change will have different mean levels or intensity of implications at the different elevation zones in these mountainous landscapes, making it difficult to take adequate steps to prevent or mitigate disasters as the need may be.

Along with the projected temperature increase, there is also the projection of increased precipitation levels over South Asia.[9] The seasonal precipitation projections show large variability with precipitation increasing in the Western Himalayas and decreased precipitation levels in general towards the Eastern Himalayas. The simulation of temperature and precipitation in Bhutan and South Asia show that until the year 2069, the temperature in both the regions will increase as much as 1.3–3.5 °C.[10] It is also expected that some regions in South Asia will experience drought with decreasing trend and some other would experience increasing trend of as much as 37% compared to the current level of precipitation. This means that some regions such as western India may experience drought, whereas some regions of South Asia such as Bhutan will have to deal with significant flooding, landslides, and erosion. On the other hand, the likelihood of heatwave and droughts is also expected to increase and such incidences will

[3] Chhogyel et al. [2].

[4] Pinninti [3].

[5] World Bank Climate Change Knowledge Portal, 'Bhutan Vulnerability' accessed on 15 September 2023.

[6] Krishnan et al. [4].

[7] Pörtner et al. [5].

[8] Eguchi[6].

[9] Ibid., 45.

[10] Chhogyel and Kumar [7], pp. 1–13, 4.

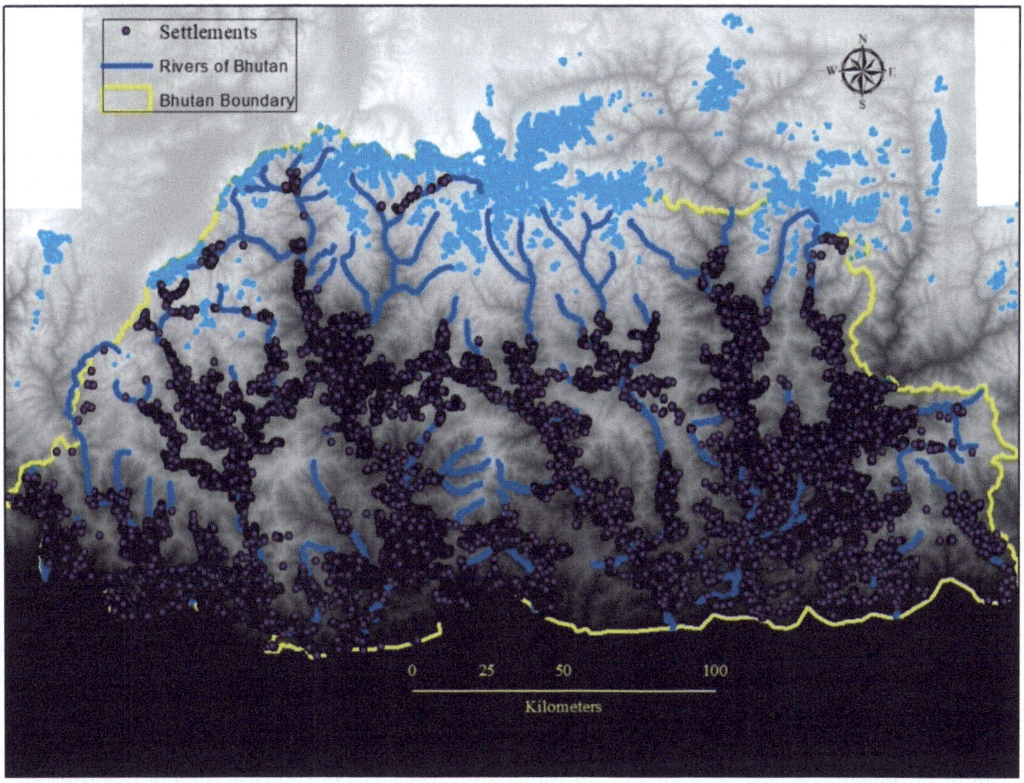

Fig. 1 Glaciers within and around Bhutan. *Source* Authors

directly affect the farming communities more than any other occupations. As a result, many biological resources will undergo significant changes ultimately affecting the socio-economic well-being of communities such as mountain farmers who depend significantly upon natural resources for their livelihood.

Bhutan is located between two areas with very contrasting climatic conditions, such as the north Indian plains with relatively higher precipitation and Tibet's dry, continental highlands which makes up the climatic complexity in the eastern to the central Himalayan region,[11] while in the southern foothills of Bhutan, precipitation is influenced by the Indian monsoon blowing against Bhutan's southern foothills and towards the high Himalayas.[12] As one moves from the southern foothills to the northern high Himalayas, the amount of precipitation declines along the altitudinal gradient. In the foothills, the average precipitation is recorded as about 3000–6000 mm per year while in the high Himalayas and central Himalayas, it is about 1500–2500 mm and 600–800 mm per year, respectively. Rainfall received in the regions of the area with windward-oriented mountain ranges is relatively elevated, even as much as double or even triple.[13] Thus the already existing climatic variability at its different locations owing to high topographic variations, by virtue of their complex biophysical characteristics coupled with the intricate network of mountains and narrow valleys, make Bhutan, one of

[11] Shah et al. [8].

[12] Hoy et al. [9].

[13] Takashi Eguchi and Pema Wangda 'Difference in Temperature Between Shallow and Deep Valleys of the Bhutan Himalaya' (2002) *J-Global* [6, 9].

the most vulnerable countries to global climatic change. Such a complexity with significant climate change poses significant risks and hazards to crop farming and human well-being.

In addition to the physical constraints due to the mountainous landscapes, glaciers account for about 3.4% of the country's total surface area and Bhutan is home to a total of 677 glaciers and 2674 glacial lakes.[14] About 66 glaciers within Bhutan (Fig. 1) recorded between the years 1963 and 1993 show an average retreat rate of 8.1%, positively correlated with temperature. Critically, those glaciers also showed a higher rate of melting leading to reduction in snow cover and in the volume of the glaciated areas.[15] This would continue to alter the environment in the coming decades faster than it did in the last several decades. For instance, significant areas of glaciers have decreased (about 11%) in the last 25 years (1990 to 2015).[16] Such melting of glaciers is compounded by the Indian monsoon during the summer, increasing the sediment loads that are discharged posing a significant danger to the infrastructure in the downstream areas. Floods particularly Glacial Lake Outburst Floods[17] (GLOFs) not only affect Bhutan but have a major impact further downstream in the Brahmaputra basin, affecting even Bangladesh, which is categorised as one of the most densely populated areas in the world.[18]

The risk of landslides in the mountainous topography of Bhutan increases manifold, especially during the monsoon rains and flash floods. Such incidences of climate events that have been occurring in the past years have shown relative increase more recently. Floods are the most common occurrence particularly affecting the river basins where most of the farmland and agricultural infrastructure are located. Increasing incidences of floods and droughts also pose a major risk to infrastructure and hydropower projects which account for more than 25% of the total revenue in the country's Gross Domestic Product, due to the significant change in the amount of water discharged.[19] In urban areas, flooding and heat stress have been pronounced already and Bhutan needs to prepare for these climate risks and hazards in order to reduce its vulnerability. Recent reports published by the UNDP[20] in Bhutan show that Bhutan is experiencing the relatively high risks and hazards emanating from climatic hazards.[21]

3 The Impact of Climate Change on Agriculture

Climate change can have a negative impact on agricultural systems such as farming in the mountainous and steep slopes which is relatively more susceptible to its impact.[22] In Bhutan, most of the farm lands are located on steep slopes with more than 50% of the farmers practising subsistence farming and most of these farm plots being rain-fed. Given that Bhutan practises subsistence crop farming, which is highly dependent on precipitation and temperature, any changes in these variables would be expected to have a significant impact on agriculture.[23] When agriculture is affected, this would certainly lead to food security issues. Consequently, there appears to be limited prospects of enhancing economic activity owing to its location and hence, as a land locked, resource poor country, Bhutan, is likely to face significant environmental and economic changes in the future. This is not only relevant for human well-being but also for natural ecosystems. As Bhutan is located in one of the biodiversity hotspot regions of the world, it would

[14] Chophel GK et al. [10].

[15] Tshering and Fujita [11].

[16] Ji et al. [12].

[17] Rinzin et al. [13].

[18] Biswas [14].

[19] Tariq et al. [15]

[20] UNDP [16].

[21] The World Bank Group and the Asian Development Bank Climate Risk Country Profile: Bhutan (2021) [13–16], https://www.adb.org/sites/default/files/publication/722636/climate-risk-country-profile-bhutan.pdf, accessed on 28 September 2023.

[22] Bhalerao et al. [17].

[23] Chogyel et al. [18].

lose a significant number of species or at least experience a distinct alteration in the behaviour of species. The following section elaborates on the specific hazards and livelihood impact with respect to climatic impact.

3.1 Impact on Crop Farming and Livestock Rearing

A large number of the Bhutanese populace is employed in the agricultural sector (more than 50%), and agriculture is extremely vulnerable to climatic conditions particularly with changing monsoon cycles and farming periods.[24] The projections until the year 2069 show that precipitation in Bhutan will increase significantly and such increasing precipitation may pose a risk to the limited agricultural land in Bhutan. This is particularly so because landslides and soil erosion are two major drivers of land degradation. Arable land in Bhutan accounts for less than three per cent of the total land area and about 30% of this land is located on slopes that are vulnerable to landslides and soil erosion.[25] With the increasing trend of precipitation levels, the already vulnerable land is likely to be degraded completely.

Changing temperatures and precipitation patterns affect crop yields as well, and as a result, its contribution to the GDP would diminish significantly. It is a fact that crop yields have already begun showing a declining trend. For instance, in the year 2000, agriculture contributed 26.8% of the total GDP, in 2011 this reduced to about 18%, and in 2022, it has accounted for a mere 12%.[26] Smaller land holdings, climate change impact, land degradation, decreasing areas for productivity (28,000 ha in 1981 to 20,000 ha in 2017), decrease in farming population, and a decrease of about four per cent in investments have deeply affected the agriculture sector.[27] While some variables may not be directly associated with climatic impact, variables such as investments are indirectly associated with climate change.

Furthermore, crops and livestock are also susceptible to various pests and diseases, and with climate change, it is argued, which will lead to new diseases among livestock as well as more pests affecting crops. The appearance of new pests and diseases are already reported in many cases such as army worms, giant African land snails, fruit fly in citrus fruits, citrus greening, late blight in potatoes, turcicum leaf blight, grey leaf spot in maize, ginger rot, cardamom rot, and maize root rot. Such pests and diseases eventually reduce crop productivity. For instance, maize harvest in the year 2021 was recorded at 30,939 metric tonnes, which is 10,026 metric tonnes less than the yield in 2020. The decline is about 24%, and 24,321 metric tonnes less than the 55,259 metric tonnes in 2018, which is believed to have associated to occurrence of several pests and diseases.[28] The increase in the pests and the diseases is found to have occurred due to changes in the temperature and precipitation patterns.

Agriculture including crop farming and animal husbandry is integral to rural development. In this context, the availability of water plays a significant role in crop and animal production. Bhutan is known to have rich water resources, with per capita mean annual flow availability at 109,000 cubic metres.[29] However, most of the farmers depend on monsoon rain, small streams, and spring water for crop cultivation. It is challenging for farmers to access perennial rivers for irrigation to their farms, owing to the topography and gradient of the landscape. A relatively small majority of the farmers has access to stream water, but recent reports have documented the drying up of the water sources in Bhutan. Consequently, as a result of increasing scarcity of water across Bhutan, farmers are experiencing an increasing trend of food shortage and incidents of conflicts concerning water sharing among farmers in the rural areas due to

[24] Ibid.

[25] Katwal et al. [19].

[26] National Statistical Bureau [20].

[27] Parker et al. [21].

[28] National Statistics Bureau, Agricultural Survey Report (2021) Royal Government of Bhutan 1–91 [20].

[29] Tariq and others (n 19) 24, 27.

a dwindling water supply are also on the rise. The cases of water related conflicts are only expected to increase in the future.[30]

In the year 2011, the Pemagatshel district in the south eastern region of Bhutan experienced a severe dry spell during the months of May and June, and in the same year, hundreds of acres of land were left fallow, as farmers were unable to cultivate any crop due to the drought during the farming season. Due to the mountainous landscape, creating watersheds is important so that spring water can be stored to provide water for farmers. This is one of the reasons why Bhutan has always protected its forests so as to protect its water sources in the mountains.

There are approximately 186 watersheds in the country that are fed by various sources such as lakes and marshes. However, it is alarming to note that the International Centre for Integrated Mountain Development states that water sources in the Himalayas are drying up at a relatively faster rate than in any other region and that Bhutan is a vulnerable area that is going to be critically affected.[31] The effect of the drying up of the water sources may become apparent with the passage of time and with the slow onset events in the long run; however, the extreme events are leading to crop loss and damage. Since water is a fundamental resource, the increasing scarcity of this resource would ultimately affect the local people forcing them to look for more viable alternatives.

Similarly, incidences of extreme weather events have significantly affected farmers as cultivated crops are damaged. The evidence reveals that the frequency of events has significantly increased in the recent decade than in the 2000s, 1990s or the 1980s.[32] Extreme events include flash floods, high-intensity rainfall, wind-storms, hailstorms, landslides, crop diseases, and pests. The loss of crops puts additional pressure on farmers in the rural area where production is already constrained by the slow onset of climatic events such as changing temperature and precipitation patterns. When the livelihood of farmers is put at risk, they have no other option but to migrate elsewhere for their safety and well-being. Climatic events not only jeopardise the livelihood of farmers but also affect culture and tradition. In Bhutan, culture, tradition, and biodiversity are closely related in the concept of *dru-na-gu*, which, in the national language, means nine basic crops that are offered to appease deities and gods during rituals. When the composition of the nine basic crops or the value of the *dru-na-gu* is incomplete, the culture and tradition are also put at risk and hence the local people's tendency to migrate elsewhere becomes strong, as the need to cling to one's ancestral place is ultimately diminished.

3.2 Livelihood Impacts

Bhutan has a shortage of agricultural land as with less than three per cent of the total area being cultivable. Further, only about seven per cent of this land is arable,[33] and most of it is located on the steep slopes of the mountains. As the crop yield has been decreasing over the years, it has led to a major food shortage among the farmers and consequently an increasing trend towards food imports.[34] Food security, therefore, has become a pressing issue in the rural areas, and with further climatic impacts, this phenomenon is likely to worsen in the

[30] Hoy and Katel [22].

[31] Tshering Dendup, Water Sources Drying up in the Himalayan Nations: ICIMOD; BBS, Water sources drying up in the Himalayan nations: ICIMOD - BBSCL, accessed on 20 September 2023.

[32] Chogyel and Kumar [7], pp 2–13 [8]; see also National Center for Hydrology and Meteorology, Record of Extreme Events in Bhutan, Annual Report, 2018, 2019, 2020 & 21, [3, 5, 6].

[33] UNDP, 'Turning Slopes, Dry Land into Viable Agricultural Land in Trongsa' https://www.undp.org/bhutan/stories/turning-slopes-dry-land-viable-agricultural-land-trongsa, accessed on 29 September 2023.

[34] Sangay Wangchuk and others, 'Rural Depopulation and Empty Rural Houses in Bhutan: How Different Stakeholders Interpret the Local Term *Gungtong*' (2023) 43 (1) *Mountain Research and Development*, 8, 9.

future. When crops fail in rural areas, farmers find themselves in difficult situations and are forced to find alternative ways to sustain their life. The failure of crops also means that farming continues to become more expensive as the output would always appear in the negative compared to the input cost for farming. In such a case, farmers migrate elsewhere especially to urban centres leaving their lands fallow. In the rural areas in Bhutan, there are significant number of abandoned houses or in Bhutan's national language what are known as *Gungtongs*.[35] Such a phenomenon of *Guntongs* also means that all the lands registered to abandoned houses remain fallow and also a lesser number of farmers engage in food production leading to a decreasing food supply to the urban areas resulting in food insecurity. As a landlocked mountainous country, with a unique culture and tradition, the issue of food security becomes crucial given that agriculture is the backbone of economic activities for a subsistence farming community.

The mountains of Bhutan are home to these subsistence farmers who are highly dependent on natural resources for their livelihood; however, these natural resources are very sensitive to climatic change. Bhutan being located in the eastern Himalayan region is also among the ten biodiversity hotspots as it is home to multiple species of flora and fauna that are in the sensitive zone particularly from the point of view of conservation and these too are vulnerable to climatic changes.[36] Therefore as a mountainous country, it faces significant challenges currently and is bound to face more in the future. Climate change also threatens the significant poverty reduction success that Bhutan has achieved in the recent decades. While facing such adaptation challenges, it is likely that people especially living in the rural areas would fall back into poverty and ultimately would be forced to migrate to the urban areas. According to the University of Notre Dame's Global Adaptation Initiative ranking index for 2020, Bhutan is ranked 94th among 182 countries with respect to the country's vulnerability to climate change.[37]

4 Human Mobility and Rural Agricultural Livelihoods

The majority of the farmers living on the steep slopes in the mountains have small landholdings and practise subsistence farmers.[38] Given the dangerous trend of the drying up of spring water, decreasing crop yield, and increasing incidences of human-wildlife conflicts, it is likely that events such as these would exacerbate rural to urban human mobility or at least internal human mobility. Although not declared officially, there is a sense that many of these migrants could be classified as a climate-displaced population.[39] Additionally, human mobility could lead to struggles for control over scarce resources, exerting greater pressure on the state.

The rate of urban expansion and human mobility from the rural areas is highly connected in Bhutan. Data reveals that in 2005, about 70% of the people lived in rural areas, and the expansion rate in the capital city was recorded as seven per cent.[40] In 2009, the internal human

[35] Ibid., 10.

[36] Philip Wester and others, *The Hindu Kush Himalaya Assessment—Mountains, Climate Change, Sustainability and People* (Springer Nature 2019) 128–157.

[37] University of Notre Dame, Global Adaptation Initiative (ND-GAIN), 'Rankings', 2022a, https://gain.nd.edu/our-work/country-index/ rankings, accessed on 24 September 2023.

[38] Pradeep Rai and others, 'Empirical Evidence of the Livelihood Vulnerability to Climate Change Impacts: A Case of Potato-based Mountain Farming Systems in Bhutan' (2022) 14 (4) *Sustainability* 14, 16.

[39] Emma Johnson, 'Climate Change and Refugees in Bhutan: The Future Impacts' in Amit Ranjan, Rajesh Kharat, and Pallavi Deka (eds) *Environment, Climate Change and Migration in South Asia* (Routledge 2023) 12,14.

[40] Mayur Gosai and Leanne Sulewski, 'Urban Attraction: Bhutanese Internal Rural–Urban Migration' (2014) 31 (1) *Asian Geographer* 1–16, 5.

mobility rate was six per cent,[41] and by 2040 or so, it is estimated that about 70% of the country's population will be living in urban areas.[42] In such a situation, there is likely to be a major labour shortage in the rural areas, ultimately exerting pressure on the socio-economic fabric of the country. The present situation also shows evidence of the fact that more and more male members (about 54%) from rural households are migrating out to urban centres leaving female members behind. However whether the reasons for this human mobility are specifically the result of climate change impact or not is not known.[43] It can be argued that when farmers leave their lands to turn fallow, such land would eventually be covered by forests. Although increasing forests can contribute to the carbon sequestration capacity of Bhutan but it is also a fact that forests cannot be a substitute for agricultural crops, in one hand, and more forests may also mean more wildlife population that may exacerbate the already existing human-wildlife conflicts, on the other, and such a shift can pose a risk to food security.

Currently, the capital city of Bhutan alone has an overall population of 13,8736, which accounts for about 20% of the country's total population. It is also a fact that a major proportion of this population consists of people who have migrated from rural areas,[44] thereby exhibiting a relatively higher net positive rate of human mobility. In fact this rate of human mobility appears to be one of the highest in South Asia.[45] Although the direct link associating human mobility with climate change as a driver is weak, it can be noted that the districts exhibiting net negative human mobility appear to have a relatively higher proportion of people engaged in agriculture and also with higher climate vulnerabilities.[46] It is difficult to determine through any official report, the extent to which climate change impact is a driver for human mobility of people in Bhutan. 'In 2017, 0.1% of the Bhutanese population—or 556 people—listed "Natural Calamities" as the reason for migrating, which could indicate a move because of an extreme weather event such as flooding or drought, events that are becoming stronger with climate change. This percentage was reported as 0.2 in 2005.'[47] Although these numbers have been obtained from the population and housing census data, it should be noted that climate change entails an extended period of time leading to different environmental shifts which therefore makes it difficult to pin down specific impacts in a census questionnaire. Thus, a scientific research specifically on climate change impact affecting mobility is necessary to generalise the claim. Human mobility has many dimensions and one of the reasons is that water is a fundamental resource required for drinking and farming. When farmers experience water scarcity, there is no better option than to migrate to a different place. For instance, as of April 2023, most of the drinking water sources (about 1856 water sources) in Bhutan have dried up.[48] More than half of the farmers who cultivate paddy in Dagana district have migrated elsewhere due to the shortage of water.[49]

This issue is part of a larger public debate in the country, with a recent newspaper article projecting that

[41] Ibid.; see also Mayur Gosai and Leanne Sulewski, 'Attraction and Detraction: Migration Drivers in Bhutan' in Irudaya Rajan (ed) *Migration in South Asia: IMISCOE Regional Reader* (Springer 2023) 111–125, 115.

[42] Dorji Yangka and others, 'Carbon Neutral Policy in Action: The Case of Bhutan' (2019) 19 *Climate Policy* 672–687, 682.

[43] Mayur Gosai and Leanne Sulewski, 'Internal Migration in Bhutan' (2020) *Internal Migration in the Countries of Asia: A Cross-National Comparison* 229–247, 237.

[44] Ibid., 241–242.

[45] Jamyang Choda, 'Rural Out-Migration Scenario in Khaling Gewog, Trashigang, Eastern-Bhutan' (2012) 6 (2) *Journal of Agroforestry and Environment* 29–32, 31.

[46] UNDP (n 20) 10–15

[47] Johnson (n 39) 84.

[48] Choki Wangmo, '*1856 Water Sources Drying Up in Bhutan*' Kuenselonline, 16 April 2023 < 1856 water sources drying up in Bhutan > accessed on 28 September 2023.

[49] Ibid.; see also '*284 Water Sources in Dagana Dried Up or Drying*', *Kuneselonline*, 11 April 2023, https://kuenselonline.com/284-water-sources-in-dagana-dried-up-or-drying/, accessed on 28 September 2023.

> ...the drying up of water sources will eventually affect the agriculture system, and it will have a greater impact on country's food security and sufficiency, ... When the land becomes uncultivated due to scarcity of water, people living in rural areas will be forced to migrate towards urban areas whereby it will lead to rural-urban migration.[50]

In Bhutan, the internal rate of human mobility is the highest among all the countries in South Asia.[51] The major proportion has moved from rural areas that are relatively more vulnerable, to the less vulnerable areas, and in many cases, migration has been from rural areas to urban areas.[52] Another notable feature is the fact that areas with relatively higher climate hazard incidents exhibited relatively higher incidences of poverty.[53] It can be deduced that there is also an overlap of incidences of poverty and vulnerability probably because farmers lose their income through various other ways. It could be that some of the most cited reasons recorded were in search of economic opportunity, land fragmentation, natural disasters, lack of capital, and lack of labour. Some of the reasons may appear to be solely economic, but it can be deduced that among subsistence farming communities, agriculture forms the major economic activity, and therefore, the reasons cited may be associated with agriculture either directly or indirectly.[54] A study involving 240 samples from six districts in western and central Bhutan indicates that to adapt to climate change impact such as erratic patterns of rainfall, unpredictable weather patterns, and drying of irrigation sources, farmers relate to untimely rainfall (57%) and droughts (45%) as the most common weather events and believe that migrating to an urban area would be the better option for their well-being.[55] It is difficult to ascertain completely, and it is solely climatic events that influence human mobility from one place to the other as there exists a knowledge gap around climate change, climate-induced, or climate-influenced human mobility in Bhutan. Thus, a proper scientific research and study to understand rural–urban dynamics are highly recommended as this will in turn help informed decision-making for policymakers and implementers of sustainable development.

5 Policies and Strategies to Address Issues of Climate Change Impacts and Human Mobility

Bhutan has recognised climate change impact as one of the serious challenges to sustainable development as it faces a series of existential challenges. At the global level, the country is committed to addressing the global agenda of sustainable development by incorporating developmental goals in the country's five-year plans where thematic areas such as poverty reduction, equitable socio-economic development, environmental sustainability, gender equity, good governance, and community participation are prioritised for implementation. These agenda points are included in Bhutan's policy document, namely Bhutan 2020,[56] National Environment Strategy,[57] and the Bhutan water vision and policy. These documents specifically highlight the limits of environmental sustainability, low-carbon and climate-resilient development, and the need for integrated water resources management, respectively. These policies and plans are formulated to enhance

[50] Jigme Wangchen, Drying Up of Water Sources, a Major concern – Business Bhutan, https://businessbhutan.bt/drying-up-of-water-sources-a-major-concern/, accessed on 5 September 2023.

[51] Gosai and Sulewski (n 40) 11–12.

[52] UNDP (n 20).

[53] Ibid.

[54] Tshewang Dorji and Rajvinder Kaur, 'Rural Out Migration of Youth in Bhutan: Issues and Challenges' (2023) 12 (5) *Eur. Chem. Bull* 2506–2517, 2513.

[55] Chogyel and others (n 23) 7–11.

[56] Royal Government of Bhutan, Bhutan 2020: A Vision for Peace, Prosperity and Happiness 1999, 1–33, 26.

[57] National Environment Commission, The Middle Path: National Environmental Strategy (2019) Royal Government of Bhutan. 125–126, 126.

self-reliance and inclusive socio-economic development in the country.

Currently, climatic events such as unpredictable growing seasons, floods, flash floods, and droughts have become a common phenomenon in the country. Therefore, strong policy interventions are required to slow or reduce these harmful patterns. Although Bhutan has enacted various climate change policies to address climate-related issues, but implementation at the local level is relatively slow. At the national level, Bhutan has made significant progress in strengthening institutional structures to address the issues of imbalanced spatial economic development and climate change. For example, a National Centre for Hydrology and Meteorology, an autonomous entity was established in 2016[58] and is mandated for weather monitoring, forecasting, and early warning. This centre is responsible for assessing and informing the nation about any hazards or risks pertaining to climate change so that appropriate action can be taken by relevant agencies.

There are some examples that merit highlighting such as the fact that Bhutan, unlike many other countries, has shown strong political commitment to conservation and sustainability with a pledge to remain carbon neutral in its energy consumption.[59] This is expected to enhance the country's capacity to respond to the uncertain environmental challenges. The other is by enacting a national climate change policy which is expected to help address the climate-related issues in the country. Thirdly, to be transparent and to support and uphold the pledge to remain a carbon neutral country, Bhutan has submitted its third national communication to the United Nations Framework Convention on Climate Change (UNFCCC). Lastly, Bhutan has formulated the National Adaptation Programme of Action (NAPA) priority based on the level of vulnerability of different sectors such as agriculture and livestock, forestry and biodiversity, health, water resources, energy, natural disasters, and infrastructure.[60] The NAPA priority is integrated into the national five-year plans so that the budget allocated in the beginning of the five-year plans can be used in an integrated manner. Accordingly, Bhutan enacted and implemented the low-carbon development policy and also incorporated many of the points from the United Nations 2030 Agenda for Sustainable Development, into its five-year development plans. This is to ensure that greenhouse gas emissions do not exceed the country's sequestration capacity and it is also to ensure that policies and institutional responses to climate change are in accordance with the country's developmental philosophy which is sustainable and climate resilient. Five-year plans were started in Bhutan in the 1960s, and currently, Bhutan is drawing towards the end of its 13th Five-Year Plans.

Bhutan's agricultural research and development policy acknowledges that limited land for agriculture, dynamics of rural–urban human mobility, decreasing investment in agriculture, and the under-developed value chains constrained by climate change[61] are the critical areas that need immediate attention, if the country needs to move forward in sustaining a self-sufficient economy. The limited land for agriculture (2.75% of the country's total land area) is a significant bottle neck that urgently requires a suitable strategy to improve crop productivity, managing drying up water sources or the utilisation of water for development along with the development and improvement of infrastructure and addressing soil and land

[58] National Center for Hydrology and Meteorology, https://www.nchm.gov.bt/home/pageMenu/10, accessed 20 November 2023.

[59] Dorji Yangka and others, 'Carbon Neutral Bhutan: Sustaining Carbon Neutral Status Under Growth Pressures' (2023) 6 (4) *Sustain Earth Reviews*, 1–16.

[60] UNDP, 'Bhutan National Adaptation Progrmme of Action', http://undp.atendesigngroup.com/projects/bhutan-national-adaptation-programme-action-napa, accessed on 3 November 2023.

[61] Lakey and Sangay Chophel. 'Agricultural Research and Development: Policy and Program Priorities in Bhutan' (2019) 72 *Agricultural Policy and Program Framework: Priority Areas for Research & Development in South Asia* 23, 29.

management. And therefore in addition to the implementation of formal policies and strategies, Bhutan has implemented flagship projects to address these recurring issues and also to pilot test the implemented activities.

One of the sectors identified is the agricultural sector that emits relatively higher levels of greenhouse gases; however, as a country embarking on carbon neutrality, it is crucial for the government to take steps to ensure that strategies are put in place to bring down these emissions. In keeping with this, the Bhutanese government has developed an agriculture land development guideline[62] for farmers to enhance land productivity through specific land use. It is expected that by this farmers would be able to enhance their land's productivity in terms of crop yield. The government has organised advocacy programmes encouraging farmers to develop entrepreneurial skills so that available land can be made more productive. One such initiative is the organic flagship programme[63] which is expected to address the employment issue and to enhance economic opportunities in rural Bhutan. The flagship programmes are created to deliver 'immediate' and 'tangible' impacts on society and economic growth, by addressing nine priority areas that require coordination and multi-sectoral interventions. The programmes include Health, Education, Digital Drukyul, Water, Organic Bhutan, Start-up, Tourism, Waste management, and Footprints of National Day, and the implementation of these programmes is monitored directly by the Prime Minister's office. For instance, the climate-resilient theme[64] programme includes the enhancement of climate smart agriculture,[65] sustainable management of lands prone to landslides, flood prevention, national capacity for weather forecasting, mobilisation of the community to develop climate-resilient farmlands, water harvesting, and the adaptation to droughts. Climate smart agriculture is expected to reduce greenhouse gases and also enhance the resilience of farming practices. These flagship projects are expected to address rural–urban human mobility by enhancing farming practices a climate resilient in the rural areas and consequently increase the food security status of the country.

Furthermore, the climate policy of the Kingdom of Bhutan reiterates that the commitment made by the country to remain carbon neutral is consistent with the Paris Agreement.[66] This is a significant milestone in keeping the country in the forefront in addressing global issues among others. Though climate change is a global issue, yet its impact is felt locally. It is vital that the vulnerabilities at the local level must be addressed. At the local level in Bhutan, very little has been achieved so far due to the lack of resources. Local governments are not mandated to address climate change issues although the five-year plans do contain specific key result areas in an integrated manner. The issue of human mobility has been significant in rural areas and that has created a major labour shortage and, in many cases, has led to feminisation of agriculture in farming; there are also cases where lands are left fallow and this phenomenon is increasing steadily.[67] It is interesting to note that despite Bhutan having several policy instruments and legal frameworks in place

[62] Karma Cheki 'Ministry Launches Guideline for Sustainable Management of Agricultural Land' 19 June 2017 https://kuenselonline.com/ministry-launches-guideline-for-sustainable-management-of-agricultural-land/ accessed on 25 August 2023.

[63] YK Poudel 'Bhutan's Organic Goal on Track?' 23 August 2023, https://kuenselonline.com/bhutans-organic-goal-on-track/ accessed on 20 September 2023.

[64] 'Flagship Programs', https://www.pmo.gov.bt/?stm_service=flagship-programs#1591395549430-7b6add26-2177, accessed on 20 November 2023.

[65] Jigmi Wangdi 'Climate Smart Agriculture: Overcoming Challenges in Bhutan's Agricultural Sector' 07 May 2023, https://kuenselonline.com/climate-smart-agriculture-overcoming-challenges-in-bhutans-agricultural-sector/, accessed on 23 September 2023.

[66] Megan Mills,Novoa and Diana Liverman, 'Nationally Determined Contributions: Material Climate Commitments and Discursive Positioning in the NDCs' (2019) 10 (5) *Wiley Interdisciplinary Reviews: Climate Change* 1589.

[67] Ibid., 29, 34.

to provide the agriculture sector an enabling environment to develop and sustain agriculture, the share of public investment in agriculture has been diminishing. Concerning the issue of rural–urban human mobility, the government has introduced an equity-based approach by investing in education, income, and health indicators in rural areas and other welfare programmes such as subsidies, transfers, enhancement of off-farm income by rural–urban connectivity, rural service provisions, connectivity to external markets, and addressing institutional and human resources requirements, although, connectivity to external markets, achieving human resource requirements are still a challenge owing to limited resources.[68] Yet, such reforms are expected to address spatial inequalities and reduce rural–urban human mobility and ultimately enhance a more balanced socio-economic development.

6 Gaps in Policy and Practices to Address Climate Change Impacts

Bhutan has made considerable investments in water resources, infrastructure development, and soil and land management yet there has been limited input to address future climate change impacts. For instance, investments have been made in irrigation systems in the country, as majority of the farmers are small landholders and practise subsistence farming requiring irrigation channels for a continuous supply of water to their farms, but these infrastructures have not been made climate resilient. This is because most irrigation schemes were constructed as earthen canals that are in the mountainous landscape where landslides and soil erosion are frequent, and these structures therefore are not particularly efficient or resilient to extreme climate events.

Climate change impact is a cross-cutting issue where stakeholder collaboration and coordination are important; however, coordination among the institutions and stakeholders appears to be relatively weak in Bhutan. Institutional linkages require strengthening collaboration and coordination for better outcomes. Managing water resources, managing land, and improving irrigation structures definitely require more than one agency to devise a sustainable solution.

Furthermore, understanding constraints concerning biophysical, financial capacity, and technical knowledge in managing water resources in a holistic manner hinders farmers in achieving food and nutrition security.

Similarly, there has been an impetus given to farm road construction in the country providing access to farmers. Indeed having a strong road network, access to hydropower as energy and telecommunications are fundamental for rural development. These services can bring significant benefits to farmers. The construction of farm roads has indeed brought notable benefits to farmers in Bhutan. However, since these roads were constructed on mountain slopes with minimal technologies, the development has also left several negative impacts such as soil erosion and landslides, as these roads were not constructed to be climate resilient. Most of these roads now require additional upfront investment. As a mountainous country, Bhutan requires a robust financial base to reduce risks that stem from climate impacts; however, as a small country with limited prospects in business, it has a small economy, limited public funds, and competing needs for other development investments. Such competition results into a series of fragmented small investments that are unsustainable in the long term and lead to a recurrent loss of funds due to repeated repair and maintenance needs and often leaving no or limited fund for technology development and research. There is also a lack of reliable and quality data because of which it is difficult to trace the historical development of the agriculture sector. Credible data is required to understand the dynamic of climate change impact especially when a sizable investment to enhance climate-resilient agriculture is undertaken. Quality and reliable climate data is equally paramount to forecast and target

[68] Gyem Dorji, 'Increasing Rural-Urban Migration: Policy Briefs on Rural Depopulation and Urbanization in Bhutan (2022) 29 (1) *Asian Journal of Education and Social Studies* 23–27, 25.

advisories to support farmers with appropriate local actions. However, technical capacities among the hydromet, agriculture, and local governments remain inadequate. Similarly, limited disaster knowledge and information on potential sources and locations of natural hazards such as landslides, risks of floods, or drought are a major challenge.

The lack of financial and technical capability is compounded by poor coordination and minimal collaboration among stakeholders which is ultimately affecting the much-needed climate initiatives and actions. It is highly recommended that any climate-resilient development plan or policy ought to adopt an integrated sectoral approach. Furthermore, while climate change dominates the discourse at the national level and often among researchers and academia, there is still a lack of awareness among policymakers and often the policymaking is not based on research results. This is often reflected in the lack of capacity at the local level with a relatively poor understanding of localised impacts of climate change and variability. Such a situation creates a vacuum at the local level in developmental planning and managing limited resources efficiently and effectively and in building resilient infrastructure that would be able to address the climate-related hazards and risks in the rural areas. The lack of facilities and desirable infrastructure in the rural areas compels people to migrate from rural to urban centres. It is seen that rural development has been lagging behind due to the lack of synergy with an effective urbanisation policy that will promote balanced growth and boost local economies and subsequently link the rural and urban populace.

7 Conclusion

The chapter and its assessment reveal that Bhutan's climate ranges from warm and humid in the south to cold climate in the north. The Indian summer monsoon and the wintery Siberian winds affect its seasonal atmospheric circulation. Temperature and precipitation changes have changed significantly in the recent decades in Bhutan, and such a development poses strong risks to the ecology, the economy, and the vulnerability of the community. Bhutanese farmers practise subsistence agriculture and farming on limited pockets of land situated on the steep slopes of a mountainous topography of Bhutan and have experienced serious climatic impacts with crop losses owing to unpredictable rainfall patterns, wind-storms, droughts and flash floods, and other calamities in the past, as well as in the more recent years, making current agricultural practices highly vulnerable to climatic change. The impact of climate change has been huge and one of the significant developments has been human mobility especially from rural to urban resulting in a shortage of labour force and thereby threatening food and nutrition security. Owing to low productivity and high cost of input, agriculture is no longer a lucrative and viable source of income any more for people living in rural Bhutan. Although the data on the impact of climate change resulting in human mobility in Bhutan is relatively difficult to ascertain, but it is interesting to note that until the current year, more than 50% of the country's population continued to be employed in agriculture and it is one of the important sectors that contributes to the Gross Domestic Product of Bhutan. It is clear that changing climatic variables has affected crop farming in the recent years and consequently food security in the rural areas. Thus, the economic activity around agriculture and human mobility can be linked and it is apparent that the same as an compounding effect on food and nutrition security. Since climate change impacts are significant, making Bhutan one of the most vulnerables in the world, it is of vital importance to delve into research that helps design effective mitigation and adaptation strategies to address climate change impacts and its influence on human mobility.

It appears that Bhutan has been putting notable efforts to reduce the vulnerability and create resilient farming practices crafting relevant policies and introducing low-carbon development strategies, including measures for supporting farmers for entrepreneurship development. However, there has been a lack of reliable data and funds for other competing developmental needs. These are further constrained by a lack of coordination and collaboration and sectoral linkages without realising that climate change is a cross-cutting issue and linkages among different stakeholders and institutions are vital. Therefore, more robust data should be generated and shared among sectors to facilitate informed decision-making. Current research shows that mitigation and adaptation efforts must go hand in hand, but how addressing agricultural productivity and climate change impacts on farming can address human mobility should be given priority, in order to ensure sustainable development. Furthermore, the procedures concerning agricultural practices and local economic development should be streamlined, reinforced, and strengthened for climate-resilient development.

References

1. Molden DJ, and others (2022) The great glacier and snow-dependent rivers of Asia and climate change: heading for troubled waters. In: Biswas A, Tortajada C (eds) Water security under climate change. Springer, pp 223–250
2. Chhogyel N, and others (2020) Perception of farmers on climate change and its impacts on agriculture across various altitudinal zones of Bhutan Himalayas. Int J Environ Sci Technol 17:3607–3620
3. Pinninti KR (2013) Climate change loss and damage: economic and legal foundations. Springer Science & Business Media, pp 18–19
4. Krishnan R, and others (2019) The Hindu Kush Himalaya assessment mountains, climate change, sustainability and people. Springer, pp 58–91
5. Pörtner H-O, and others (2022) Climate change; impacts, adaptation and vulnerability. In: Summary for policymakers a report of working group II of the IPCC 2022, pp 3–118, 44
6. Eguchi T (1997) Regional and temporal variations in precipitation in the Eastern Himalayas. Faculty of Humanities and Economics, Kochi University, Japan, pp 55–82, 57
7. Chhogyel N, Kumar L (2018) Climate change and potential impacts on agriculture in Bhutan: a discussion of pertinent issues. Agric Food Secur 7(79):1–13, 4
8. Shah SK, and others (2022) Potential utility of Himalayan tree-ring to reveal spatial patterns of past drought variability—its assessments and implications. In: Kumaran N, Padmalal S (eds) Holocene climate change and environment. Elsevier, pp 265–293
9. Hoy A, and others (2016) Climatic changes and their impact on socio-economic sectors in the Bhutan Himalayas: an implementation strategy. Reg Environ Change 16:1401–1415
10. Chophel GK, and others (2011) Securing the natural freshwater systems of the Bhutan Himalayas: climate change and adaptation measures on water resources in Bhutan. In: A climate summit for a living Himalaya, vol 6, p 9
11. Tshering P, Fujita K (2016) First in-situ record of decadal glacier mass balance (2003–2014) from the Bhutan Himalaya. Ann Glaciol 57(71):289–294, 292
12. Ji Q, and others (2022) Variations in glacier coverage in the Himalayas based on optical satellite data over the past 25 years. Catena 214:106240
13. Rinzin S, and others (2023) GLOF hazard, exposure, vulnerability, and risk assessment of potentially dangerous glacial lakes in the Bhutan Himalaya. J Hydrol 619:8
14. Biswas A (2011) Cooperation or conflict in transboundary water management: case study of South Asia. Hydrol Sci J 8(1):662–670, 662
15. Tariq MAUR, and others (2021) A critical review of water resources and their management in Bhutan. Hydrology 8(1):1–24, 14
16. UNDP (2022) Climate change vulnerability analyses and mapping for national adaptation plan (NAP) formulation process in Bhutan. https://www.undp.org/bhutan/publications/climate-change-vulnerability-analyses-and-mapping-national-adaptation-plan-nap-formulation-process-bhutan. Accessed on 09 June 2023
17. Bhalerao AK, and others (2022) Sustainable agriculture in Northeastern India: how do tribal farmers perceive and respond to climate change? Int J Sustain Dev World Ecol 29(4):291–302, 299
18. Chogyel N, and others (2020) Consequences of climate change impacts and incidences of extreme weather events in relation to crop production in Bhutan. Sustainability 12(10):4319
19. Katwal TB, and others (2015) Community perspectives on the on-farm diversity of six major cereals and climate change in Bhutan. Agriculture 5(1):2–16
20. National Statistical Bureau (2022) Royal Government of Bhutan 1–395:88–92
21. Parker L, and others (2017) Climate change impacts in Bhutan: challenges and opportunities for the agricultural sector. CCAFS Working Paper. https://ccafs.cgiar.org/resources/publications/climate-change-impacts-bhutan-challenges-and-opportunities-agricultural. Accessed 6 Oct 2023
22. Hoy A, Katel O (2019) Status of climate change and implications to ecology and community livelihoods in the Bhutan Himalaya. In: Saikia, Thapa P (eds) Environmental change in the Himalayan Region: twelve case studies. Springer, pp 23–45, 37

23. Wangchuk S, and others (2023) Rural depopulation and empty rural houses in Bhutan: how different stakeholders interpret the local term *Gungtong*. Mount Res Dev 43(1):8, 9
24. Wester P, and others (2019) The Hindu Kush Himalaya assessment—mountains, climate change, sustainability and people. Springer Nature, pp 128–157
25. Rai P, and others (2022) Empirical evidence of the livelihood vulnerability to climate change impacts: a case of potato-based mountain farming systems in Bhutan. Sustainability 114(4):14, 16
26. Johnson E (2023) Climate change and refugees in Bhutan: the future impacts. In: Ranjan A, Kharat R, Deka P (eds) Environment, climate change and migration in South Asia. Routledge, pp 12, 14
27. Gosai M, Sulewski L (2014) Urban attraction: Bhutanese internal rural–urban migration. Asian Geogr 31(1):1–16, 5
28. Gosai M, Sulewski L (2023) Attraction and detraction: migration drivers in Bhutan. In: Rajan I (ed) Migration in South Asia: IMISCOE regional reader. Springer, pp 111–125, 115
29. Yangka D, others (2019) Carbon neutral policy in action: the case of Bhutan. Clim Policy 19:672–687, 682
30. Gosai M, Sulewski L (2020) Internal migration in Bhutan. In: Internal migration in the countries of Asia: a cross-national comparison, pp 229–247, 237
31. Choda J (2012) Rural out-migration scenario in Khaling Gewog, Trashigang, Eastern-Bhutan. J Agrofor Environ 6(2):29–32, 31
32. Dorji T, Kaur R (2023) Rural out migration of youth in Bhutan: issues and challenges. Eur Chem Bull 12(5):2506–2517, 2513
33. Royal Government of Bhutan (1999) Bhutan 2020: a vision for peace, prosperity and happiness, vol 1–33, p 26
34. National Environment Commission, The Middle Path: National Environmental Strategy (2019) Royal Government of Bhutan, pp 125–126, 126
35. Yangka D, and others (2023) Carbon neutral Bhutan: sustaining carbon neutral status under growth pressures. Sustain Earth Rev 6(4):1–16
36. Lakey L, Chophel S (2019) Agricultural research and development: policy and program priorities in Bhutan. In: Agricultural policy and program framework: priority areas for research & development in South Asia, vol 72, pp 23, 29
37. Cheki K (2017) Ministry launches guideline for sustainable management of agricultural land, 19 June 2017. https://kuenselonline.com/ministry-launches-guideline-for-sustainable-management-of-agricultural-land, Accessed on 25 Aug 2023
38. Poudel YK (2023) Bhutan's organic goal on track?, 23 Aug 2023. https://kuenselonline.com/bhutans-organic-goal-on-track/. Accessed on 20 Sept 2023
39. Wangdi J (2023) Climate-smart agriculture: overcoming challenges in Bhutan's agricultural sector, 07 May 2023. https://kuenselonline.com/climate-smart-agriculture-overcoming-challenges-in-bhutans-agricultural-sector/. Accessed on 23 Sept 2023
40. Mills-Novoa M, Liverman DM (2019) Nationally Determined contributions: material climate commitments and discursive positioning in the NDCs. Wiley Interdisc Rev Clim Change 10(5):1589
41. Dorji G (2022) Increasing rural-urban migration: policy briefs on rural depopulation and urbanization in Bhutan. Asian J Educ Soc Stud 29(1):23–27, 25

Dr. Om N. Katel holds a Ph.D. and is a faculty member at the Department of Environment and Climate Studies, College of Natural Resources, Royal University of Bhutan (RUB), Lobesa, Bhutan. Dr. Katel has 21 years of teaching experience and taught courses such as Environmental Governance, Natural Resources Management, Climate Change Vulnerabilities and Adaptation, and Integrated Watershed Management. Dr. Katel is an alumnus of Youth Encounter on Sustainability (YES), Intensive Program on Sustainability (IPoS); International Center for Climate Change and Development (ICCCAD); Alliance for Global Sustainability (AGS); and South Asian Network for the Development and Environmental Economics (SANDEE). Dr Katel is a focal person from RUB for Least Developed Countries University's Consortium on Climate Change, RUB representative for SWISSnex India, and country's representative to IPCCC. Dr. Katel is a post-doctoral research fellow at the Grassroots Institute Canada.

Anooja Nair joined the College of Natural Resources, Royal University of Bhutan, in 2012 and since then working at the Department of Food Science and Technology. She holds a Master of Science Degree in Food Science and Technology, from Wageningen University form the Netherlands and Bachelor of Science degree from India. Since joining the College, she shouldered the responsibility as program leader to develop the courses for Food Science and Technology program and subsequently served as the head of the Department. Anooja is interested in urban climate resilience, food security and safety, bee keeping and climate change, food production, packaging and distribution, and sustainable development through food innovation in the context of climate change.

Ugyen Yangchen is a faculty member at the Department of Sustainable Development, College of Natural Resources, Royal University of Bhutan, Lobesa, Punakha, Bhutan. She holds a Master's degree in Sustainable Rural Development and Bachelor of Science in Life Science. She has more than ten years of teaching and research experience in agriculture and development studies. She joined the College in 2007 and since then supervised numerous undergraduate and master thesis research. Yangchen's secured and managed projects and grants for capacity development and research in agriculture and development from within Bhutan and international donors. Yangchen is also a member of

Least Developed Countries University Consortium on Climate Change (LUCCC). Her research interests are but not limited to plant protection and human health, sustainable agriculture, food security, and sustainable rural development.

Chogyel Wangmo has an M.Sc degree in Environment and Forest from AgroParisTech and University of Lorraine, France. She has post graduate diploma in higher education from Samtse College of Education, Samtse, Bhutan and B.Sc in environmental science from Asian University of Women, Chittagong, Bangladesh. She joined the College of Natural Resources under RUB as a Lecturer in the Environment and Climate Science department from 2014. She is currently the programme leader of BSc Environment and Climate Studies. Chogyel is a technical member of Sustainable and Environmental technical committee of Bhutan Standard Bureau and managing editor of Bhutan Journal of Natural Resource and Development. She is actively involved in climate and environment related projects, training farmers, local leaders as well as field staff and extension workers of different government agencies.

Climate Change-Induced Disaster Displacement and Law in India: Positioning the Operationalization of Artificial Intelligence for Protecting Human Rights

Nafees Ahmad

Abstract

Artificial Intelligence (AI) is playing an increasing role in humanitarian space requiring the integration of human rights into AI and human rights-based approach (HRBA) into law and policy relating to the Climate Change-induced Disaster Displacement (CiDD) in India. AI has the potential to significantly impact low-emission, climate-resilient development as well as CiDD. The UN Climate Change Initiative on AI for Climate Action investigates how AI might be used as a potent instrument to advance and intensify climate action that is revolutionary and pragmatic. Therefore, the CiDD crisis must also be re-imagined and operationalised for protecting human rights instead of posing risks to inherent human dignity and human protection at the intersection of National Disaster Laws (NDL), National Climate Law Frameworks (NCLF) and International Human Rights Law (IHRL). The institutional capabilities of the AI-driven systems pose challenges to the human rights of the CiDD population that has become a threat multiplier. Therefore, AI technologies must be designed and developed to incorporate diagnostic, formative, and summative digital assessments of the CiDD situations. The digital assessments must be integrated with NDL, NCLF, and IHRL frameworks to facilitate the protection of human rights. In protecting human rights, the changing climate conditions have necessitated a specific legally binding international Human Rights Protection Framework (HRPF) during CiDD situations. Therefore, India can create AI-driven automatic decision-making (ADM) systems in collaboration with Asia–Pacific countries or otherwise to protect human rights in CiDD scenarios. Thus, in India and elsewhere, AI technologies, AI policies, and accountability systems operationalise human rights protection and contribute to attaining Sustainable Development Goals (SDGs) in climate statecraft and beyond. In this chapter, I explore the space to integrate AI into NDL, NCLF, and IHRL frameworks with the human rights-based approach (HRBA). I investigate the questions around legal and policy considerations, such as the divergences and convergences of AI norms between NDL on CiDD and the protection of human rights in India. Can AI help national governments better prepare for CiDD migration? What are the specific AI tools and AI best practices that India can adopt for operationalizing the ADM systems and integrating

N. Ahmad (✉)
Faculty of Legal Studies, South Asian University, New Delhi, India
e-mail: drnafeesahmad@sau.ac.in

S. Jolly et al. (eds.), *Climate-Related Human Mobility in Asia and the Pacific*, Sustainable Development Goals Series, https://doi.org/10.1007/978-981-97-3234-0_8

them with the NDL and IHRL framework? Can AI be used to direct and operationalize the IHRL framework in the CiDD crisis? Should AI Ethics be invoked to plug gaps in CiDD emergency and raise standards above the minimum requirements of IHRL obligations in India? Would the age of algorithmic humanitarianism be sufficient for protecting human rights in CiDD prosceniums?

1 Introduction

Artificial Intelligence (AI) technology is permeating more areas of everyday life and every aspect of human life will be changed by AI-driven machines: driverless cars will dominate the roads, AI-powered robots will staff medical and manufacturing floors, AI will control war and weaponry, and AI will prognosticate the Climate Change-induced Disaster Displacement (CiDD) crisis in advance. The term "CiDD" is used to describe the potential for widespread disasters due to rising sea levels, violent storms, wildfires, and other extreme weather phenomena, which are worsened by global warming. These factors can make physical spaces more vulnerable, endangering interactions between people and places[1] and potentially causing widespread CiDD. Despite the potential for disasters, survival remains possible. The number of displaced persons in the Global South is rising; projections indicate that by 2050, there may be as many as 250 million[2] of them. In addition, they are causing death of people—more to CiDD than to war and bloodshed. Weather-related factors accounted for more than 97% of displacement and relocation caused by CiDD in 2022.[3] Vulnerable populations may discover that moving away from their current location becomes the only practical choice as the effects of climate change worsen. Indeed, in response, individuals have relocated or withdrew throughout human history due to the CiDD. Administrators and analysts worry about the effects of such AI evolutions[4] and innovations and the harm that will be done to people and businesses. They contend that governments must assist in controlling the costs of progress. These discussions often approach technological progress as a force of nature that must be accommodated, crashing inescapably into the profound customs and presumptions of the rule of law, justice, and human rights.

AI and its deployment and research are all advancing quickly and bolstering[5] the world economy. AI-related software, hardware, services, and sales are expected to generate $900 billion in revenue globally by 2026, up from $318 billion in 2020, at a rate of approximately 19% annually. Some projections state that by 2030, Artificial Intelligence would boost the world economy by more than $15 trillion.[6] Improvements in creativity, services, safety, and lifestyles are some of the notable advantages of AI out of an invisible assortment of benefits. Still, it also generates a lot of worries and concerns due to its potential adverse effects on human autonomy, privacy, and fundamental freedoms. The chapter advocates for the use of AI in disaster management due to the combined effects of disasters and climate change. AI can identify susceptible locations, simulate future climate scenarios, and evaluate risks and opportunities for businesses and infrastructure. It can also process vast amounts of data from satellite

[1] Reference [1].

[2] United Nations, 'Land—the Planet's Carbon Sink' https://www.un.org/en/climatechange/science/climate-issues/land. Accessed 30 October 2023.

[3] Internal Displacement Monitoring Centre, The Global Report on Internal Displacement (2023), https://www.internal-displacement.org/sites/default/files/2023-05/IDMC_GRID_2023_Global_Report_on_Internal_Displacement_HQ.pdf. Accessed 30 October 2023.

[4] Reference [2].

[5] Reference [3].

[6] BANK OF AMERICA, 'Artificial intelligence: A real game changer' Businesses & Institutions https://business.bofa.com/en-us/content/economic-impact-of-ai.html#:~:text=Global%20revenue%20associated%20with%20AI,with%20%24318%20billion%20in%202020.&text=According%20to%20some%20estimates%2C%20AI,the%20global%20economy%20by%202030. Accessed 25 October 2023.

imagery and climate models, providing insightful information for informed decision-making and preparedness for climate change's effects. This technology is crucial for informed disaster management in the face of climate change.

AI technologies must be designed and developed to incorporate diagnostic, formative, and summative digital assessments of the CiDD situations. The digital assessments must be integrated with NDL, NCLF, and IHRL frameworks to facilitate the protection of human rights as India experiences each year some of the highest rates of Internal Displacement[7] worldwide, most of which are caused by climate catastrophes. India's vast population, socioeconomic drawbacks, and exposure to severe and frequent climate calamities all contribute to the scale of international displacement. The latest climate change calamity Biporjoy[8]—a challenging tropical cyclonic storm on June 15, 2023, left a trail of destruction through Gujarat's Kutch area and continued towards Rajasthan. Fortunately, no one is recorded as having been hurt by the cyclone except, according to media[9] sources, a father and son goatherd pair lost their lives trying to save their flock from a flooded ravine in the Saurashtra region's Bhavnagar district. According to climate experts, Biporjoy made landfall around 6.30 pm on the evening of June 15, 2023, after becoming the longest-lived cyclone in both the pre-and post-monsoon Arabian Sea. On June 16, 2023, reports[10] stated that the cyclone's eye had moved to Pakistan after passing through Kutch. The maximum wind speed was 108 kms per hour (kmph), and the average wind speed was 70 kmph. In the city of Dwarka, it was 60 kmph. In 940 communities, 1000 villages were devastated, hundreds of houses were damaged, and more than 20 electricity poles were smashed. A total of 22 persons were hurt. Twenty-three cattle perished, 524 trees were felled due to the strong winds and downpours, and people were evacuated from low-lying locations. The electricity supply has been entirely out of sync ever since the hurricane hit, and the essential requirements of life derailed in the middle of the rescue operations. In the post-cyclone scenario, all sorts of diseases like anaplasmosis, anthrax, antibiotic-resistant infections, dengue, fungal diseases, and valley fever are collateral challenges requiring immediate attention.

In 2022, there were 2.5 million[11] fewer CiDDs than in 2021, a nearly 50% decrease,[12] and 96% resulted from flooding. The normal or below-average rainfall and flooding, with some regions of India reporting 2022 as their lowest July rainfall in 122 years, explain the downward trend. Assam was the state most severely impacted, with the same districts experiencing flooding in May and June. As their homes were devastated, the CiDD people fled in improvised boats, many of whom ended up in camps with deficient access to drinking water and sanitation. The Kopili River in Nagaon district touched an unprecedented level of more than a metre above the danger limit. Between the middle of May and the middle of July, the pre-monsoon season saw an estimated 742,000 flood displacements.[13] A significantly lower cyclone season in 2022 led to fewer displacements being reported, totaling 95,000. Cyclone Sitrang, which struck the states of Odisha and West Bengal in October 2021, caused over 66,000[14] of this CiDD. Other incidents were cyclones Asani and Mandous, which in May and December, respectively, caused about 1500 and 9500 moves in Andhra Pradesh, Tamil Nadu, and Puducherry.

There is no systematic data on displacement related to conflict and violence in India. A little

[7] Reference [4].

[8] Reference [5]. In the Arabian Sea, a powerful cyclonic storm known as Biparjoy has formed. According to the Indian Meteorological Department (IMD), it has already made landfall along Gujarat's coastline with wind speed between 145 and 155 kmph. The IMD initially did not anticipate any big effects on nations like India, Oman, Iran, and Pakistan.

[9] Reference [6].

[10] Reference [7].

[11] Reference [8].

[12] Ibid.

[13] Ibid.

[14] Ibid.

over 1000 cases of inter-communal violence were reported in 2022 due to isolated clashes. Almost 631,000 persons[15] were displaced as of the end of 2022. Many of them were displaced by conflicts and insurgency in the Northeastern states of Assam, Mizoram, and Tripura decades ago, and Jammu and Kashmir have not made any headway towards finding lasting peace. However, despite responding to persons harmed by conflict, many humanitarian actors continue to view CiDD and those displaced by them as peripheral to the core mission of humanitarian intervention.[16] International cooperation is crucial for preventing and addressing the issue of Internal Displacement (IDD) in disasters. Humanitarian intervention can mitigate the negative effects of CiDD, but collaboration with other actors is essential for longer-term development. Coordination initiatives involving AI technologies and local communities can reduce persistent data gaps. The Guiding Principles on Internal Displacement, adopted for natural disasters, can also be applied to those displaced in disaster contexts.[17] GPID principles created the basis for the recognition of the CiDD[18] as one of the primary causes of displacement.

2 AI and the Law on CiDD in India: Scale, Impact, Exposure, Vulnerability, Harm, and Beyond

The CiDD has raised concerns about the international human rights order and global climate stability. Nation-states are becoming more aware of the problems posed by CiDD, environmental degradation, and global warming. The relationship between International Climate Change Law (ICCL) and International Human Rights Law (IHRL) is often cordial, but this perspective has obscured the divergences between the two. This chapter examines the operationalisation of AI for protecting human rights in socio-political and eco-cultural transformations orchestrated by CiDD in India and elsewhere. The CiDD has changed the geopolitics of migration and facilitated the super-diversity[19] that has been producing the super-diverse nations worldwide, including India. Super-diversity refers to the unique socio-political demographics and environmental security dynamics arising from the convergent intersection of various categories, such as nationality, race, ethnicity, gender, age, immigration status, religion, and language, during the process of CiDD migration, resettlement, and transnational practices.

Presently, there are three types of transitions, "revolution in mobility, the revolution in energy systems, and revolution in connectivity such as AI, Robotics, and Data,"[20] that need "acts of imagination[21] to determine moments of power and distinguish it from the force" to fill the gaps between disaster displacement and AI for protecting the human rights of CiDD people. The stable climate has been essential for civilisation's development, especially since we have already left the stable climate and entered the Anthropocene or Capitalocene. Such a transition requires National Climate Law Frameworks (NCLFs) in India and elsewhere for sustainable AI technology-triggered automatic decision-making (ADM) to respond to the CiDD crisis while incorporating the AI and its collateral instruments to protect human rights during the CiDD calamity. These incorporations must be considered in the national climate policy of India in the decades ahead. It must find its space in National Disaster Law (NDL, International Change Law (ICCL), and NCLF. The legal ramifications of sustainable transition, specifically within the NDL, NCLF, and ICCL, are bound to fill the void regarding the positioning of the operationalisation of AI for protecting the human rights in CiDD situations.

[15] Ibid.

[16] Reference [9].

[17] Reference [10].

[18] Reference [11].

[19] Reference [12].

[20] Reference [13].

[21] Ibid.

For more than 20 years, India has been at the forefront of highlighting climate concerns.[22] The CiDD conditions in the fields of human rights protection are forcing international players, including India, to broaden their conflict prevention tools, including AI applications, and reassess current NDL norms and policies in the light of new realities. India also strives to rethink the relationship between climate change and human rights. India must set ambitious goals for its future foreign and climate policies, following in the footsteps of initiatives like the Strategic Compass,[23] the European Green Deal,[24] and the concept for an HRBA for climate change and human security, the climate change and defence roadmap, and others. The climate change and defence plan received high marks from India's Parliamentary Committee on Foreign Affairs in a report.[25] Human dignity is at stake in the operationalization of AI to defend human rights. Therefore, an HRBA provides actors positioned to operationalize AI aspirational and normative direction to defend human dignity and the intrinsic value of every person, regardless of nation or jurisdiction. The operationalization of AI to protect human rights during CiDD conditions can aid in identifying and prognosticating some of the worst and best social impacts of ADM technologies.[26]

Technology can aid climate change adaptation and risk reduction by generating data and insights that enhance responsive policies. There is a need to develop an AI tool to map disaster-prone areas using satellite images and combine this information with community-generated data on migration and displacement, access to water, power, and transportation, and climate risks. A compendium of disaster management laws[27] in India applicable to the transdisciplinary sectors and industries created a formidable infrastructure and machinery that provide aid and assistance in CiDD crises beyond national boundaries. However, India has adopted a national AI strategy for achieving the country's total capacity to facilitate AI development. India has identified its focus areas such as health care, education, agriculture, smart cities and infrastructure, smart mobility, and transportation for applying AI. India used remote sensing technology due to its experience with flooding brought on by the monsoon. The government can anticipate flood-prone areas and prepare evacuation plans by using satellite data, meteorological predictions, and Geographic Information Systems (GISs). For example, real-time water level monitoring during the 2018 Kerala floods was made possible using remote sensing, which aided in resource allocation and rescue efforts. Consequently, laws and regulations are the safety nets that guarantee AI technology's effective and ethical use. But, unfortunately, there is no integration of AI technologies into NDL to deal with CiDD situations that remain limited only to AI policies and practices.

India is susceptible to various climate calamities due to its location. India's total area is in Seismic Zones III to V, which puts 55% at risk for earthquakes. Additionally, 68% of net sown land is at risk for drought, 40 million hectares of land are at risk for flooding, the sub-Himalayan/Western Ghats is at risk for landslides, and the coastal states are at risk for cyclones.[28] The nation is exposed to manufactured calamities in addition to natural disasters. There is a strong movement to do away with the concept of "natural" disasters because it hides the social dimension, as evidenced by the Sendai Framework.[29] In India, disaster is defined as "a catastrophe,

[22] Reference [14].

[23] Reference [15].

[24] Reference [16].

[25] Ministry of External Affairs-New Delhi (The 19th Annual Report of Parliamentary Committee-2022–23).

[26] Reference [17].

[27] Reference [18].

[28] The Government of India-UNDP Disaster Risk Management Programme, Ministry of Home Affairs National Disaster Management Division, Government of India https://ndmindia.mha.gov.in/images/public-awareness/Primer%20for%20Parliamentarians.pdf. Accessed 24 October 2023.

[29] Denis McClean, 'SENDAI FRAMEWORK 6th ANNIVERSARY: Time to recognize there is no such thing as a natural disaster—we're doing it to ourselves' United Nations Office for Disaster Risk Reduction March 17, (2021) https://www.undrr.org/news/sendai-framework-6th-anniversary-time-recognize-there-no-such-thing-natural-disaster-were. Accessed 27 October 2023.

mishap, calamity or grave occurrence in any area, arising from natural or man-made causes, or by accident or negligence which results in substantial loss of life or human suffering or damage to, and destruction of, property, or damage to, or degradation of, environment, and is of such a nature or magnitude as to be beyond the coping capacity of the community of the affected area."[30] Nansen Initiative Protection Agenda[31] refers to disasters as disturbances brought on by or associated with geophysical, hydrometeorological, and climatological natural disasters, particularly those related to anthropogenic global warming.

However, in the context of AI risks that are involved when climate lawyers employ AI without understanding its limitations, especially the Large Language Models (LLMs) or Big Parlance Models (BPMs) like Open AI's ChatGPT, Bing Chat, and Google Bard. However, just because technology can be abused does not mean it should be disregarded. These tools have sparked a multitude of legal technology start-ups and scholarly contributions because they have the potential to revolutionise the NDL and IHRL frameworks for addressing the integration of AI into them. However, in NDL, NCLF, and IHRL, the potential of BPMs has largely gone unrealized. The capacity of these tools to navigate NDL's special characteristics will determine how it is changed to respond to the CiDD crisis.

2.1 The State, Status, and Strategies of AI in India

India's AI strategy is firmly rooted in ethical AI concepts and ensures that comparable provisions are made in international frameworks. This is demonstrated by the national programme on AI that is in place, the National Data Governance Framework (NDGF) policy and one of the most extensive publicly accessible datasets' programme currently under development.[32] India is one of the founding members of the Global Partnership on Artificial Intelligence (GPAI), and it intends to work closely with other GPAI members to create a framework that would encourage AI use that is safe, accountable, open, and trustworthy. In the province called Tamil Nadu in India adopted the Tamil Nadu Safe and Ethical Artificial Intelligence Policy 2020 that offers a roadmap for the state's policymakers to adopt AI-based solutions in particular sectors and recommends a framework for evaluating AI-based systems and the same may be a guiding factor in the context of CiDD crisis at national in India. The latest data on facial recognition technology is included in a series of publications on Responsible Artificial Intelligence (RAI) issued by the think tank of the Indian government, The National Institution for Transforming India (NITI) Aayog. Compared to other countries (like the UK), India has a comprehensive AI strategy covering a wide range of subjects. But, unfortunately, it does not adequately address climate change. An increased need to collaborate with stakeholders to improve government readiness for extreme weather events and to access datasets of extreme occurrences in order to build a robust climate data ecosystem consisting of data from satellites and weather, demographics, losses and damages, public infrastructure access, and government physical reaction are the five categories of datasets. With so many datasets, a deployable platform for disaster management is required, together with an Intelligent Data Solution for Disaster Risk Reduction (DRR). Using such dynamic data, India benefits from AI practices and initiatives to meet societal health care, education, agriculture, smart cities, and infrastructure demands, including smart mobility and transportation.

[30] Section 2, Disaster Management Act, 2005.

[31] Nansen Initiative Protection Agenda https://disasterdisplacement.org/wp-content/uploads/2015/02/PROTECTION-AGENDA-VOLUME-1.pdf. Accessed 27 October 2023.

[32] Draft National Data Governance Framework Policy https://www.meity.gov.in/writereaddata/files/National-Data-Governance-Framework-Policy.pdf. Accessed 27 October 2023.

The Government of India (GoI) has focused on developing cutting-edge technology as part of the Digital Government Mission[33] to enhance the provision of citizen services in India. According to recent findings,[34] India has carved out its conspicuous presence among the top ten nations in the world in terms of technological developments and discoveries in AI. Regarding public expenditures and investments in AI made by governmental programmes, corporate institutions, and organisations, India is ranked sixth. It is significant to note that while the USA, China, the UK, France, Japan, and Germany are ahead of India in AI, other nations like Canada, South Korea, and Italy are behind it. Therefore, technology advances rapidly and there is a need to ensure that India remains among the technology savvy countries in acquiring AI technologies and applying them to protect the human rights of CiDD populations at the right moment. Currently, there are around 14 million[35] CiDD people in India, and by 2050, climatic catastrophes in India alone will cause 45 million people to leave their homes[36]; this is three times the number of people displaced due to extreme weather events. Over 3 million people were forced to flee their homes in 2020–2021, making India the fourth most affected country in the world by CiDD migration,[37] according to the "State of India's Environment-2022" study. Therefore, India is set to take advantage of the prospects of the impending fourth Industrial Revolution.

To advance AI research and development for the benefit of society, NITI[38] Aayog introduced the National Programme for AI. When India's largest High-Performance Computing (HPC) AI supercomputer, PARAM SIDDHI AI, commissioned by C-DAC, was placed 62 on the top in the list of 500 supercomputers in November 2020, and it signified a significant advancement towards this aim. India is establishing its dominance as the world's top AI destination, with AIRAWAT[39] listed among the top 500 supercomputers worldwide and an AI System called PARAM SIDDHI[40] that is among the top 100 supercomputers globally. The AI market is expected to multiply at a five-year multifactorial yearly growth rate (CAGR) of 20.2% and reach total revenues of USD 7.8 billion[41] by 2025.

India is a multi-cultural, multi-lingual, and multi-religious country that generates a wealth of data, and the magnitude of AI deployment, India, will have to adopt an HRBA to democratise AI technology than the rest of the globe. According to HRBA, discrimination of any kind in the pursuit of rights must be outlawed, stopped, and eradicated. It also implies that those who are most disadvantaged or marginalised and who have the greatest obstacles to achieving their rights should be given precedence. An HRBA involves integrating human rights into climate policy and initiatives. The broad digital tsunami the pandemic heralded has influenced AI adoption and awareness. NDL, NCLF, IHRL, and International Organisations (IOs) must use AI with a democratic vision, cutting-edge computing, real-time tracking, and Industrial Revolution 4.0 to produce innovative and sophisticated solutions for India's CiDD population and beyond.

---

[33] AI and Emerging Technologies Division, 'Promoting the Adoption of Cutting-edge Technologies to create Significant Economic and Societal Impact' Ministry of Electronics and Information Technology, Government of India (2023) https://www.meity.gov.in/emerging-technologies-division. accessed 02 November 2023.

[34] Reference [19].

[35] Internal Displacement Monitoring Centre (2022) https://www.internal-displacement.org/countries/india#:~:text=As%20of%20the%20end%20of,people%20were%20living%20in%20displacement. Accessed 02 November 2023.

[36] Reference [20].

[37] CENTRE FOR SCIENCE AND ENVIRONMENT, 'State of India's Environment-(2022)' https://www.cseindia.org/static/page/CSE-annual-report-2022.pdf. Accessed 04 November 2023.

[38] Reference [21].

[39] Reference [22].

[40] Ibid.

[41] Special Correspondent, 'India's AI market to reach $7.8 billion by 2025: IDC' *The Hindu*, October 05, (2021) https://www.thehindu.com/business/indias-ai-market-to-reach-78-billion-by-2025-idc/article36846106.ece. Accessed 06 November 2023.

2.2 Re-Imagining the AI Interface Between Climate Protection and Human Rights Protection of the CiDD

The climate change crisis is a global human rights crisis, causing forced migration and displacement. India experiences both voluntary and involuntary internal migrations due to conflicts, development, and natural disasters. The destruction of semi-urban villages, livelihoods, and lush terrain is causing immense suffering, particularly in rural areas. India faces environmental challenges, with floods affecting 1.5 million in Assam in 2012, and water shortages forcing farmers and fishermen to migrate to cities, highlighting the urgent need for climate change mitigation.[42] India has a legal framework in place to safeguard displaced individuals brought about by war and development, but it ignores IDPs caused by CiDD. India is used to IDPs, which are eligible for protection under the 2005 Indian Disaster Management Act (DMA) and its implementing guidelines. Unfortunately, however, there is no functioning national protection framework. AI can provide insightful information for decision-making and prepare for climate change effects. However, ethical questions and potential biases must be addressed to ensure open and equitable implementation. Recent advancements highlight the potential benefits and ethical issues of AI-enabled techniques for addressing the CiDD crisis, aiming for a more resilient, sustainable, and fair future for everyone. Unfortunately, AI is neither assisting CiDD migration nor the AI application for climate protection. In theory, this might be plausible, but not in practice; the use of AI in CiDD situations has been highly unsettling thus far. It is improbable that applying AI explicitly to climate change-related mobility will result in compassionate solutions to the problem. AI is widely used to put RAMS[43] (Refugees, Asylum-seekers, Migrants, and Stateless) under surveillance[44] and incarcerate them. It is now a significant component of border militarization.[45]

Conflicts or natural disasters often lead to similar human experiences, including unequal access to aid, discrimination, forced relocation, sexual and gender-based violence, loss of documentation, kidnapping, unsafe resettlement, and property restitution issues. Despite these challenges, most remain within their nation's borders, receiving protection, aid, and long-term relocation solutions. The Guiding Principles on Internal Displacement acknowledge natural disasters as a significant factor in displacement, emphasising the importance of respecting human rights of those affected by floods, hurricanes, and earthquakes. The international community's response to CiDD should focus on the Operational Guidelines on Human Rights and Natural Disasters,[46] considering the human rights of those displaced by conflict and natural disasters.

However, many people in society are currently unsure of the pragmatic impacts of AI systems on the CiDD people. There is evidence[47] that some AI systems are already infringing upon fundamental rights and freedoms, despite promises that AI can contribute to the global good. We may rely on the operationalisation of AI for protecting human rights to help chart the future of CiDD populations. A significant class of dangers and harms can be identified, prevented, and remedied with the help of integrating AI applications into the NDL, NCLF, and IHRL framework. Those engaged in commercial AI research may question the importance of human rights. Stakeholders are increasingly looking to the private sector to defend their rights.[48] The UN Guiding Principles

[42] Reference [23].

[43] Reference [24].

[44] Reference [25].

[45] Ibid.

[46] IASC Operational Guidelines on the Protection of Persons in Situations of Natural Disasters (The Brookings—Bern Project on Internal Displacement 2011).

[47] Reference [26].

[48] Reference [27], See: Reference [28].

on Business and Human Rights (GPBHR), an important document published by the UN in 2011, encourages businesses to respect, defend, and provide for human rights.[49] On the intersection of AI and human rights, AI systems lack the human traits of empathy, intuition, and emotional intelligence, even while they can automate work and increase efficiency. However, it will take a lot of work to create best practices for integrating HRBA into AI algorithms and to take into account the proper scope, structure, scalability, timeliness, and administrative overhead of incorporating human rights aspects into AI for their application in a range of CiDD scenarios.

The GPBHR represents a beginning point for implementing human rights in the AI industry. Working together, civil society and AI developers will assist in identifying risk areas and anticipate the requirements of the people in situations of vulnerability. AI systems cannot protect human rights during CiDD challenges like climate disaster preparedness, disaster prevention, multi-hazard early warning system, and improving weather prediction accuracy to lower the likelihood of disasters unless stakeholders collaborate across silos to prevent harm. AI-driven projects integrate into national and local response strategies by focusing on high-risk locations. For instance, mapping can assist local authorities in developing and implementing sustainable development strategies, lowering hazards, and guaranteeing the safety of citizens in high-risk circumstances in areas prone to landslides. Recent work by the World Meteorological Organisation (WMO), UN Environment Programme (UNEP), and International Telecommunication Union (ITU) recognised related advances in robotics and AI as among the instruments. The UN Secretary-General's ground-breaking Early Warnings for All plans includes utilising AI's advantages. By the end of 2027, the action plan, which was introduced earlier in 2023, is intended to guarantee that everyone on Earth is shielded from dangerous weather, water, or climate disasters by early warning systems.

Governments and people can utilise AI to assist them become ready for the CiDD migration. A new group of people known as "climate migrants" or "climatically displaced persons"[50] (CDMs) was developed as a result of the CiDD, one of the gravest problems of our time, which has caused climate migration and frequently led to human rights transgressions. These people cross municipal and state borders to avoid the devastation that climate change is causing to their homes and neighbourhoods. Over the next three decades, the International Organisation for Migration (IOM) predicts that up to 2 billion people will migrate due to the CiDD crises.[51] The world population will be 9.8 in 2050; by 2100, this estimate will be 11.2 billion.[52] Although these numbers are worrisome, the state of the environment makes them predictable. Thus, there is a pertinent question of how we can conceptualise the future of climate change migration using AI.

The 1951 UN Convention relating to the Status of Refugees (UNCSR)[53] describes refugees as being compelled to leave their home country because they face an immediate risk of persecution and human rights abuses. States are legally required to review refugee situations and provide necessary protections. Refugees have a

[49] The Guiding Principles on Business and Human Rights, United Nations, June 16, (2011), https://ww.ohchr.org/Documents/Publications/GuidingPrinciplesBusinessHR_EN.pdf. Accessed 09 November 2023.

[50] Reference [29].

[51] Think about Tomorrow, Act Today: The Future of Mobility and Climate Change (IDM Second Session 2023, IOM UN Migration 05 October (2023) https://www.iom.int/international-dialogue-migration-2023-think-about-tomorrow-act-today. Accessed 11 November 2023.

[52] UN Department of Economic and Social Affairs, 'World population projected to reach 9.8 billion in 2050, and 11.2 billion in 2100' https://www.un.org/en/desa/world-population-projected-reach-98-billion-2050-and-112-billion-2100. Accessed 12 November 2023.

[53] UN General Assembly, Convention Relating to the Status of Refugees, 28 July 1951, United Nations, Treaty Series, vol. 189, p. 137, https://www.refworld.org/docid/3be01b964.html. Accessed 12 November 2023.

legal right[54] to request international protection. However, the category of migrants is broader and lacks a clear definition. Economic migrants emigrate for employment, education, and family opportunities. Climate migrants, particularly those in immediate danger, are not classified as refugees. International organisations hesitate to provide additional safeguards for climate migrants. However, by improving disaster relief assistance and raising public awareness, the UN, other organisations, and countries are increasingly addressing humanitarian crises like the CiDD. Despite the tremendous progress in extending safeguards for climate migrants, it is critical to consider how we may manage climate migrants' increasing numbers more effectively and efficiently in India and elsewhere, I think AI might hold the key to the CiDD. India is experimenting with AI to improve weather forecasting due to severe weather system conflicts caused by global warming. The Indian Meteorological Department uses supercomputers to produce mathematical models, and an extended observation network combined with AI could provide better prediction data at a lower cost. The government plans to expand weather observatories and provide higher-resolution data for predictions. AI models can function on high-end desktop computers without the overhead of running a supercomputer. However, better data is necessary for location-specific amplification of current model projections.

3 Digital Assessments of the CiDD Situations: Integrating AI Application, Tools, and Practices into NDL, NCLF, and IHRL for Protecting the Human Rights

The global canvas of human rights, peace, security, and development is being redefined and altered by CiDD crisis. It requires actionable information and analysis about the risks related to CiDD that must be produced with accuracy and in a timely manner. AI technologies have the potential and can provide such information with precision and implementability. Therefore, AI can contribute in preserving the human rights of CiDD populations in India and elsewhere for their adaptation, mitigation, and planned relocation. Digital assessments can expedite the protection response for the CiDD people. There are AI tools and applications[55] to design and develop robust digital evaluations while considering the human rights[56] ramifications in the CiDD trajectories. Much has changed since the term "artificial intelligence" was first used in 1956. Today, we can access various AI tools and technologies, such as computer vision, deep learning (DL), machine learning (ML), and natural language processing. New breeds of intelligent systems that can carry out jobs long regarded as the sole preserve of humans are being powered by these technologies.

Digital assessments occupy a central space in the CiDD repercussions. In addition, climate change impact assessments must be digitalized and duly integrated with the legal frameworks. AI is a crucial tool, and AI algorithms are improving predictions[57] of the short- and long-term implications of climate change, including those related to weather patterns, floods, droughts, and human migration patterns. However, it is unclear whether India is genuinely gaining from these AI technologies, especially given that they were created mainly by Global North nations utilising datasets that are distinctive to those same nations. Therefore, it is crucial to describe the initiatives being made to employ AI in a way that uniquely helps India with climate change adaption. According to the IPCC's Second Assessment Report (WG II

[54] Reference [30].

[55] Reference [31].

[56] Aizenberg, E., & van den Hoven, J., 'Designing for human rights in AI' 7 *Big Data & Society* 2(2020) https://doi.org/10.1177/2053951720949566. Accessed 13 November 2023.

[57] Reference [32].

contribution) published in 2022, India is among the nations most "economically harmed" by climate change. It outlines the possible applications of AI for climate action in water management and transportation and evaluates the degree or kind of effect associated with each. Projects that are physically carried out in India and those that run in other nations must be the guiding force for the CiDD population in India.

Integrating AI applications into the NDL, NCLF,[58] and IHRL[59] for protecting human rights offers a unique perspective into the CiDD migration that can benefit climate migrants and states alike consistent with the Global Platform for Disaster Risk Reduction[60] (DRR) strategies that take into account regional and transnational perspectives and incorporate provisions that aim to prevent the CiDD and decrease disaster displacement risk,[61] respond to the protection requirements of CiDD people, and promote durable solutions to displacement. The Platform on Disaster Displacement[62] (PDD) encourages states to develop and incorporate provisions for human mobility[63] and the CiDD into their national and regional DRR programmes. There is a need to consider improved data-gathering methods for increasing AI efforts to protect the human rights of the CDMs. National authorities, NGOs, IGOs, and administrative data sources, like numbers for humanitarian visas, are some current data sources, including other data originating from databases developed by institutions like the IOM's Displacement Tracking Matrix (DTM),[64] which tracks and monitors the CiDD people. By utilising AI, India and other nations will be able to anticipate substantial population displacement and, where possible, allocate the necessary resources for assistance. Some organisations are also utilising AI to build resilient infrastructure[65] for climate change. Google uses ML to combat riverine floods, a major natural disaster. The Google Flood Forecasting Initiative, an AI-based system, helps send flood alerts earlier than conventional methods. The system was launched in Bihar, India, and allows the government to evaluate potential floods. Google continues to develop AI by providing vast amounts of data. However, internet connectivity remains a challenge in many areas. By doing this, climate calamities will be prevented before they can have disastrous repercussions.

AI technologies are often seen as vulnerable social actors, and design decisions, such as gender, can significantly impact people's engagement with them, affecting their perception of competence and autonomy. Because of this, it is critical for practitioners creating AI experiences to recognise that people engage with AI socially. Individuals react to AI socially and unintentionally mistake it for humanity. These social dynamics have significant design consequences that should be well researched and thought through. Ultimately, our ability to comprehend this particular aspect of human–AI relations will determine how successful AI innovation and adoption are. AI tool designers must be aware of the trade-offs inherent in these choices. Researchers must look into how different design choices affect CiDD people's perceptions and social reactions. By comprehending these elements, AI applications can incorporate data analysis from satellites and sensors, enhancing renewable energy systems,

[58] Reference [33].

[59] Reference [34].

[60] Reference [35].

[61] Reference [36].

[62] Daria Mokhnacheva, 'Implementing The Commitments Related To Addressing Human Mobility In The Context Of Disasters, Climate Change And Environmental Degradation: A Baseline Analysis Report Under the Global Compact for Safe, Orderly and Regular Migration' Platform on Disaster Displacement (2022) https://environmentalmigration.iom.int/sites/g/files/tmzbdl1411/files/inline-files/17052022_PDD_Baseline_Mapping_Report_final_compressed_0.pdf. Accessed 18 November 2023.

[63] IOM, 'Mapping Human Mobility (Migration, Displacement and Planned Relocation) and Climate Change in International Processes, Policies and Legal Frameworks' (2018) Analysis Report, August (2018).

[64] IOM Displacement Tracking Matrix (2023) https://dtm.iom.int/about/infosheets. Accessed 18 November 2023.

[65] Reference [37].

maximising agriculture, minimising waste, and tracking environmental conditions into the NDL, NCLF, and IHRL to safeguard the human rights of CiDD populations and provide more useful, approachable AI solutions that are appropriate for CiDD circumstances. AI usage, however, is not without danger. Some of these hazards include the possibility of increasing resource exploitation, a worsening of environmental injustices, and job displacement. Guaranteeing that AI is created and applied in ways that prioritise sustainability, fairness, and justice involves interacting with CiDD people.

4 Algorithmic Humanitarianism: Positioning AI's Operationalisation for Human Rights Protection in CiDD Prosceniums

Operationalisation of AI can be defined as the deployment and utilisation of algorithmic technologies to predict CiDD situations and protect their human rights by sharing information regarding legal processes, fairness, and integration into the host societies while providing them migration models such as adaptation, mitigation, rehabilitation, relocation, resettlement, and resilience within the gamut of algorithmic humanitarianism[66] in which AI applications and ADM technologies are programmed with ethics, justice, and fair play. Public authorities increasingly use AI technology to assess people's personalities or skills, allocate resources, monitor activities of people on social media and digital platforms, and make other decisions that could have real and severe repercussions for people's human rights. However, finding the ideal balance between technological advancement and human rights protection is a pressing issue, as emphasised by the Office of the High Commissioner for Human Rights (OHCHR) in a Human Rights Comment.[67] The OHCHR made recommendations on AI and human rights following its mandate to promote the awareness of, practical observance of, and full enjoyment of human rights in all Council of Europe[68] member states and provide advice and information on protecting human rights. There is no established definition of "artificial intelligence" at the moment. However, for this recommendation, the word "AI" refers broadly to a collection of disciplines, theories, and methods aimed at enhancing the capacity of machines to perform tasks requiring intelligence.

AI perceives real or virtual surroundings and first uses machine or human-based inputs. AI machines either manually or automatically abstract these perceptions into models. Finally, AI machines, either manually or automatically, derives outputs from these models through recommendations, projections, or judgments. AI has potential and risks, but it must not violate human rights. Earth and climate sciences are among the numerous industries where AI is reputed to be helpful. Currently, AI and big data are being used in climate and sustainability projects[69]; for example, to prevent forest fires and monitor wildlife, Microsoft's[70] AI for Earth grants is being given out to solve climate issues. The Snow Leopard Trust (SLT) has obtained funding from Microsoft to monitor and count snow leopards in the wild. AI can classify thousands of photos in a few seconds or minutes. The same big data scale classification could need hundreds of "human" hours to duplicate.

---

[66] See: Reference [17].

[67] Reference [38].

[68] Council of Europe Directorate General, 'Human Rights and Rule of Law: Human rights and the environment—a priority for the Council of Europe' https://www.coe.int/en/web/human-rights-rule-of-law/humanrights-environment. Accessed 22 November 2023.

[69] Reference [39].

[70] Lukas Velush, 'Tackling environmental sustainability from the inside out at Microsoft, *Microsoft-Inside Track, August 13(2019)* https://www.microsoft.com/insidetrack/blog/tackling-environmental-sustainability-from-the-inside-out-at-microsoft/. Accessed 22 November 2023.

4.1 Future AI Techniques for Protecting the CiDD People

Future AI techniques might develop a digital console for the planet Earth enabling global monitoring, modelling, prognosis, and management[71] of climate processes. Monitoring CO2 levels, carbon emissions, ocean curves, wildlife movement, deforestation, illicit actions, pollution, and improving climate catastrophe prediction are just a few examples. Time and resources are running out worldwide, so this strategy must be implemented immediately to improve CiDD.[72] For the benefit of the CiDD people and the quality of life in the future, a worldwide partnership involving the Government of India and its research organisations, businesses, industries, and NGOs must begin to accomplish AI-powered solutions to transform the CiDD calamity into the opportunity for human integrity.

By effectively tracking climate change permutations, AI can reduce CiDD incidences. Additionally, AI helps to position its operationalisation to protect human rights and ADM technologies. However, AI may be less effective due to its high energy needs and restricted accessibility. An overreliance on AI may also hamper the implementation of critical systemic reforms in the AI technologies. AI's capacity to process and analyse massive volumes of data fast and precisely makes it possible to more pragmatically monitor CiDD indicators like temperature, rising sea levels, and carbon emissions. AI can also reduce waste and optimise energy systems, reducing carbon footprints. The evidence is increasing that AI is crucial in protecting the CiDD people as the most countries are addressing climate change, whether it is quicker catastrophe mitigation from floods and fires or creating a cleaner electricity grid. Humanity will require all the assistances it can obtain. Thanks to scientific evidence, the need to minimise global carbon footprint is more evident than ever.[73] Otherwise, we risk making the Earth uninhabitable. However, AI has the potential to be a very beneficial tool for improving both our ability to respond to extreme weather occurrences and the resilience of our infrastructure. A predictive algorithm that analyses a large quantity of data to forecast what will happen next is often the core of an AI-powered solution. Implementing an AI-driven "smart grid"[74] solution that manages electricity from various renewable resources more effectively to control power distribution could be a significant step towards reaching low or zero-carbon energy usage.[75] Data is abundant in many electricity networks, and the sector has started to develop next-generation systems (smart grids) powered by AI and ML.

There is software that monitors flood conditions and warns individuals in CiDD risk locations, such as Google FloodHub,[76] to help with CiDD mitigation and adaptation. It displays maps of inundated areas and measurements of floodwaters concerning the height of an average adult body. FloodHub is now only accessible in a few locations in Bangladesh, India, South Asia, and South America, but the business wants to make it available everywhere. The Global South[77] and its cities are typically characterised

[71] UNEP, 'How artificial intelligence is helping tackle environmental challenges' Climate Action November 07(2022) https://www.unep.org/news-and-stories/story/how-artificial-intelligence-helping-tackle-environmental-challenges. Accessed 24 November 2023.

[72] Reference [40].

[73] Climate Change 2022: Mitigation of Climate Change. Working Group III Contribution to the IPCC Sixth Assessment Report.

[74] Entefy, 'The smarter grid: AI fuels transformation in the energy sector' November 15, (2022) https://www.entefy.com/blog/the-smarter-grid-ai-fuels-transformation-in-the-energy-sector/. Accessed 24 November 2023.

[75] Reference [41].

[76] Google Flood Hub https://sites.research.google/floods/l/0/0/3?lat=16.130262012034756&lng=2.8125&zoom=3.6.

[77] Solarz, Marcin Wojciech, 'The Global North–South Atlas: Mapping Global Change' *Routledge* p 2. (2019). ISBN 9780429959684. Accessed 25 November 2023.

by stable informal economies, slow economic expansion, fast population growth, and deteriorating infrastructure, which set them apart from the usual, open spaces of cities in developed or Global North countries with a high risk of being damaged by CiDD events and few resources[78] to deal with them, will benefit significantly from these technologies. For instance, a SWAMP project used the internet of things (IoT) to control water[79] use in Brazil, Italy, and Spain and cut use from 18 to 38%. That might enable nations affected by drought to increase their water supplies. Some climate-related applications of AI serve only to raise awareness. For instance, the climate issue is digitally brought to people's doorsteps with a AI tool called "This Climate Does Not Exist,"[80] which simulates imagery of floods, smog, or flames for whatever address is entered into to see how CiDD could impact familiar locations.

Therefore, climate change AI can be a tremendous assistance, even though it is not a miracle cure. It is a component of the AI labyrinth that got the capability to offer more sustainable modes of prognosis about the CiDD crisis along with addressing the collateral sectors like eco-compliant transportation and renewable energy sources and can redesign AI-powered infrastructure to help build a more climate-friendly and sustainable society.

4.2 AI Ethics and Reconciliation of Tensions Between CiDD and AI Applications: Raising Constitutional Standards and IHRL Obligations for Fostering Human Dignity in the CiDD Circumstances

The Constitution of India protects the right to dignity, diversity, and development, along with an assortment of human rights. Article 21 protects the right to life and personal liberty that has been bridging the legal, social, and policy gaps in the law. India is a party to the majority of the IHRL instruments,[81] but Article 21 of Indian constitution remains the potential instrument to position the operationalisation of AI-powered solutions for CiDD communities. In India, the legislative attempt follows years of worry over AI deployments, ranging from the national security implications of AI technology to the profiling of individuals for social security purposes.[82] The effects of AI on a range of human rights of the CiDD people such as the right to normal life, housing, health care, education, culture, diversity, identity, access to financial systems, human resources (recruitment and hiring), community development, good governance, and the right not to be displaced must be appreciated.

Historically, technology has aided in advancing human rights and harmed people unanticipatedly. It was nearly impossible to foresee many of the distressing ways that internet search tools, social media,[83] and mobile technology became drivers and multipliers of human rights abuses[84] worldwide when they were first released and as they grew in widespread adoption and accessibility. The use of nearly undetectable spyware to turn mobile devices into 24-h surveillance machines[85] used against journalists and human rights defenders, for example, Facebook's role[86] in the 2017 ethnic cleansing of the Rohingya[87] in Myanmar, are both effects of the introduction of disruptive[88] technologies whose social and political implications had

[78] Reference [42].

[79] Reference [43].

[80] Reference [44].

[81] National Human Rights Commission, 'A Handbook on International Human Rights Conventions'. https://nhrc.nic.in/sites/default/files/A_Handbook_on_International_HR_Conventions.pdf. Accessed 28 November 2023.

[82] Reference [45].

[83] Reference [46].

[84] Reference [47].

[85] Reference [48].

[86] Reference [49].

[87] Reference [50].

[88] Reference [51].

not been given serious consideration. The ethical compass is more critical in AI than in any other technology. These all-purpose technologies change how we work, engage, and live.[89] The ethical ramifications and moral issues raised by the creation and application of AI technologies are crucial in the CiDD circumstances to examine the policies and frameworks India and other nations have developed to deal with them. The powerful frameworks with ethical concerns acknowledged the gaps in the mechanisms for equitable benefit sharing, responsibility distribution, worker exploitation, energy demands in the light of CiDD treatment, and more nuanced and less specific implications of AI, such as those about human relationships. Global businesses now need to guarantee the ethical growth of AI technologies.[90]

Moreover, these businesses are in a position to ensure that AI either helps humanity or, at the very least, does no harm. The human rights community has created a pragmatic human rights due diligence framework (HRDF) that derives its legitimacy from the IHRL regime and the same must be applied to businesses in identifying, preventing, and minimising sufferings of the CiDD communities. Before it is too late, the industries creating AI technologies must have the HRDF procedures and protocols in place. Such algorithms use training data from enormous real-world datasets to produce outputs that are frequently impossible to distinguish from "actual" data, making it difficult to determine whether a human or an algorithmic application had a piece of content.

Tools like ChatGPT, which produce text; tools like Dall-E, Midjourney, and Stable Diffusion,[91] which generate images; and tools like Codex and Copilot, which generate compute code, are the three primary types of Generative AI applications available to date. Therefore, applying an HRBA to Generative AI is crucial, but it might be challenging to integrate AI tools with HRDF to protect the human rights of CiDD communities. To better understand the potential effects of these new tools, active interaction with CiDD stakeholders must be a part of the application design and policy formulation processes. Therefore, human rights organisations should lead the way in doing AI Ethics research, advocating for change, and engaging in other activities that foresee the revolutionary potential of AI technologies. It is in everyone's interest that this omnipotent new technology is applied with HRBA to advance the rights of CiDD humanity.

5 Conclusion

AI is revolutionising the collection and processing of data in CiDD situations, enabling quick and efficient decision-making. Its data processing techniques, including machine learning, real-time data analysis, and automation, enable the extraction of intelligence for informed decision-making. AI has the potential to significantly change the way we manage CiDD and is a potent ally for local and global security. Integrating AI applications into the NDL, NCLF, and IHRL can provide a novel perspective on CiDD migration, benefiting both states and climate migrants. AI can collect data on actual and fictitious CiDD and accurately predict climate occurrences using satellite imagery and local information. India, with a regional focus, needs to use AI to create climate change-resistant infrastructure. The ethical use of AI can significantly improve climate resilience and accessibility measures for governments and communities affected by climate change. By acknowledging and openly discussing concerns, AI systems can be developed that are fair and just. Future AI technology could create AI-powered consoles for monitoring, modelling, prognosis, and management of climate processes. The human rights community has established a pragmatic HRDF to help businesses identify,

[89] Reference [52].

[90] Reference [53].

[91] James Myers, '7 Alternatives to Midjourney for Creating AI-Generated Images (Including Examples)' Bootcamp. https://bootcamp.uxdesign.cc/7-alternatives-to-midjourney-for-creating-ai-generated-images-including-examples-35c4a4c3d5ee. 04 December 2023.

prevent, and minimise damage to CiDD communities. However, integrating AI tools with HRDF may be challenging to protect human rights. Businesses creating Generative AI tools must implement a stringent and ethically oriented HRDF process in line with the UN-GPBHR to uphold their obligation to respect human rights.

AI has the potential to help address CiDD crises, as the climate change sector seeks innovative methods to support individuals. While the security risks of CiDD may be real, they are not part of a securitisation narrative that could perpetuate harm. Human security and dignity are affected areas, crucial for building ecologically sustainable nation-states. To address the CiDD threat, India and other countries need to adopt CiDD-oriented laws and AI policies. The operationalisation of AI in response to the CiDD crisis requires rewriting laws in India and abroad to incorporate AI-powered strategies, protocols, methods, procedures, tools, and technology. AI should be used to enhance diagnostic, formative, and summative digital assessments of CiDD, linking them to NDL, NCLF, and IHRL frameworks based on the centrality of human survival with human dignity and AI Ethics. This will help protect human rights and speed up the protective response for the CiDD population, ensuring a more effective response to climate disasters. Internet constitutionalism in India aims to protect the human rights of CiDD communities with the operationalisation of AI. However, India has shut down the internet more frequently than any other country over the past five years, often as a collective punishment for disrupting access to critical services. To defend fundamental human rights and the lives of CiDD people, "Digital India" must guarantee an internet accessibility and affordability consistent with internet constitutionalism. The Constitution of India protects the right to dignity, diversity, development, and various human rights, with Article 21 bridging legal, social, and policy gaps. Article 21 remains the potential instrument to position AI-powered solutions for CiDD communities.

AI operationalisation is not sufficient to address the immediate and future issues related to the CiDD crisis. Research should focus on how AI policy and practice can apply a human rights perspective in institutional transformation. AI can analyse risks, identify susceptible locations, and model future climate scenarios using data from satellite images and climate models. This information can guide decision-making and prepare for climate change effects while addressing ethical questions and biases. Satellite imagery can track the movements of CiDD persons, monitor internal mobility, and identify potential dangers. AI algorithms can identify patterns, pinpoint conflict zones, and support early warning systems. Governments and organisations can use satellite images and AI algorithms to collect data, track climate change patterns, forecast extreme weather, improve agricultural practices, and enhance disaster response and recovery operations.

References

1. Mach KJ, Siders AR (2021) Reframing strategic, managed retreat for transformative climate adaptation. Science 372(6548):1294–1299. https://doi.org/10.1126/science.abh189
2. Cockburn IM, Henderson R, Stern S (2018) The impact of artificial intelligence on innovation. Working Paper 24449. National Bureau of Economic Research. http://www.nber.org/papers/w24449. Accessed 30 Oct 2023
3. Feijóo C et al (2020) Harnessing artificial intelligence (AI) to increase wellbeing for all: The case for a new technology diplomacy. Telecommunications Policy 44(6):1–14
4. Ahmad N (2017) Internally displaced persons and international refugee law: protection gaps, challenges and implementation in practice. The Kings' Students Law Rev 8(2):94–117
5. Krishnamurthy R (2023) Biporjoy: Saurashtra, Kutch highly vulnerable to cyclones. Down To Earth June 14. https://www.downtoearth.org.in/news/natural-disasters/biporjoy-saurashtra-kutch-highly-vulnerable-to-cyclones-90029. Accessed on 20 Oct 2023
6. Dighe S (2023) Army teams ready for cyclone Biporjoy relief ops in Gujarat. The Times of India June 15. https://timesofindia.indiatimes.com/city/pune/army-teams-ready-for-relief-ops-in-guj/articleshow/101004828.cms. Accessed 21 Oct 2023
7. Bhayani H (2023) Cyclone Biporjoy causes large-scale property destruction in Kutch; moves towards Rajasthan. Down To Earth June 16. https://www.downtoearth.org.in/news/natural-disasters/

cyclone-biporjoy-causes-large-scale-property-destruction-in-kutch-moves-towards-rajasthan-90075. Accessed 21 Oct 2023
8. IDMC, 'Internal Displacement Monitoring Centre Report 2022' (2022) Internal displacement data updates. https://www.internaldisplacement.org/countries/india#:~:text=The%20scale%20of%20displacement%20is,other%20forms%20of%20violence%2C%2013%2C000. Accessed 21 Octo 2023
9. Ferris E (2008) Displacement, natural disasters, and human rights. Brookings Friday, October 17. https://www.brookings.edu/on-the-record/displacement-natural-disasters-and-human-rights/. Accessed 22 Oct 2023
10. Scott M (2020) The role of national law and policy in addressing displacement in the context of disasters and climate change in Asia and the Pacific. In: climate change, disasters, and internal displacement in Asia and the Pacific. Routledge 37. eBook ISBN9781003015062
11. Ahmad N (2023) Disaster displacement and international refugee law: locating legal protections in the context of climate change migration. In: International handbook on disaster research. Springer
12. Vertovec S (2023) 'SUPERDIVERSITY: migration and social complexity. Routledge, pp 159–201
13. Gandhi R (2023) Bharat Jodo Yatra—the power of truth. The Speech at Stanford University on 03 June. USA YouTube video at 15.26 minutes https://www.youtube.com/watch?v=xnpWtq6FsiE. Accessed 22 Oct 2023
14. Gupta A, Ghosh S (2023) India's climate research agenda: 2030 and beyond. Department of Science and Technology, Ministry of Earth Sciences, Government of India. https://dst.gov.in/sites/default/files/India%27s%20Climate%20Research%20Agenda%202030%20and%20beyond.pdf. Accessed 22 Oct 2023
15. Council of the EU Press Release (2022) A Strategic Compass for a stronger EU security and defence in the next decade, March 21. European Council. https://www.consilium.europa.eu/en/press/press-releases/2022/03/21/a-strategic-compass-for-a-stronger-eu-security-and-defence-in-the-next-decade/. Accessed 23 Oct 2023
16. The EU SWITCH to Green Flagship Initiative (2023) The EU green deal—a roadmap to sustainable economies February 14. https://www.switchtogreen.eu/the-eu-green-deal-promoting-a-green-notable-circular-economy/. Accessed 23 Oct 2023
17. Ahmad N (2021) Refugees and algorithmic humanitarianism: applying artificial intelligence to RSD procedures and immigration decisions and making global human rights obligations relevant to AI governance. 28 Int J Minority Group Rights 3:367–435
18. National Disaster Management Authority Government of India (2015) Compendium of laws on disaster management, January. https://ndma.gov.in/sites/default/files/PDF/COMPENDIUM-OF-LAWS-ON-DISASTER-MANAGEMENT.pdf. Accessed 24 Oct 2023
19. Maslej N,Fattorini L, Brynjolfsson E, Etchemendy J, Ligett K, Lyons T, Manyika J, Ngo H, Niebles JC, Parli V, Shoham Y, Wald R, Clark J, Perrault R (2023) The AI index 2023 annual report. AI Index Steering Committee, Institute for Human-Centered AI, Stanford University, Stanford, CA, April. https://aiindex.stanford.edu/wp-content/uploads/2023/04/HAI_AI-Index-Report_2023.pdf. Accessed 02 Nov 2023
20. Krishnan M (2023) India: migration from climate change getting worse (DW 19 April 2023). https://www.dw.com/en/india-migration-from-climate-change-getting-worse/a-65369043#:~:text=India%20alone%20will%20see%2045,result%20of%20extreme%20weather%20events. Accessed 04 Nov 2023
21. NITI Aayog (2018) National strategy for artificial intelligence. https://niti.gov.in/sites/default/files/2019-01/NationalStrategy-for-AI-Discussion-Paper.pdf. Accessed 06 Nov 2023
22. PIB Delhi (2023) AI Supercomputer 'AIRAWAT' puts India among top supercomputing league 'AIRAWAT' ranks 75th in top 500 Supercomputing List, May 24. https://pib.gov.in/PressReleaseIframePage.aspx?PRID=1926942. Accessed 06 Nov 2023
23. Kugelman M (2020) Climate-induced displacement: South Asia's clear and present danger. 30 Sept 2020. Wilson Center. https://www.wilsoncenter.org/article/climate-induced-displacement-south-asias-clear-and-present-danger#:~:text=In%202012%2C%20floods%20displaced%201.5,livelihoods%20to%20migrate%20to%20cities. Accessed on 06 Nov 2023
24. Ahmad N (2023) Artificial intelligence and international refugee law, 05 Sept. Modern Diplomacy. https://moderndiplomacy.eu/2021/09/05/artificial-intelligence-and-international-refugee-law/. 07 Nov 2023
25. Ozkul D (2023) Automating immigration and Asylum: the uses of new technologies in migration and Asylum governance in Europe. Algorithmic Fairness and Asylum Seekers and Refugees (AFAR) Project and Oxford Refugee Studies Centre. University of Oxford. https://www.hertie-school.org/fileadmin/2_Research/1_About_our_research/2_Research_centres/Centre_for_Fundamental_Rights/AFAR/automating-immigration-and-asylum_final_afar.pdf. Accessed 07 Nov 2023
26. Rodrigues R (2020) Legal and human rights issues of AI: gaps, challenges and vulnerabilities. 4 J Responsible Technol. https://doi.org/10.1016/j.jrt.2020.100005. 09 Nov 2023
27. Latonero M (2018) Governing artificial intelligence: upholding human rights and dignity. Data Soc

28. Droubi S (2016) Transnational corporations and international human rights law. Notre Dame J Int Comparative Law 6(1)
29. Jolly S, Ahmad N (2019) Climate refugees in South Asia: protection under international legal standards and state practices in South Asia. Springer
30. Goodwin-Gill GS, 'The International Law of Refugee Protection' (2014) The Oxford handbook of refugee and forced migration studies. In: Fiddian-Qasmiyeh E, Loescher G, Long K, Sigona N (eds). https://doi.org/10.1093/oxfordhb/9780199652433.013.0021. Accessed 12 Nov 2023
31. Mantelero A, 'Human Rights Impact Assessment and AI' (2022) In: Beyond data, information technology and law series, vol 36. T.M.C. Asser Press, The Hague. https://doi.org/10.1007/978-94-6265-531-7_2 Accessed 13 Nov 2023
32. Rutenberg I, Gwagwa A, Omino M (2021) Use and impact of artificial intelligence on climate change adaptation in Africa. In: Oguge N, Ayal D, Adeleke L, da Silva I (eds) African handbook of climate change adaptation. Springer, Cham. https://doi.org/10.1007/978-3-030-45106-6_80. Accessed 15 Nov 2023
33. Ferris E (2008) Natural disasters, human rights, and the role of national human rights institutions. 25 Oct. Brookings. https://www.brookings.edu/articles/natural-disasters-human-rights-and-the-role-of-national-human-rights-institutions/. Accessed 15 Nov 2023
34. Scott M et al (2020)FIRE: a framework for integrating human rights and gender equality in disaster risk reduction and climate change adaptation. Clim Dev. https://doi.org/10.1080/17565529.2022.2138694. Accessed 16 Nov 2023
35. Lassa JA et al (2019) Measuring political will: an index of commitment to disaster risk reduction. 34 Int J Disaster Risk Reduction 64–74
36. Sommario E, Venier S (2018) Human rights law and disaster risk reduction QIL 49 29–47
37. Yabe T et al (2022) Toward data-driven, dynamical complex systems approaches to disaster resilience PNAS Perspective 8(119). https://doi.org/10.1073/pnas.211199711. Accessed 18 Nov 2023
38. Khalili-Borna CA (2007) Technological advancement and international human rights: is science improving human life or perpetuating human rights violations? MICH. J INT'L L 29:95–125
39. Galaz V et al (2022) Artificial intelligence, systemic risks, and sustainability. 67 Technol Soc 101741. https://doi.org/10.1016/j.techsoc.2021.101741. Accessed 22 Nov 2023
40. Kaplan S (2023) World is on brink of catastrophic warming, U.N. climate change report says. The Washington Post, 20 Mar. https://www.washingtonpost.com/climate-environment/2023/03/20/climate-change-ipcc-report-15/. Accessed 24 Nov 2023
41. Liu Z et al (2022) An artificial intelligence-based electric multiple units using a smart power grid system. Energy Rep 8:13376–13388. https://doi.org/10.1016/j.egyr.2022.09.138. Accessed 25 Nov 2023
42. Arbab P (2019) Global and globalizing cities from the global south: multiple realities and pathways to form a new order. 18 Perspect Global Dev Technol 3:327–337. https://doi.org/10.1163/15691497-12341518. S2CID191718311. Accessed 27 Nov 2023
43. Aparicio CB, Ofa SV (2021) Digital technologies for climate change adaptation in Asia and the Pacific. Working Paper Series December ESCAPc. https://reliefweb.int/report/world/digital-technologies-climate-change-adaptation-asia-and-pacific. Accessed 27 Nov 2023
44. Bogna J (2023) Can AI help us save the planet from ourselves? PC Magazine April 6. https://www.pcmag.com/news/disaster-prevention-and-climate-resilience-how-ai-helps-us-save-the-planet. Accessed 28 Nov 2023
45. Marda V (2018) Artificial intelligence policy in India: a framework for engaging the limits of data-driven decision-making. Phil Trans R Soc. A.37620180087 20180087 https://doi.org/10.1098/rsta.2018.0087. Accessed 28 Nov 2023
46. Kapoor KK, Tamilmani K, Rana NP et al (2018) Advances in social media research: past, present and future. Inf Syst Front 20:531–558. https://doi.org/10.1007/s10796-017-9810-y. Accessed 29 Nov 2023
47. Campbell E, Kleinman M (2023) AI must not become a driver of human rights abuses. Al Jazeera January 13. https://www.aljazeera.com/opinions/2023/6/13/ai-must-not-become-a-driver-of-human-rights-abuses. Accessed 29 Nov 2023
48. Horak G (2023) Personal details exposed: spyware and human rights in the middle east and North Africa. Master's thesis, Harvard University Division of Continuing Education
49. Mozur P (2018) A genocide incited on facebook, with posts from Myanmar's Military. The New York Times October 15. https://www.nytimes.com/2018/10/15/technology/myanmar-facebook-genocide.html. Accessed 30 Nov 2023
50. Akinwotu E (2021) Facebook's role in Myanmar and Ethiopia under new scrutiny. The Guardian October 07. https://www.theguardian.com/technology/2021/oct/07/facebooks-role-in-myanmar-and-ethiopia-under-new-scrutiny. Accessed 30 Nov 2023
51. Guzman CD (2022) Meta's facebook algorithms 'Proactively' promoted violence against the Rohingya, New Amnesty International Report Asserts. TIME01 December 2023. https://time.com/6217730/myanmar-meta-rohingya-facebook/. 01 Dec 2023
52. Hughes L (2022) Metaverse beyond the hype: multidisciplinary perspectives on emerging challenges, opportunities, and agenda for research, practice and policy. 66 Int J Inf Manag 102542 https://doi.org/10.1016/j.ijinfomgt.2022.102542. 01 Dec 2023
53. Hunkenschroer AL, Kriebitz A (2023) Is AI recruiting (un)ethical? A human rights perspective on the use of AI for hiring. AI Ethics 3:199–213. https://doi.org/10.1007/s43681-022-00166-4. 02 Dec 2023

Nafees Ahmad is Associate Professor at the Faculty of Legal Studies, South Asian University (SAU), New Delhi. He holds a doctorate in international refugee law and human rights. His scholarship focuses on RAMS (refugees, asylum-seekers, migrants, stateless) and the role of artificial intelligence (AI) in their protection, global forced displacement, global circumstantial migration (GCM) governance, and climate refugees in South Asia. He also addresses the international politics of asylum, refugee policy paradigms, invisible frames of asylum, disconnects of durable solutions, and SAARC connects and contexts of refugee protection. He conceived and introduced a new SAARC-specific Program in 2011 called Comparative Constitutional Law of SAARC Nations (CCLSAARCN) at the LLM level. He publishes inter-alia in the *International Handbook on Disaster Research* (Springer 2023), *The Asian Yearbook of Human Rights and Humanitarian Law* (2023), *Groningen Journal of International Law* (The Netherlands), *Harvard International Law Journal* (Harvard), *Asia-Pacific Journal of Human Rights and Law, International Journal of Environment and Waste Management International Journal on Minority and Group Rights, Kings' Student Law Review* (KCL-London), *ISIL Year Book on International Humanitarian Law and Refugee Law, ELCOP Year Book of Human Rights* and *NUJS International Journal of Legal Studies and Research* (IJLSR) etc. Dr. Ahmad has co-authored a book on 'Climate Refugees in South Asia' (Springer 2019). He is a member of Editorial Advisory Board of *Iranian Journal of International and Comparative Law*. Dr. Ahmad has been a Resource Person and External Reviewer for the Ministry of Law; Government of India-sponsored Research Project on "Judicial Reforms since June 2016" at the Indian Institute of Management (IIM-Kashipur) and accomplished Research Projects from the ICSSR and RWI (Sweden) on "The Muncipal Solid Waste Management in Delhi: A Socio-Legal Study of Okhla Lanfill" and "Climate-related Human Mobility in Asia and the Pacific: Interdisciplinary Rights-based Approaches" respectively. Further, he has been a visiting professor at the Indian Society of International Law (ISIL), Jamia Milia Islamia-New Delhi, and Judicial Academies of various states of India.

Humanitarian Aid Distribution in the Context of Human Rights-Based Approach Among Vulnerable Communities: Flash Floods and Climate Change in North Luwu, Indonesia

Dina Ruslanjari, Cahyadi Ramadhan,
Inayah Bastin Al Hakim and Feby Aulia Marsida

Abstract

North Luwu Regency in South Sulawesi Province, Indonesia, was struck by a flash flood in 2020, resulting in 38 reported deaths and 106 people missing. Additionally, 20,562 residents were forced to evacuate and were referred to as internally displaced persons (IDPs). The handling and protection of the affected community or IDPs in an emergency and a post-emergency period requires proper, integrated, and measurable management. The research done in North Luwu Regency, employed a qualitative approach and was conducted through observation, in-depth interviews, and focus group discussions (FGDs) with heads of villages and the officials, religious and local traditional leaders, NGOs, related agencies, and the IDPs. The findings of this study indicate that the policies regarding aid provision during flash floods are considered quite comprehensive, given that the region experiences similar disasters frequently. The findings also suggest that the implementation of IDP protection is comprehensive but encounters some challenges, including hindered aid distribution due to difficult access to disaster-stricken areas, non-targeted and non-centralised aid distribution, as well as the ineffectiveness of the random system for temporary housing for IDPs. In terms of perception, the community expresses satisfaction with the government's performance in providing humanitarian aid and logistics, resulting in tangible impacts. Furthermore, no discriminatory practices based on religion, ethnicity/race, gender, background, or economic status were found in the aid distribution.

D. Ruslanjari (✉) · C. Ramadhan · I. B. Al Hakim · F. A. Marsida
Masters in Disaster Management, Universitas Gadjah Mada, Yogyakarta, Indonesia
e-mail: dienarus@ugm.ac.id

C. Ramadhan
e-mail: cahyadi.ramadhan@mail.ugm.ac.id

I. B. Al Hakim
e-mail: inayah.bastin.a@mail.ugm.ac.id

F. A. Marsida
e-mail: febyaulia97@mail.ugm.ac.id

1 Introduction

Climate change increases the risk of disasters.[1] It expands the distribution of intense rainfall, increases the variability of extreme rainfall, and

[1] Reference [1].

S. Jolly et al. (eds.), *Climate-Related Human Mobility in Asia and the Pacific*, Sustainable Development Goals Series, https://doi.org/10.1007/978-981-97-3234-0_9

accelerates the hydrological cycle,[2] so that its amplification increases the intensity and frequency of floods.[3] Flash floods are generally caused by heavy or extreme rain that can be localised, affecting large areas in one to several basins, especially in urbanisation zones and upstream bare areas.[4]

Flash floods are categorised as hazards that damage human life and are triggered by an anomaly of rainfall intensity in mountainous areas.[5] It is strongly influenced by the geological conditions of slope-forming rocks or soil, vegetation or trees, slope, land use, and geological structure.[6] Flash floods occur when spillage flows out of river channels carrying debris (mixed with mud, gravel, boulders, and wood waste). Such a catastrophic event can destroy urban settlements, causing death and damage to infrastructure.[7]

Climate change can trigger the higher intensity of flood events, including flash floods and urban floods, caused by heavy rainfall.[8] This arises from the fact that climate change encompasses alterations in precipitation, as well as temperature, wind dynamics, atmospheric pressure, cloud cover, and moisture content due to short-duration extreme precipitation.[9] Thus, with climate change, the hydrological cycle has been expectedly increased along with the existence of global warming.[10]

Flash floods have occurred in many places in Indonesia, especially in several watersheds formed from hilly valleys with steep slopes and abundant water sources. One area that has a high exposure to flash floods is North Luwu Regency in South Sulawesi Province. The area's exposure is due to the predominance of a steep topography with an elevation of 1200–2000 m. Most of the area is surrounded by various water points (Malacca Straits, Pacific Ocean, Maluku Sea, Bone Bay), which are ideal and complex terrains for flash floods.[11] During the last 23 years, there have been multiple hydrometeorological disasters triggered by floods, landslides, tornadoes, and droughts in North Luwu Regency. Floods were the most dominant trigger of disaster in the area. Climate change is expected to exacerbate these threats. Flash floods hit North Luwu Regency on 13 July 2020 following the landslides in the upper reaches of the Masamba, Rongkong, and Radda rivers.[12] The affected Regencies include Masamba, Sabbang, Baebunta, South Baebunta, Malangke, and West Malangke. The leading cause of the flash floods in North Luwu Regency was the high rainfall pattern that had occurred gradually in the previous months. Heavy rainfall can be one of the main factors that triggers flash floods.[13] In this case, it is unusually high as flooding is frequently caused by prolonged periods of heavy rainfall, a phenomenon that occurs naturally and raises the amount of moist air in the atmosphere, leading to heavy rainfall over most of Indonesia. The angle slope or local terrain has an impact on flooding as well. A flash flood will be more destructive the steeper the slope as then the faster the water will flow. There are numerous ways to reduce the likelihood of landslides and floods. Among these are afforestation which requires the growing and upkeep of a certain number of trees in the watersheds to restore the ability of forests to store water effectively. Additionally, we must keep an eye out for the dangers posed by mining and logging operations that invade watershed areas. Before the flash

[2] Reference [2]. See also: References [3, 4].

[3] Reference [5].

[4] Reference [6].

[5] Reference [7].

[6] Disaster Risk Index in Indonesia, (National Agency for Disaster Management 2022) 1.

[7] Reference [8], See also: References [9, 10]

[8] Reference [11]. See Also: Climate Change Guide, 'Flooding and Climate Change: Everything You Need to Know' (3 November 2023) https://www.climate-change-guide.com/floods-and-climate-change.html accessed on 3 June 2023.

[9] Reference [12].

[10] Reference [13].

[11] Reference [14].

[12] Reference [15].

[13] Reference [16].

floods on 13 July 2020, rainfall was expected to be around 50 mm over 10 days.[14]

In contrast to the flash floods that occurred in other areas, the material in the flash floods in North Luwu was mainly sand and a little mud.[15] The Communication and Information Office of North Luwu noted that the impact of the flood led to the death of 38 people and 106 were injured. In addition, 1600 houses were heavily damaged, 136 moderately damaged, and 2384 slightly damaged. The flash floods also damaged public facilities, 3 units of health facilities, 25 units of educational facilities, and 25 units of houses of worship. Furthermore, an estimated 20,562 residents were forced to flee from 75 Masamba, Baebunta, and Sabbang Districts posts.

The established regency is home to a population of 327,820 residing across 15 sub-districts, with the reported poverty rate at 13.22 per cent in 2022, showing a decrease from the previous year's 13.59 per cent.[16] In terms of disaster vulnerability, this area is classified as highly susceptible to hydrometeorological and climate-related events, with a low level of community resilience against such occurrences, and the population exposed to specific hazards being notably high; for instance, 268,345 individuals live in flood-prone areas.[17] North Luwu is the regency with the highest flash flood hazard area in the South Sulawesi Province, covering 48,083 hectares.[18] The local government has recognised this issue and is committed to safeguarding its vulnerable residents through a human rights-based approach. This dedication is exemplified by the accolade bestowed upon the North Luwu local government by the Ministry of Laws and Human Rights (Kemenkumham) in 2022.[19] The award is a testament to the various indicators addressing community needs, including the right to legal assistance, access to information, promotion of diversity and pluralism, healthcare access, educational opportunities, protection of women and children's rights, population rights, and ensuring a healthy environment and adequate housing.

Handling the community affected by flash floods in North Luwu Regency required appropriate, integrated, and measurable actions, especially during the COVID-19 pandemic. Managing humanitarian assistance was a multidimensional challenge as it had various problems, such as aid distribution, accessibility, and meeting the needs of displaced populations. These problems are often presented during the emergency response and post-disaster phases, where survivors need temporary and permanent shelters to start life again after a disaster. Further challenges were the unavailability of land or space, the distribution of permanent shelters, and the building process.

In this background, the study aimed to understand the policies, practices, and public perceptions regarding managing internally displaced persons (IDPs) during emergencies and post-emergencies from a human rights perspective. This study specifically aimed to (1) examine the protection of displaced populations as reflected in logistics management policies in humanitarian assistance; (2) review the implementation of IDP protection in the management of humanitarian assistance and logistics; and (3) examine the perceptions of the affected community towards fulfilling the needs of displaced populations and management of humanitarian logistical assistance.

[14] Deti Mega Purnamasari and Krisiandi, 'BMKG Mentions the Rainfall Intensity Triggers the Flash Flood in North Luwu' (Kompas, 19 July 2020). https://nasional.kompas.com/read/2020/07/19/16052071/bmkg-sebut-curah-hujan-tinggi-jadi-penyebab-banjir-bandang-luwu-utara (in Indonesian).

[15] Reference [17].

[16] BPS North Luwu, 'North Luwu Regency in Figures' (Central Bureau of Statistics of North Luwu 2023) 53, 141 (in Indonesian).

[17] See Footnote 16.

[18] BPBD South Sulawesi *'Disaster Risk Study Document of South Sulawesi Province 2022–2026'* (Regional Disaster Authority of South Sulawesi Province 2022) (in Indonesian).

[19] IKP, 'North Luwu Receives an Award from Ministry of Law and Human Rights of the Republic of Indonesia' (The Government of North Luwu Regency, 2022).

2 Research Methodology

This research was conducted in North Luwu Regency, South Sulawesi Province. This location was chosen by purposive sampling because floods often occur for a long duration, causing many fatalities and had damaging effects on 20,562 residents. From this category, there were 15 affected sub-districts in North Luwu Regency. However, in this research, the researchers only focused on the two districts, i.e. Masamba and Baebunta, which were mostly affected by the flash floods that occurred on 13 July 2020. The field study was carried out within four days from 4 to 8 October 2021 to collect the data.

Masamba is known for its strategic location and serves as the regency's capital.[20] The region is characterised by its diverse landscape, including plains and hills, and is an essential economic hub for the surrounding areas. Baebunta, on the other hand, is recognised for its unique geographical features and plays a role in the economic and social life of the regency. Both districts are part of the larger North Luwu Regency, contributing to the cultural and economic diversity of the region. The communities in these districts are often engaged in agricultural activities, and the areas face specific challenges, such as the risk of flash floods, given their geographical characteristics. These two districts are traversed by the Masamba River, the Radda River (a tributary of the Masamba River), and the Rongkong River,[21] which often causes flash floods.

This research uses a qualitative approach with documentation, observation, interviews, and focus group discussions (FGD). Documentation was carried out to identify policies and examine the perception of displaced populations on the humanitarian and logistical assistance delivered by aid providers. The initial observation of this research was carried out to find out the socio-cultural characteristics to the distribution of displaced populations in the study location. The distribution of displaced populations was analysed using spatial tools such as ArcGIS and Google Earth, to simplify the data collection scheme that had to be carried out optimally to divide the number of respondents for sampling. Data collection was carried out using structured and in-depth interview techniques with 32 displaced respondents, 16 each from Baebunta sub-district, and 16 from Masamba sub-district. The FGDs involved key informants, including regional/village heads and their officials, local traditional and religious leaders, NGOs, related agencies, and temporary and permanent housing displaced populations, totalling 14 people at the meeting in Baebunta sub-district and 13 people at the meeting in Masamba sub-district.

Before the flash flood occurred, the community depended on the agricultural sector as paddy field farmers, meeting their livelihood needs through harvest yields. Moreover, they were able to provide for their children's education without difficulty. However, following the flash floods, many from the agricultural community lost their steady employment because the land they had cultivated for so long was buried under materials such as mud and rocks. This has significant economic implications, particularly for those farmers with low incomes and limited education, making it challenging for them to find alternative employment to sustain the economic and educational needs of their children. Consequently, the community had to undergo relocation and a shift in their means of livelihood. Formerly employed as paddy field farmers on their own lands, the community found themselves working as daily wage labourers on others lands.[22]

---

[20] BPS North Luwu Regency *'Masamba in Figures'* (Central Bureau of Statistic of North Luwu Regency 2021) (in Indonesian).

[21] Ministry of Public Works and Housing of The Republic of Indonesia, *'Brief Description: Construction Of Sediment Control for the Radda River in North Luwu Regency 2023'* (in Indonesian).

[22] Reference [18].

This research is expected to be a reference in designing effective strategies to improve the management of humanitarian assistance in areas affected by disasters such as flash floods due to climate change and incorporate issues of prioritising a human rights perspective. With this aim, the research results are divided into four sections to explain related logistics management policies in existing humanitarian assistance, the implementation of IDP protection in humanitarian assistance and logistics management, and the related perceptions of displaced populations on the fulfilment of protection and management needs of humanitarian logistics assistance that has been carried out.

3 Disaster Management in Indonesia and the Regulations on Humanitarian Aid Management

The regulation and the implementation of disaster risk reduction activities has changed over time in the world's biggest archipelagic country. Following the 2004 earthquake and tsunami that killed 160,000 people in Indonesia alone,[23] improvement of regulation and practice, in the disaster risk reduction (DRR) context, has taken place in the nation. The catastrophic event triggered the government, community, and stakeholders to be more aware of disasters, not only just in terms of the action in emergency response, and there appears to be significant progress in all phases of disaster that include pre-disaster, during disaster, and post-disaster. The emergence of the Law on Disaster Management 2007 and the establishment of the National Disaster Management Authority (BNPB) and the Regional Disaster Management Authority (BPBD) prove that the Government of Indonesia (GoI) is seriously committed to disaster issues.

While the government emphasises decentralisation, key areas such as foreign policy, defence, legal systems, and monetary policy remain under national government jurisdiction. In the context of DRR, the BPBD are entrusted with DRR responsibilities within their respective domains, while the BNPB as the authority in the national level acts as the central agency for national DRR policies.[24] The issues cover all areas of DRR practices, including humanitarian aid management that should be prepared right from inception when there is a likely potential of hazard. It refers to how the government with the participation of academics, business/private sector, community, media, and other stakeholders, can collaboratively conduct displacement and humanitarian assistance management, particularly in the prone-areas. The objective is to reduce the risk and fatalities in the upcoming disasters. These processes could either be meticulously organised and officially endorsed, or they may arise spontaneously from affected populations in accordance with their own understanding and assessment of risks.[25]

3.1 Humanitarian Aid Management Policy at the National Level

Disaster Management Law in the Republic of Indonesia[26] is the foundation of all the derivative regulations (the derivative regulations can be presidential decree, regional regulation, etc.) and community implementation in DRR. Disaster management handling should be based on humanity; justice; equality before the law and government administration; balance, harmony, and congruence; order and legal certainty; togetherness; environmental conservation; and

[23] Reference [19].

[24] UNDRR, 'Disaster Risk Reduction in Indonesia: Status Report 2020' (United Nations Office for Disaster Risk Reduction 2020).

[25] UNISDR and IDMC, 'Displacement in the Greater Horn of Africa: A Disaster Risk Reduction Perspective' (United Nations Office for Disaster Risk Reduction and Internal Displacement Monitoring Center 2017).

[26] 'Law of The Republic of Indonesia Number 24 Year 2007 on Disaster Management' article 3 para 6.

science and technology.[27] These humanitarian principles mandate the State's obligation to protect and respect every Indonesian with human rights and human dignity. Specifically, the 'principle of justice'[28] means that every implementation of DRR reflects justice for every citizen without exception, regardless of background, including religion, ethnicity, race, class, gender, or social status. The status of 'citizen' refers to individuals of the indigenous Indonesian population and individuals of other nationalities officially recognised as citizens by law,[29] and residents are Indonesian citizens and foreigners residing in Indonesia.[30] Thus, disaster management based on humanitarian principles has a strong statutory basis to protect people in disaster prone-areas in Indonesia territory.

The BNPB has established regulations for further streamlining the disaster management process, particularly focusing on basic needs assistance.[31] It gives a message to the government and the stakeholders involved, to ensure the fulfilment of the rights of disaster-displaced populations and the affected community, especially in providing essential services. The BNPB states that the principle of assistance must be carried out based on gender equality and a human rights approach.[32] Moreover, the agency is also concerned with gender equality in disaster, followed by the regulation mentioning gender mainstreaming.[33] This regulation provides guidelines for implementing Gender Responsive Planning and Budgeting (PPRG) in the scope of disaster management, as well as gender mainstreaming at the pre-disaster, disaster emergency, and post-disaster phases. In addition, the regulation also indicates that assistance should be based on the principle of sustainability, local wisdom, and participation between stakeholders and the affected community.These regulations at the national level are supplemented by the regional-level policies, described below.

3.2 Humanitarian Aid Management Policy at the Regional Level

The Minister of Social Affairs of the Republic of Indonesia issued a regulation concerning social assistance in catastrophic events,[34] which is and becomes one of the legal bases for managing humanitarian assistance at the regional level. In the regulation, it suggests that direct distribution of aid in the form of clothing, food and boards, health services, provision of temporary shelters, psychosocial therapy services in shelters, cash, waivers on the cost of processing population and ownership documents, condition of low-cost necessities, provision of public kitchens, clean water, and healthy sanitation, and provision of funerals for disaster victims are carried out by officers in the provincial and regency/municipal levels, involving sub-district or village officials or other similar names in their places of residence.[35] It is also complemented by the BNPB Regulation concerning Guidelines for Logistical Assistance.[36] Logistical assistance is needed for disaster management activities, primarily when a disaster occurs, covering the community's daily needs, such as food, clothing and its complementary, water sanitation, sleeping bags, and so forth.

27 Ibid, article 2 para 5.

28 Ibid, article 2 para 4.

29 'Constitution of The Republic of Indonesia 1945' article 26(1).

30 Ibid, article 26(2).

31 'Head of BNPB Regulation Number 18 Year 2009 on Guidelines for Basic Needs Assistance' article 1.

32 'Head of BNPB Regulation Number 3 Year 2018 on Displacement Management in Emergency Responses' article 2.

33 'Head of BNPB Regulation Number 13 Year 2014 on Gender Mainstreaming in Disaster Management' article 1.

34 'Ministerial Regulation of Social Affairs Number 1 Year 2013 on Social Assistance for Disaster Affected Community' article 5.

35 Head of BNPB Regulation Number 13 Year 2014 on Gender Mainstreaming in Disaster Management' (n 26) article 8.

36 'Head of BNPB Regulation Number 4 Year 2009 on Guidelines for Logistic Assistance' article 1.

In addition, the Government of North Luwu Regency has a series of regulations related to the management of social assistance that can be funded from the revenue budget and regional expenditures.[37] The North Luwu Regency has also provided space for efforts to implement community-based disaster risk reduction (CBDRR),[38] which is based on the objective to involve community participation in DRR efforts. People are also enabled to use their resources so as to prevent, reduce, avoid, and recover from the effects of disasters. More than that, the implementation of CBDRR also addresses the obligation of the government to protect the community as part of human rights.[39]

The local government has implemented various regulations as part of its emergency response, rehabilitation, and reconstruction strategies in the light of the flash floods. Most recently, the government has identified recipients, from stimulus budgets, to aid affected communities in repairing moderately and lightly damaged houses.[40] Emergency expenditure budgets have been distributed to the IDPs for logistic assistance in the affected districts.[41] In addition, another policy outlines technical guidance for the development and reconstruction of houses impacted by the flash floods.[42] The authorities are committed to ensuring that communities receive funding assistance for Phase II,[43] the construction of permanent shelters.[44] These measures collectively represent a comprehensive approach to addressing the challenges posed by the flash floods.

Policies at the Ministerial and national levels, and their derivatives, as well as the Regent Decree on flash flood disaster, show that there was a firm reference or legal basis for the management of humanitarian assistance for the protection of displaced populations caused by disaster. The national/regional/local policies in response to managing humanitarian aid to protect IDPs due to flash floods in the North Luwu Regency were comprehensive. These policies aimed to protect communities affected by disasters and prevent various possibilities of conflict. However, there were some differences between the existing policies and their implementation at the disaster site. Flash flood management policies in North Luwu Regency were not optimal due to ineffective government communication.[45] The following section enumerates the field study conducted by the authors in the North Luwu Regency to understand the effectiveness of disaster assistance following the flash flood tragedy.

4 Implementation of Humanitarian Aid Distribution on Flash Floods

The impact of the flash floods in North Luwu Regency resulted in socio-economic changes for the affected people.[46] Various parties

[37] 'Regent of North Luwu Regulation Number 13 Year 2021 on Procedures for Budgeting, Implementation and Administration, Accountability and Reporting as well as Monitoring and Evaluation of Grants and Social Assistance Sourced from Regional Revenue and Expenditure Budgets' article 1.

[38] 'Regent of North Luwu Regulation Number 28 Year 2017 on Community-Based Disaster Management Implementation'.

[39] Ibid.

[40] 'Regent Decree Number 480 Year 2021 on The Determination of Recipients from Stimultant Fund to Repair Moderately Damaged House and Slightly Damaged House Due to Flash Flood and Landslide in North Luwu'

[41] Regent Decree Number 425 Year 2021 on Emergency Expenditure Budget for Logistic Assistance to Flash Flood Affected Community in Malangke, West Malangke, South Baebunta, Mappedeceng and Sukamaju Districts'.

[42] 'Regent Decree Number 411 Year 2021 on The Establishment of Technical Guidelines for Implementing the Construction of Severely Damaged Houses, Repair of Moderately Damaged Houses, and Minor Repairs, as well as Provision of Transitional Housing Funds during the Emergency to Recovery Phase of Flash Flood in North Luwu Regency'.

[43] Regent Decree Number 194 Year 2021 on The Determination of Life Security Assistance (Phase II) Recipients for Flash Flood Affected Community'.

[44] Ibid.

[45] Reference [20].

[46] Reference [21].

participated in disaster management efforts from the first critical period (Golden Hour) to the rehabilitation and reconstruction. Humanitarian assistance for flash flood-affected displaced populations came from various parties, including Regional Disaster Management Agency (BPBD), Social Service Organisations, Disaster Voluntary Group (*Tagana)*, Civil Service Police Unit (*Satpol PP*), National Search and Rescue Agency (Basarnas), NGOs, academics, business institutions, religious institutions, as well as families and the surrounding community.[47] Multi-sectoral involvement in the distribution of humanitarian aid during the flash flood event in North Luwu Regency at the various phases of disaster assistance is described as follows.

4.1 Distribution of Humanitarian Aid During the Golden Hour

The first 72 h after a disaster are referred to as the '*Golden Hour',* considering the fact that these times are very crucial for people to survive.[48] Flash floods in Luwu in 2020 occurred at night, damaging many buildings, houses and public facilities, and causing road access and main telecommunications to be cut off. Information on the occurrence of flash floods was constrained by the disconnection of two main communication accesses. As a result, the flow of emergency response coordination and aid distribution during the Golden Hour did not run optimally. The distribution of humanitarian aid to the Baebunta sub-districts on the first day could not be carried out because the land access connecting the two districts was damaged. Most of the humanitarian assistance in Masamba district was provided by the closest family, from the Indonesian Red Cross, and the nearest sub-district.

4.2 Distribution of Humanitarian Assistance During the Emergency Response Period

The emergency response period for the North Luwu flash floods covered a span of 60 days. During the first 30 days, several agencies such as the BPBD, Social Services, and government agencies from other districts and provinces began to provide humanitarian and logistical assistance as well as evacuation assistance, especially in the Masamba district and the Baebunta district. Meanwhile, based on findings in the field during the emergency response period, several NGOs or humanitarian agencies also participated. In addition, academics, scouts, and the Cakra community were active in providing logistics and psychosocial services, and the Indonesian Red Cross and *Aksi Cepat Tanggap* (ACT) a humanitarian organisation provided logistics and clean water in the Baebunta district.

During the extension of the emergency response period to an additional 30 days, all agencies and NGOs provided and distributed logistical and psychosocial assistance, clean water supply, and other public facilities in the two sub-districts. These activities began to involve many parties, including foreign parties such as the International Organisation for Migration (IOM), which initially focused only on the Baebunta district, and later began to enter the Masamba district and significantly impacted the handling of displaced populations, especially in the provision of logistics and the provision of medicines, tents, and psychosocial services (see Table 1).

The terms 'disaster relief' and 'humanitarian response' encompass the initial response and early recovery phases, with the foremost priority being the provision of essential necessities (such as food, water and sanitation, clothing, health service, and shelter) to the individuals most severely affected in the immediate aftermath of a disaster.[49] The distribution of humanitarian

[47] Reference [22].

[48] Kominfo, 'Why Phone Communication is Still in Trouble after Earthquake and Tsunami Event?' (2 October 2018) Kominfo 2 Oct. 2018 <https://www.kominfo.go.id/content/detail/14823/usai-gempa-dan-tsunami-di-palu-kenapa-komunikasi-ponsel-bermasalah/0/sorotan_media. Accessed on 3 July 2023.

[49] 'Head of BNPB Regulation Number 7 Year 2008 on Guidelines for Social Assistance of Basic Needs' article 4.

Table 1 Distribution of humanitarian aid in Baebunta and Masamba district during the emergency response

Area	Disaster relief provider	Type of disaster relief	Time
Baebunta	Anonymous volunteer	Logistics	On the day of the disaster (13 July 2020)
Baebunta	Community organisation *Masyarakat Rongkong Akar Bersatu*	Logistics	
Baebunta, Masamba	North Luwu Red Cross	Clean water, psychosocial, drugs, tarp, tent	During the emergency response (14 July–12 September 2021)
Baebunta, Masamba	Scout	Logistics, psychosocial	
Baebunta, Masamba	Organisation *Aksi Cepat Tanggap* (ACT)	Logistics	
Baebunta, Masamba	Community organisation *Rumah Kemanusiaan Chakra Abhipraya Responsif*	Logistics, psychosocial	
Baebunta	Islamic Community Organisation *Peduli Dakwah*	Logistics	
Masamba	Mosque youth community	Logistics	
Baebunta	Academics	Logistics, psychosocial	
Baebunta, Masamba	Nearest District	Logistics	
Masamba	Nearest Province	Logistics	

assistance (Table 1) prioritised the basic needs when the flash floods had just hit the area. The form of assistance can change over times following the needs of the community in each phase of disaster.[50] Nevertheless, pre-disaster evacuation plans were not practised in the case of flash floods in North Luwu Regency. There were disparities in emergency response actions during the floods that occurred in Masamba and Baebunta. The community in the Masamba district, which serves as the capital of North Luwu Regency, received a one-hour flash flood warning, and the local government promptly directed the evacuation of residents to the Regent's Office. Meanwhile, there was no warning in the Baebunta district, resulting in a lack of preparedness among the residents when the flash floods occurred.

There are persisting deficiencies within the distribution framework. The BPBD of North Luwu, as the primary disaster response entity, bears the responsibility of coordinating and overseeing humanitarian aid efforts during emergency situations such as flash floods. The agency is tasked with procuring aid from external sources, encompassing both organisations and individuals, subsequent to a swift quantitative needs assessment, and subsequently disseminating it to the affected community.[51] This procedural approach is designed to optimise the precision and effectiveness of aid allocation. Unfortunately, not all stakeholders possess a comprehensive understanding of the associated regulations. Some donors opt to directly disburse social assistance to the community, permitting their involvement in emergency scenarios, albeit potentially leading to suboptimal distribution outcomes. Moreover, each engaged donor pursues its own agenda and carries distinct beliefs and systems.[52] Consequently, instead of being active participants in development, vulnerable individuals have been relegated to the status of objects. The conflicting ideologies, systems, and approaches among various actors have further exacerbated this problem. Furthermore, the BPBD faces internal challenges in this context, including insufficient human resources, logistical and equipment

[50] Reference [23].

[51] Reference [24].

[52] Reference [25].

limitations, budgetary constraints, and the logistical complexities of reaching isolated areas.

4.3 Distribution of Humanitarian Assistance During the Rehabilitation and Reconstruction Period

The types of assistance distributed during the rehabilitation and reconstruction phase focused on increasing the welfare and capacity of the community, including the provision of a DTH (House Waiting Fund) of USD 32/family/month for six months by the BPBD and the Ministry of Social Affairs (see Table 2). The organisations and the individuals provided the advanced assistance when the essential aids were no longer needed, such as the training for the affected community to build resilience and the life security for administrative purposes. Displaced populations also received permanent housing assistance of USD 37/person. Nonetheless, numerous inhabitants of Bone Tua Village in Masamba expressed dissatisfaction, asserting that the aid provided did not align with the commitments made by the government.

Distribution of humanitarian assistance for the flash flood-affected community in North Luwu Regency has been carried out in a multisectoral manner from the government and non-government sectors. A number of policies related to displaced populations have been well implemented, starting from the establishment of the Indonesian National Disaster Management Agency that has the role to lead and coordinate DRR practices in Indonesia.[53] The Government of Indonesia has also issued the regulation to emphasise that the importance of DRR action, in this case is an emergency response that focuses on the services to the affected community or IDPs at the shelters.[54] Specifically, the BNPB has mandated the policy in regard to logistical assistance guidance which proposes the objectives: logistic assistance planning in disaster response should be in accordance with needs; the procurement of logistics aid in disaster response adheres to applicable regulations concerning the quantity, type, and quality of assistance; the distribution of logistic aid in disaster response is effective, efficient, targeted, and accountable.[55] Communication between the local government and institutions, organisations or associations that provide logistical assistance to the community has not been well developed[56] and is causing the distribution of logistical assistance to be unreliable. It is not in line with the government regulations that emphasise the principal of coordinative and supportive communication.[57] Data and information management in handling IDPs should be integrated into the Emergency Response Management Command System.[58]

Based on the regional level, the regulation of Regent Decree Number 290/2021 concerning the Determination of Recipients of Stimulant Funds for the Repair of Heavily Damaged Houses Due to Flash Floods and Landslides is also considered to be ineffective. Most IDPs complained about the inconsistency between what was announced by the government and what was distributed to them. In addition, the allocation system (by lottery) for the distribution of temporary and permanent housing for IDPs is also considered inaccurate, unfair, and unobjective.

---

[53] 'Presidential Regulation Number 8 Year 2008 on National Disaster Management Agency' article 1.

[54] 'Government Regulation Number 21 Year 2008 on Disaster Management Implementation' article 40.

[55] 'Head of BNPB Regulation Number 4 Year 2008 on Logistic Assistance Guidance' article 1.

[56] Reference [20].

[57] Head of BNPB Regulation Number 4 Year 2008 on Logistic Assistance Guidance' (n 47) article 10.

[58] Head of BNPB Regulation Number 3 Year 2018 on Displacement Management in an Emergency Response.

Table 2 Distribution of humanitarian aid in Baebunta and Masamba district during the rehabilitation and reconstruction

Area	Disaster relief provider	Type of disaster relief	Time
Baebunta	Training Centre *Balai Latihan Kerja (BLK)* North Luwu	Training such as sewing and cooking for women and mechanical engineering for men	During the transitional period of recovery and reconstruction rehabilitation (at the shelter)
Baebunta	International Organisation for Migration (IOM)	Lighting (street lights), boreholes, widened septic tank, water tank	
Baebunta	North Luwu Women and Children Empowerment Service	Training such as sewing and cooking for women	
Baebunta	Red Cross	Water tank	
Baebunta	Dompet Dhuafa	Water tank	
Baebunta	Baznas	Prayer room	
Baebunta, Masamba	BPBD	Residential waiting fund (DTH) per family for six months	
Baebunta, Masamba	Ministry of Social Affairs	Life guarantee for two months (for administrative arrangements such as a complete KTP and family card)	
Masamba	South Sulawesi Provincial Government	Permanent residents whose distribution was based on a lottery	

5 Fulfilment of Basic Needs and Perception of Displaced Populations in Humanitarian Aid Distribution

5.1 Protection of Displaced Populations and Fulfilment of Basic Needs

During the evacuation, displaced populations have the right to receive protection in accordance with human rights and humanitarian law standards governing IDPs. One of the rights that displaced populations must obtain is fulfilling their basic needs.[59] Every individual who is affected by a catastrophic event has the right to receive assistance to meet their basic needs.[60] Furthermore, the assistance for basic needs intended for disaster-affected people is in the form of temporary housing, food assistance, clothing, and clean water and sanitation, and health services.[61] Disaster emergency assistance is provided by considering the minimum standard of basic needs and prioritising vulnerable groups, such as older people, women, persons with disabilities and children.[62]

Based on the research results on displaced populations, perceptions of protection and fulfilment of their basic needs during flash flood events in Luwu District, the researchers found that most of the disaster-affected community was quite satisfied. The disaster management that was carried out was very good. This perception was taken from the percentage of displaced populations already in permanent housing, where they had gone through the Golden Hour, and the emergency response, rehabilitation, and reconstruction. The qualification for permanent housing recipients were those who had experienced housing damages due to flash floods. When our team was collecting the data, the distribution was still in progress. The community was considered to have a sufficient understanding of the protection and fulfilment of the basic needs of displaced populations and was open to share about their experiences in receiving the aid (see Table 3). It should be noted that the total respondents were 32, including 5 pregnant and breastfeeding women, 4 people with disability and 3 senior citizens from the population. The table shows the response of all respondents.

Protection and fulfilment of basic needs such as clean water, health services, clothing, food, and vulnerable groups were considered very good by the displaced populations. The practice is based on the mandate from the BNPB Regulation to provide basic needs as primary assistance in a disaster event.[63] Clean water was always available thanks to the Indonesian Red Cross and the state-owned enterprise ‘PDAM’ that provided water tanks at all evacuation points. Quality of health services during the evacuation were considered reasonable by the displaced populations. Many volunteers helped with the health services. The Indonesian Red Cross assisted by providing various medicines to help the community. According to the displaced populations, health workers served them well even though they had mild symptoms such as headaches.

Many volunteers came to the evacuation point to provide logistical assistance in the form of rice, fast food, and various clothes. Most displaced populations consider that the provision of clothing and shelter was good enough. However, some considered the realisation of the assistance poor, especially for evacuation locations that were very far away and difficult for volunteers to reach. Meanwhile, some people thought sanitation was inadequate because trash bins were not provided.

[59] Head of BNPB Regulation Number 13 Year 2014 on Gender mainstreaming in Disaster Management’ (n 26) article 3 para 7.

[60] Ibid, article 5 para 14.

[61] ‘Government Regulation Number 22 Year 2008 on Funding and Management of Disaster Aid’ article 4 para 13.

[62] Ibid.

[63] ‘Head of BNPB Regulation Number 13 Year 2014 on Gender mainstreaming in Disaster Management’ (n 26) article 7 para 24.

Table 3 Perception of displaced populations on the fulfilment of basic needs

Perception	Very good (%)	Good (%)	Fair (%)	Poor (%)	Very poor
Fulfilment of food aid during the evacuation	59	16	25	–	–
Fulfilment of clothing assistance during the evacuation	61	29	10	–	–
Fulfilment of temporary shelter during the evacuation	37	47	13	3	–
Fulfilment of clean water during the evacuation	72	22	6	–	–
Fulfilment of sanitation during the evacuation	39	32	26	3	–
Meeting the needs of children (vulnerable groups) during the evacuation	59	41	–	–	–
Meeting the needs of pregnant or breastfeeding women (vulnerable groups) during the evacuation	44	44	11	–	–
Meeting the needs of people with disabilities (vulnerable group) during the evacuation	46	42	8	4	–
Meeting the needs of sick people (vulnerable groups) during the evacuation	55	29	16	–	–
Meeting the needs of older people (vulnerable group) during the evacuation	57	37	13	3	–
Fulfilment of health services during the evacuation	63	31	6	–	–

In general, the satisfaction in meeting the needs of vulnerable groups was deemed satisfactory. Children's necessities, such as diapers, warm clothing, blankets, and specialised food, were adequately provided for displaced populations. Pregnant and lactating women also reported their various needs being adequately met. Assistance for pregnant women came in the form of packages comprising milk, vitamins, clothing, masks, and food. Aid providers visited evacuation tents to meet pregnant and lactating mothers to ensure targeted assistance. Furthermore, special food assistance for the celebration of Eid al-Adha was also provided to this group as it coincided with the Muslim community's celebration. The support for older individuals was also noted as being fairly robust, with the head of the North Luwu area promptly providing canes for those in need.

Additionally, a majority of the requirements for persons with disabilities were addressed. Those with hearing impairments received hearing aids, and individuals with mobility challenges were supplied with wheelchairs courtesy of the North Luwu Regency government. Nonetheless, some of the displaced populations with disabilities and older individuals expressed that their needs were minimally met. Upon the arrival of volunteers at the evacuation site, able-bodied community members swiftly sought to meet their basic needs, inadvertently causing some difficulties for people with disabilities and older individuals in accessing the necessary support. In this phenomenon, community awareness to prioritise vulnerable groups is still low. The government should also take the action concerned regarding the challenge in order to actualise the regulations that provide the first assistance to these vulnerable groups.

5.2 Perception of Fulfilment of Basic Needs

The basic needs of the displaced populations from the North Luwu flash flood disaster was achieved according to the demands of the situation. The basic requirements for vulnerable groups have also fulfilled the priority elements. Some displaced populations considered that the coordination between the government and volunteers was ineffective due to the lack of information about the presence of disabled and older people at the evacuation points. Therefore, the volunteers were not aware of it. From this experience, it is evident that the communication procedure is not well implemented. The Emergency Disaster Command System (SKD) is a standardised protocol for handling disaster emergencies, utilised across all agencies and institutions. It involves the coordinated mobilisation of facilities, equipment, personnel, procedures, and communication channels within a unified organisational framework under a single command structure. This system is implemented by the BNPB/BPBD to effectively deploy resources from various agencies, institutions, businesses, and the community for an efficient response to disaster emergencies.

Assistance for displaced populations and disaster-affected people had been carried out in an open, accountable manner, involving various parties and without the propagation of a particular religion or belief. The perception elements of the community (see Table 4) are in accordance with the principles of the BNPB policy framework,[64] which consists of responsiveness and accuracy, priority, coordination and integration, efficiency, transparency and accountability, partnership, empowerment, non-discriminative, and no religious matters. As part of the research, our team explained to the respondents about each perception in detail to make sure they understood. The elements of 'partnership' or 'coordination and integration' were how the respondents saw the solidity of various stakeholders in handling aid distribution.

The evaluation carried out by the community regarding the implementation of basic needs assistance practices in the aftermath of the flash flood event in North Luwu showcased a commendable level of responsiveness and accuracy. A substantial 53 per cent of the respondents expressed their utmost satisfaction with the assistance received, underscoring the effectiveness of the initiatives. The local government played a pivotal role in exhibiting responsiveness, ensuring that the community received timely and appropriate disaster aid, especially in addressing basic needs during the emergency response. According to the FGD result, the government, through agencies such as the BPBD and the Dinsos (Department of Social Affairs), has commenced active provision of humanitarian aid and logistics, as well as managing the evacuation of disaster victims (establishment of shelters/emergency tents). Thus, the efficiency of the local government's efforts was reflected in the high satisfaction rates reported by the surveyed individuals.

The community's perception reflects a positive assessment of the provision of assistance needs in the aftermath of the flash floods. With a combined percentage of 100 per cent (Very good: 35per cent and Good: 65per cent), the majority of the respondents view humanitarian aid positively, indicating a general satisfaction with the assistance provided. The aid efforts seem to have successfully targeted the priority elements outlined in the disaster response plan, aligning with the community's immediate needs. It reveals that the distribution reaches the vulnerable groups, such as pregnant and breastfeeding women, as well as older people and those with disabilities.

The community expressed gratitude for the assistance provided by donors from diverse sectors through the elements of coordination and integration. The collaboration between the local government and external donors proved instrumental in aiding the survivors of the flash floods in Masamba and Baebunta. This collaborative approach not only highlighted the effectiveness of partnerships in disaster response but also showcased the collective commitment to supporting affected communities in times of crisis. For instance, the FGD result revealed that several NGOs and other humanitarian organisations were also actively participating, such as the PMI

[64] 'Head of BNPB Regulation Number 12 Year 2010 on Guideline for Social Assistance Mechanism in Emergency' article 5.

Table 4 Perceptions of protection and fulfilment of basic needs

Perception	Very good (%)	Good (%)	Fair (%)	Poor (%)	Very poor
The provision of assistance needs has met the elements of speed and accuracy	53	47	–	–	–
The provision of assistance needs has fulfilled the priority elements	35	65	–	–	–
The provision of assistance needs has fulfilled the elements of coordination and integration	19	59	19	3	–
The provision of assistance needs has fulfilled the elements of being efficient and effective	22	63	16	–	–
The provision of assistance needs has met the elements of transparency and accountability	9	41	50	–	–
The provision of assistance needs has fulfilled the element of partnership	16	71	13	–	–
Providing assistance needs has fulfilled the element of empowerment	13	50	38	–	–
The provision of assistance needs has met the elements of non-discrimination	72	16	9	3	–
The provision of assistance needs that doesn't include the religious matters	72	22	6	–	–

(Indonesian Red Cross) and the ACT, in providing logistics and clean water. Additionally, the Scouts Movement (Pramuka) and the Cakra community were involved in supplying logistics and psychosocial support. Overall, the combined efforts of the local government and external donors contributed significantly to meeting the immediate basic needs of the flash flood survivors and facilitating a more coordinated emergency response.

The analysis of the community perception regarding the distribution of humanitarian aid in response to the flash floods in Masamba and Baebunta reveals a generally positive outlook. A combined 85 per cent of the respondents rated the provision of assistance needs as either Very Good (22 per cent) or Good (63per cent), indicating a predominant satisfaction with the efficiency and effectiveness of the aid distribution. This positive perception suggests that the aid delivery process successfully met the immediate needs of the affected community, demonstrating a level of coordination and responsiveness. While a small percentage (16per cent) rated it as fair, further investigation is warranted to understand the specific concerns or areas for improvement highlighted by this group. On the whole, the survey underscores the overall success in achieving efficiency and effectiveness in the distribution of humanitarian aid in response to the flash floods.

The examination of community perception highlights a notable focus on transparency and accountability. While a majority of respondents (91 per cent) acknowledged that the provision of assistance needs met acceptable standards, there is room for improvement, with 9 per cent rating it as Very Good. The dominant perception of fairness (50 per cent) suggests that community members perceive the aid distribution process as transparent, and the providers are held accountable for their actions. The emphasis on fairness indicates that the community values open communication and responsible handling of resources. The survey results underscore the importance of continuous efforts to enhance transparency and accountability in humanitarian aid distribution, ensuring community trust and satisfaction in future disaster response initiatives.

A significant majority of respondents, comprising 87 per cent (16 per cent Very Good and 71 per cent Good), expressed satisfaction with the provision of assistance needs, emphasising the collaborative efforts and partnerships involved in the aid distribution process. This positive perception underscores the effectiveness of cooperation among various stakeholders. The high percentage of positive responses suggests that the community values and appreciates the synergistic approach taken during the distribution of humanitarian aid, reflecting a successful collaborative effort in addressing the needs arising from the flash floods.

The community's perception regarding the fulfilment of the basic needs of the flash flood affected in Masamba and Baebunta reveals a positive acknowledgment of the element of empowerment. A substantial 63 per cent of the respondents, comprising 13 per cent Very Good and 50 per cent Good, expressed satisfaction with the provision of assistance needs, indicating that the aid distribution process contributed to empowering the affected community. This positive perception suggests that the assistance provided not only addressed immediate needs but also empowered individuals and groups to actively participate in their recovery. While 38 per cent rated it as fair, further investigation is needed to understand specific nuances or areas where empowerment could be enhanced. Overall, the survey results highlight the importance of considering empowerment as a key aspect of humanitarian aid distribution and the potential for continuous improvement in fostering community resilience. In regard to this, the FGD respondents mentioned that the BLK (Training and Skills Centre) and the Department of Women's Empowerment and Child Protection had started organising various training sessions at evacuation centres, such as culinary training.

In terms of the element of non-discrimination, an overwhelming 88 per cent of the respondents, comprising 72 per cent Very Good and 16 per cent Good, expressed satisfaction with the provision of assistance needs, suggesting that the aid distribution process was perceived as fair and impartial. The minimal percentages of fair (9 per cent) and poor (3 per cent) ratings indicate a high degree of confidence in the equitable treatment of individuals or groups, irrespective of background or demographics. This overwhelmingly positive perception underscores the success of efforts to ensure non-discrimination in the distribution of humanitarian aid, reflecting a commitment to fairness and inclusivity in addressing the needs arising from the flash flood.

A significant 94 per cent of respondents, comprising 72 per cent Very Good and 22 per cent Good, expressed satisfaction with the provision of assistance needs, emphasising that the aid distribution process was perceived as impartial and devoid of religious considerations. The minimal fair rating of 6 per cent suggests that the majority of the community views the assistance as free from any religious biases, contributing to a sense of inclusivity and neutrality in addressing the needs arising from the flash flood. This overwhelmingly positive perception underscores the success of efforts to ensure that humanitarian aid is distributed in a manner that respects and upholds the principle of religious non-involvement.

6 Conclusion

Flash floods in Masamba and Baebunta subdistricts, North Luwu Regency, in mid-July 2020 caused fatalities, damaged basic infrastructure, disrupted "residents' economic and socioeconomic activities, and destroyed their homes. Based on findings in the field and discussions with relevant stakeholders and residents affected by flash floods, the following conclusions and suggestions are made:

a. The policy for handling humanitarian aid and logistics during a disaster crisis was considered comprehensive because floods frequently hit North Luwu Regency. Flash floods in July 2020 occurred due to hydrometeorological conditions.
b. Policies and regulations related to disaster management are considered to be comprehensive. On the other hand, implementing disaster management still faces many challenges, such as the distribution system and logistics coordination due to limited accessibility to disaster-affected areas. The

fulfilment of the basic needs of the IDPs was very good while they were in the evacuation camps but they began to face problems when they moved to temporary shelters, such as the random allocation system of temporary shelters for IDPs which was considered inaccurate, unfair, and unobjective. Lessons learned from handling the disaster in North Luwu included the following: efforts to deal with the crisis were carried out in a multi-stakeholder manner, involving, the regional government of North Luwu, other regional governments, volunteers, business institutions, non-governmental organisations, and social organisations. In addition, there was a distribution of assistance carried out privately but not integrated through the government, so some assistance to certain areas was late or was not received at all. This consequence arose because of the mechanism of distributing temporary housing randomly or by lottery, and therefore some affected residents who needed houses did not get them. On the other hand, residents who had built homes elsewhere received homes.

c. Community affected by flash floods in North Luwu were quite satisfied with the government's performance in providing humanitarian and logistical assistance during and after the emergency response. Coordination between the government and other parties in handling the 2020 flash disaster was also considered good, so the evacuees felt many positive impacts. In addition, there was no discrimination in distinguishing one religion from another in the distribution or distribution of aid.

References

1. Legionosuko T, Madjid A, Asmoro N, Samudro E (2019) Indonesia's position and strategy in facing climate change to support national resilience. Jurnal Ketahanan Nasional 25(3):295–299
2. Hosseninzadehtalaei P, Tabari H, Willems P (2020) Climate change impact on short-duration extreme participation and intensity-duration-frequency curves over Europe. J Hydrol 590(14)
3. Camarasa-Belmonte AM (2021) Flash-flooding of ephermal streams in the context of climate change. Cuadernos de Investigación Geográfica 47(1):121–142
4. Wu J, Zhang Y, Zhou H (2020) Groundwater chemistry and groundwater quality index incorporating health risk weighting in Dingbian county, Ordos Basin of Northwest China. Geochemistry 80(4):1–2
5. Tabari H (2020) Climate change impact on flood and extreme precipitation increases with water availability. Scientific 10(1):1–10
6. Llasat MC (2021) Floods evolution in the Mediterranean Region in a context of climate and environmental change. Cuadernos de Investigación Geográfica 47(1):13–32
7. Sene K (2012) Flash floods: forecasting and warning. Springer Science & Business Media
8. Mulyanto HR, Nunus Aris Parikesit R, Utomo H (2012) Guidelines and mitigation system on flash flood events. Ministry of Public Works
9. Raharjanto (2012) Flash floods mixed with sediment in the Situbondo Area, East Java. In: Symposium proceedings national sediment disaster prevention Yogyakarta 12–13 Mar 2022
10. Aldosary AS, Md Nahiduzzaman K, Reza I (2016) Vulnerability of flash flooding in Riyadh, Saudi Arabia. Nat Hazards 84:1807–1830
11. Kundzewicz ZW et al (2012) Flood risk and climate change: global and regional perspectives. Hydrol Sci J 59(1):1–28
12. Prein AF, Rasmussen RM, Ikeda K, Liu C, Clark MP, Holland GJ (2016) The future intensification of hourly precipitation extremes. Nat Clim Change 7(1):48–52
13. Tabari H (2020) Climate change impact on flood and extreme precipitation increases with water availability. Sci Rep 10:1–10
14. Yulihastin E, Nuryanto DE, Trismidianto, Muharsyah R (2021) Characteristics of mesoscale convective complexes that triggered heavy rainfall related to severe flash flood in Luwu, Sulawesi, Indonesia. Atmos Sci Meteorol 3
15. Paski JAI, Makmur EES, Permana DS, Nurahhmar MH, Praja AS, Riama NF, Fitria W, Hartanto (2021) Analysis of multi-scale hydrometeorological triggering flash flood event of the 13 July 2020 in North Luwu, South Sulawesi. IOP Conf Ser Earth Environ Sci 893(1):1–10
16. Hand WH, Fox NI, Collier CG (2004) A study of twentieth-century extreme rainfall events in the United Kingdom with implications for forecasting. Meteorol Appl 11:15–31
17. Guswantoro T, Suppa R, Azahra SF, Sulaiman L (2021) The characteristics of surface topography and element of sand from flash floods in North Luwu regency. In: Proceedings of the 3rd international conference of education and science, 17–18 Nov 2021, Jakarta, Indonesia, 348

18. Jamin NH, Risfaisal (2021) Economic and social changes in the farming community following the flash flood in Meli Village, Baebunta Subdistrict, North Luwu Regency. *Aksiologi: Jurnal Pendidikan dan Ilmu Sosial* 2(1):32–39 (in Indonesian)
19. Frankenberg E, Gillespie T, Preston S, SikokiB, Thomas D (2011) Mortality, the family and the Indian Ocean Tsunami. Econ J 121(554):162–182
20. Darmadi D (2021) Government communication on flash flood crisis in Masamba, North Luwu Regency. Jurnal Ilmu Komunikasi 11(1):48
21. Rahmansyah Abdullah MT, Cangara H, Arianto (2021) The analysis of communication strategies of North Luwu regency government in socio-economic recovery after landslides and flash flood. In: 2nd International conference on science, technology, and modern society (ICSTMS 2020). Atlantis Press 453–461
22. Kastono, MM, Muhibuddin A, Surihardjo HS, Hasbi, AM, Sudirman SW, Abdulbar F (2022) Mitigation and public coordination for flood disaster risk reduction (FDRR) in the implementation of North Luwu sustainable development. IOP Conf Ser Earth Environ Sci 1109(1):012018
23. Wisetjindawata W, Itob H, Fujitaa M, Eizoa H (2014) Planning disaster relief operations. Procedia Soc Behav Sci 412–421
24. Drolet J (2015) Disasters in social, cultural and political context. Int Encycl Soc Behav Sci 478–484
25. Susilo A (2010) The ineffectiveness of aid in Aceh redevelopment projects. Media Journal Global dan Strategi 3(1) (in Indonesian)

Dina Ruslanjari is an Indonesian disaster management expert. She completed her highest education at Gadjah Mada University, specializing in Environmental Science. Her areas of expertise encompass disaster risk management, community empowerment, and capacity development. Currently, she serves as the Head of the Master's Program in Disaster Management at the Graduate School of Universitas Gadjah Mada, Indonesia. As an academician, she actively engages in the Three Pillars of Higher Education, which include education and teaching, research and development, and community service, especially in remote areas facing disaster risks. With 20 years of experience in the field of disaster management, Dr. Dina has played pivotal roles as a consultant, coordinator, and expert involving stakeholders at local, national, and international levels. Moreover, she is also prolific in producing scholarly publications, with dozens of works featured in various reputable journals. Her aspiration is to establish a disaster-resilient Indonesia by prioritizing local wisdom and engaging government, academics, private sector, NGOs, and the community itself.

Cahyadi Ramadhan a driven young researcher hailing from Sulawesi, possesses an unwavering passion for developing disaster risk reduction strategies in Indonesia. He holds a Bachelor of Science degree in Environmental Geography and a Master of Science degree in Disaster Management, both from Universitas Gadjah Mada. His specific areas of interest encompass hydrology, environmental economics, forestry, GIS, and global environmental issues, including climate change adaptation and disaster management. Since his university days, Cahyadi has been actively engaged in scientific publications, achieving the prestigious 1st place in national essay competitions in both 2018 and 2019. Recently, he and his colleagues have published two original research papers in the esteemed journal, International Journal of Disaster Risk Reduction in 2023. Despite originating from a remote area, Cahyadi possesses a resolute determination, ambition, and a deep-seated aspiration to make significant strides in his field through continuous learning and practical application.

Inayah Bastin Al Hakim holds a Bachelor of Arts degree in English Literature. Following the completion of her undergraduate studies, she chose to embark on an academic journey in Disaster Management for her master's program. Her keen interest in disaster risk issues stems from the stark reality of Indonesia's vulnerability to multiple disasters. As a nature enthusiast, she advocates for the concept of "Living in Harmony with Disaster," emphasizing the importance of recognizing and adapting to surrounding risks. Inayah has actively participated in various collaborative projects involving national and global stakeholders, particularly focusing on disaster management, climate change, community empowerment, and sustainable development. Proficient in writing, translating, external communication with stakeholders, and manuscript editing. She has devoted her time, efforts, and expertise to empowering communities to build resilience against various risks and crises. Her goal is to contribute to the realization of the United Nations' Sustainable Development Goals (SDGs).

Feby Aulia Marsida comes from South Sulawesi, Indonesia. Her specific areas of interest is Earth science which consist of geology, oceanography, meteorology, and climatology. She holds a Bachelor of Science degree in geography at Makassar State University after research about potential of Micro hydro power plant as a renewable energy. Currently, she is on-going to complete her master degree in Disaster Management at Gadjah Mada University. At the same time, she is currently working in the largest start-up company in Indonesia who focuses on education-based services as a geography master teacher since 2020. Nonetheless, she kept on fire for enter a field of disaster management.

Planned-Relocation, Resettlement, and State Responsibility

Climate Change-Related Displacement: Inter-Island and Rural–Urban Migration in the Solomon Islands: Options for a Viable National Resettlement Plan

Lovelyn Otoiasi

Abstract

Global environmental change has been projected to have a greater impact on human mobility across the globe. In the recent past, the movement of people between islands in the Solomon Islands was voluntary, either through marriage or other social interactions. However, recently climate change-related impacts have begun to force people to move, particularly from the low-lying atolls. Although there is a drive from the international community to address climate change-related resettlement and relocation, there is an urgent need to better understand local migration intervening factors and address these intervening obstacles to the resettlement programme in the Solomon Islands. This chapter discusses the intervening aspects of climate change-related migration and rural–urban migration to Honiara City, the capital of Solomon Islands, the options to address resettlement challenges, and makes recommendations for institutional, legal, and financial mechanisms for the nation's Resettlement Programme. It is hoped that discussions from this chapter will provide the needed insights and a better understanding of various social, economic, and cultural attributes of climate-related human mobility in relation to Honiara and thereby inform the development and implementation of sound and viable legal and policy measures.

1 Introduction

Global climate change is expected to have a greater impact on human mobility around the world.[1] To the east and south of Papua New Guinea, in the southwest Pacific are the Solomon Islands (Fig. 1). It consists of a double chain of 997 islands, including 6 large ones (Choiseul, Santa Isabel, New Georgia, Guadalcanal, Malaita, and San Cristobal-Makira) as well as numerous smaller ones.[2] Its topographical features serve as a metaphor for the difficulties of migration brought on by climate change. Until recently, movement between islands in the Solomon Islands was voluntary, either through marriage or other social interactions.[3] However, social, economic, and climate change-related impacts have begun to force people into migration, in particular towards Honiara

[1] Tabe [1], Manou and Mihr [2], Klepp [3], David and Jennifer [4], Adoho and Wodon [5].

[2] Population & Housing Census [6].

[3] Monson and Foukona [7]; See also Tabe [1, 1–2], Klepp [3, 6–7].

L. Otoiasi (✉)
Environment Department, Faculty of Agriculture, Fisheries and Forestry, Solomon Islands National University, Honiara, Solomon Islands
e-mail: Lovelyn.Otoiasi@sinu.edu.sb

S. Jolly et al. (eds.), *Climate-Related Human Mobility in Asia and the Pacific*, Sustainable Development Goals Series, https://doi.org/10.1007/978-981-97-3234-0_10

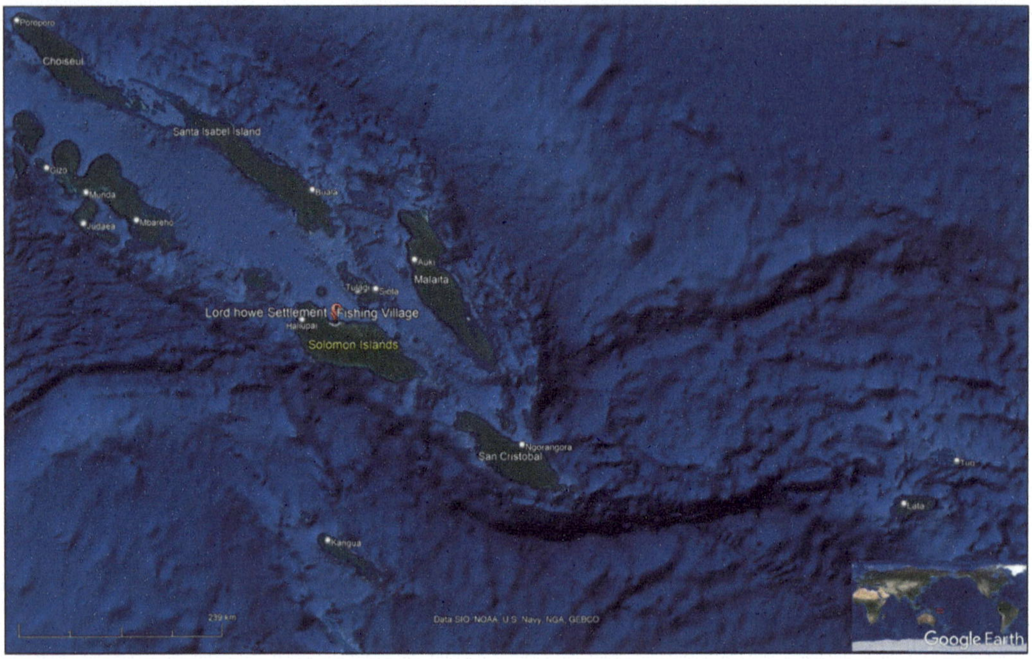

Fig. 1 Location of Solomon Islands—study site. Map shows the location of the study sites-Lord Howe settlement and the fishing village. *Map Source* Google Earth Pro

city, the capital of Solomon Islands. An increase from 68,298 to 87,633 in the total number of migrants who had moved between provinces, was recorded between the 1999 and 2009 census records.[4] Rural–urban migration can also be seen happening and was validated in the latest census of Honiara in 2019 with the fastest annual growth rate of 5.9% and an increase of 57,033 since 2009, or 5700 people per year.[5] Although there is an indication that migration is becoming more evident through these census reports and urban settlement expansions, limited studies have been conducted to determine the driving factor behind the decision to move from provinces into Honiara's peripheral zone. The literature available has evaluated the connection between migration and climate change on a global scale, particularly as it relates to the Pacific islands.[6] However, there is scant academic literature that addresses this issue locally in the Solomon Islands. Over the decades, governments have ignored the need to consider the critical linkage of climate change-induced human mobility as a priority subject in their policies. Recently, the Solomon Islands government adopted a plan for relocation guidelines in 2022 (hereinafter the SI Planned Relocation Guidelines), prepared by the International Organisation for Migration (IOM).[7] Will this Planned Relocation Guidelines 2022 be fully

[4] Solomon Island National Statistic Office [8]. Due to lack of updated data on Rural–Urban migration available in the country the latest 2019 census report is presented here in the paper.

[5] Solomon Island National Statistic Office [9].

[6] Campbell and Warrick [10], IOM Outlook on Environment Migration and Climate Change (International Organisation for Migration 2014), 16–19; Foresight Migration and Global Environmental Change - Final Report, (The Government Office for Science, London 2011), 5-6, Warn and Adamo [11], Klepp [3, 1–2], Campbell [12].

[7] Planned Relocation Guidelines, Solomon Islands Government 2022. The SI National Planned Relocation Guidelines 2022 was developed as an obligation towards international agreement by the Government. Standard Operating Procedures are yet to be developed.

supported by the government and feasible to implement by responsible authorities?

This chapter presents the results of a study conducted using mixed quantitative and qualitative methods using semi-structured, open-ended questionnaires carried out at selected communities and government offices, particularly the Ministry of Environment, Climate Change, Disaster, and Meteorology (MECDM) and the Ministry of Lands, Housing, and Survey, the two key focal organisations that will be implementing the SI Planned Relocation Guidelines 2022. A total of 38 respondents were interviewed from the two settlements. The two selected communities were the Lord Howe settlement and the Fishing Village, situated in the east to central of Honiara City, the capital of Solomon Islands that is located in the Guadalcanal Provinces. Selection of these two sites was based on the place of origin of the settlers which was the most vulnerable location of climate change impacts in the Solomon Islands.[8] The interview results show that most respondents originated from the Malaita province. Furthermore, these locations were depicted as the most vulnerable to disasters and climate-related natural hazards in Honiara City. There were 15 persons who were interviewed in the Lord Howe settlement and 23 from the Fishing Village settlement. Government officials were consulted through email and face-to-face interview conversations. The interviews took place over two months between mid-March and April 2023. Most of the secondary data was obtained from existing literature on the research topic.

This chapter discusses the intervening factors of climate change-related migration, rural–urban migration to Honiara City, and the options to address resettlement challenges and make recommendations for institutional, legal, and financial mechanisms for the nation's resettlement programme. It is hoped that discussions from this chapter will provide insights for a better understanding of the different social, economic, and cultural attributes of climate-related human mobility in relation to Honiara City and thereby inform the development and implementation of sound and viable decisions to address climate change-related planned relocation and rural–urban migration in the Solomon Islands. Also, this could point out some improvements to the provisions of the existing Land and Titles Act for relocation and resettlement in Solomon Islands. With these research objectives, the chapter is divided into five sections. Following this introduction, the second section describes features of climate change-induced human mobility in the Solomon Islands. Section 3 elaborates on the government's policy framework on climate change, migration, and planned relocation, and Sect. 4 provides critical recommendations for planned relocation. The chapter concludes with recommendations for research and policy development.

2 Intervening Factors in Climate Change-Induced Rural–Urban Migration in Solomon Islands

This section outlines the relationship between climate change and rural–urban migration in the Solomon Islands taking into account environmental, social, and economic drivers. Pacific islands consist of over 22 countries and territories accommodating more than 10 million people. The smaller islands in the Pacific are characterised by high levels of exposure to climate-related dangers and a limited ability to adapt. In this situation, people from Pacific islands may be uprooted or decide to migrate in search of safer living conditions.[9] The decision to move is still uncertain as the population in most Pacific islands has a strong connect to culture, land, and family.[10] Most Pacific island communities are in a crisis due to climate change impacts and are faced with livelihood, food, and

[8] For More Information on these Settlements refer to Sect. 2.1 and Fig. 3.

[9] Oakes [13].

[10] Ibid.

water security problems, increasing salination and droughts.[11] In addition to increasing salination, droughts and sea level rise, climate-induced coastal inundation could severely impact coastal livelihoods, thus reducing their food security status. This will affect most coastal and low-lying atolls resulting in a significant loss of livelihood security, island habitat degradation, and land loss that triggers relocation of affected communities.[12] These climate-induced impacts in the Pacific islands have become more pointed and present themselves in three aspects that add weight to the migration perspective and are presented by Campbell in Table 1.[13]

Since climate-induced problems have been occurring and are projected to get worse, relocation is considered a possibility for Pacific island groups. Migration or long-term relocation is not something new in the Pacific; as this has already been implemented recently by neighbouring countries like Fiji, which has for instance relocated two communities of Denimanu and Vunidogoloa villages. In addition, relocation has occurred across international borders as well; for instance, the I-Kiribati resettlement in the Solomon Islands at Wagina and Titiana; the Tuvalu settlement of Kioa in Fiji.[14] As households are rendered unable to meet their daily subsistence needs and wants, they will tend to decide to migrate to more favourable localities with better services and livelihood security opportunities. Urban centres or central capital towns are centres of economic development, have agglomeration effects, and provide more job opportunities, and therefore, these are becoming the prospective location for some migrants.[15]

Climate change and worldwide migration have been widely acknowledged and recorded.[16] However, as noted earlier, in the local context, limited academic literature is documented. The present chapter aims to fill this gap by investigating the link between climate change and migration and its impact on the livelihood of migrants in the local context of the Solomon Islands by highlighting the lived experiences of the people. The following section illustrates the findings of the study investigating the factors leading to migration to Honiara from other provinces.

A. Factors Leading to Migration to Honiara City

As noted in the earlier section, the study deployed semi-structured questionnaires and interviewed 38 respondents at the Lord Howe and the Fishing Village settlements of Honiara. Table 2 below shows the responses of the sample population from the two study sites.

The Fishing Village settlement was established in the 1950s, after World War II. People from this community originated from the Lau Lagoon artificial islands located in the North Malaita province of Solomon Islands. The settlement is situated within Vura ward along the Kukum Highway, adjacent to the Solomon Islands National University, and close to the coastline. The settlement is zero metres above sea level and has a relatively flat terrain. As a result, it is particularly exposed to coastal flooding and storm surges. As of the present, 200 people have been living in this community. Due to the increased population and a shortage of space inside the designated land border of the Fishing Village, a majority of their members were voluntarily relocated to Lio Creek, which is located behind the Solomon Islands National University. It is unclear who arranged

[11] Campbell and Warrick [10, 15], Michelle et al. [14], Ackerly et al. [15], Brown [16].

[12] Campbell and Warrick [10, 10–18], see Michelle et al. [14, 2049–2069].

[13] Campbell and Warrick [10, 15], see Michelle et al. [14, 2068].

[14] Ibid (n14), Piggott-Mckellar et al. [17], Campbell and Bedford [18].

[15] Ranaa and Ilina [19]. The Global Urban Economic Dialogue Series The Economic Role of Cities (UN-HABITAT 2011), 1–43.

[16] Campbell [12, 4–8], Brown [16, 19–23], International Organisation for Migration,(IOM) https://publications.iom.int/books/iom-outlook-migration-environment-and-climate-change. Accessed 16 August 2023, Campbell [12], Laczko and Piguet [20], Gray et al. [21]. Foresight Migration and Global Environmental Change—Final Report (n6) 67–119.

Table 1 Links between loss of land, livelihood, and habitat security and migration

FORCED MIGRATION →			← INDUCED (VOLUNTARY MIGRATION)			
Lost land security	Lost livelihood security	Habitat security severely reduced	Land security partly lost	Livelihood security was severely disrupted	Livelihood security was significantly but not severely disrupted	Land and livelihood security are disrupted periodically
Community relocation	**Community relocation**	**Community relocation or individual migration**	**Community reloca-tion or individual migration**	**Individual migration**	**Individual migration**	**Short-term migration**
Atoll submerged and or eroded Coastal sites submerged and/ or eroded Delta sites eroded and/ or inundated (coastal and/or river bank)	Salinization • Water polluted • Crop failure Persistent Drought • Water reduced • Crop failure	Changes in disease vectors such as malaria, dengue, and ciguatera Changes in waterborne incidence Temperature-related illnesses	Severely but not completely reduced land available for settlement or livelihoods	Severe reductions in • Water supply • Food production	Moderate reductions in: • Water supply • Food production Moderately but completely reduced the land available for settlement or livelihood	Increased frequency and or magnitude of climatic extremes Atolls/coastal land temporarily inundate

Table 2 Sample population interviewed

Location	Settlement	No. of respondents	Gender	
			M	F
Site 1	Lord Howe-Mamana water	15	7	8
Site 2	Fishing village	23	17	6
	Total	38	24	14

Source Author's survey

the relocation of these Fishing Village settlers who had illegally settled on government land but were later given land titles when those areas were formally legalised and registered by the Ministry of Lands. Most houses in the Fishing Village are permanent structures, with a few semi-temporary structures on a registered property. Livelihood strategies of households differ by gender, with the majority of the male population engaged in fishing and the majority of the female population involved in selling cooked food. Sometimes root crops and vegetables are bought from farm producers from around the Guadalcanal Plains and resold at the local market.[17] Some people were employed in the public and private sectors and received salaries and wages to sustain their family's needs and wants.

The Lord Howe settlement[18] is located at the Mataniko riverfront, geographically extending south to the Mataniko Highway and west to the National Referral Hospital in Mataniko ward. The Lord Howe settlement was originally owned by settlers from the Lord Howe atolls in the Malaita province. It is one of the most densely populated settlements in Honiara. With its geographical location, this settlement is exposed and vulnerable to flash floods and sea level rise. This settlement has suffered loss and damages due to the flash flood in 2014.[19]

[17] Interview with Interviewee Number 35, Interviewed by Lovelyn Otoiasi (16 March 2023).

[18] Lord Howe settlement [also known as Mamana water] commonly known name used by local settlers. See Fig. 3—Google aerial location of this settlement.

[19] See also Foukona [22], who describes displacement and subsequent relocation of people who had been living in the Koa Hill informal settlement in Honiara, located further inland in Honiara along the Mataniko river.

People explained that they moved to Honiara because of anticipated socio-economic benefits, and due to environmental factors. The first settlers to the Fishing Village in this case were forced to migrate due to social difficulties such as conflicts in their native land with other tribes.[20] Other relatives joined their cousins in moving to the Fishing Village to live in peace and secure land for future generations. The survey emphasised that the decision of young people to migrate is influenced by pull factors such as economic opportunities, and push factors such as an expanding population, a lack of jobs in rural areas, and risks to their livelihoods.

The study illustrated five important reasons for migration. On hearing the accounts of various interviewees about their decision to move, it is clear that they were based on the following: (1) climate change, agricultural deficit, and water scarcity that are examples of environmental challenges; (2) employment; (3) education; and (4) family, including marriage. Figure 2 demonstrates that 70% of the population interviewed, migrated from their area of origin due to serious environmental challenges and sought employment opportunities to pursue their livelihood and aspirations, 19% came to Honiara to pursue tertiary education, and 6% went for family reasons like marriage or parents working in Honiara City. The findings from the data collected show that rural–urban migration is triggered by accumulated environmental issues that threaten a household's livelihood status. Thus, the decision to move occurs when livelihood

[20] Interview with Interviewee Number 17. Interviewed by Lovelyn Otoiasi (16 March 2023).

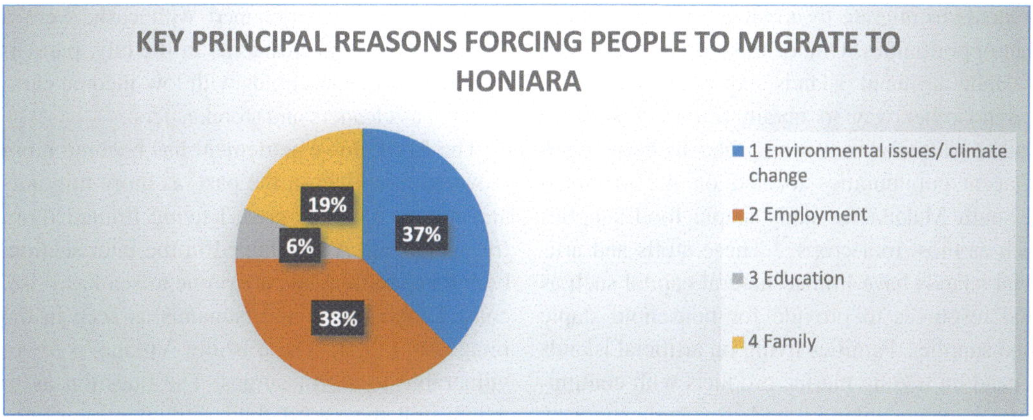

Fig. 2 Rural–urban migration factors stated by respondents

(a) Fishing Village map extracted from Google Earth Pro

(b) Lord Howe Map extracted from Google Earth Pro

Fig. 3 Location of **a** fishing village and **b** Lord Howe settlement

strategies and asset capital in the place of origin are dismissed and threatened by population growth putting more pressure on the limited resources of island habitats.[21]

In addition, it was noted that most respondents from both the Fishing Village and the Lord Howe settlements migrated from Malaita province's low-lying and artificial islands, such as Ontong Java, Lau Lagoon, Langalanga, and Walande, where people struggle with food production, availability and access owing to rising sea levels, saltwater intrusion, and dwindling water supply. Agricultural yields have decreased in atoll islands where each household was assigned a small patch of land to cultivate swamp taro.[22]

Artificial islands, such as the Walande[23] and other regions in Malaita province, have already been seriously impacted by sea level rise, forcing

[21] Interviewed by Lovelyn Otoiasi (15 March, and 16 April 2023).

[22] Interviewed by Lovelyn Otoiasi (15 March, and 16 April 2023).

[23] Walande region in the south Malaita region and Lau Lagoon in Malaita province are places of origin for some fishing village populations.

residents to migrate to urban centres with abundant opportunities. One of the interviewees from Walande artificial islands expressed that there was no other way to obtain resources such as land to cultivate root crops than to barter with adjacent communities located on the mainland of south Malaita for daily staple food supplies such as most root crops.[24] These atolls and artificial islands have limited natural capital such as land resources to provide for household staple food supplies. Families living on artificial islands depend on trading marine products with communities on the mainland islands for staple supplies, building materials, and necessities that cannot be obtained from their customary territories.[25] From their perspective, moving to Honiara City will provide access to resources, employment, and easy access to everyday necessities and wants, instead of living on their man-made islands.

The majority of respondents, in the Fishing Village and Lord Howe settlements, had daily sustenance access from Honiara's central marketplaces. Many of them were working in private and government organisations and had acquired tertiary education. They received salaries to sustain their livelihood. A number of them were self-employed or private entrepreneurs and market vendors. Some households in the Fishing Village did gardening in the backyard of their home to supplement their incomes and wages received. Some who did not own enough space around their homes, bought green vegetables, root crops, and firewood from farmers nearby Honiara city and resold it at the local markets. In the Lord Howe settlement, most respondents emphasised that there was no space to grow small green gardens, in their existing settlements. Most green staple products such as fruits, green vegetables, and root crops were purchased from the local central market, and protein was purchased from the shops. Some respondents were employed and some were market vendors who sold cooked food, mostly by women as a means to generate income for their family. However, certain items that could only be obtained with cash, were a challenge for some residents in the city, particularly for those households with low-income earners such as cleaners and storekeepers.

The Lord Howe settlement has become more crowded now than in the past, as more migrants are moving into the area, leaving limited room for expansion. As explained in the later section, both these settlements are prone to sea level rise, coastal inundation, and tsunamis as seen in the image in Fig. 3. The Fishing Village is more vulnerable to storm surges. The question as to where will they go if their settlement gets submerged by the rise in sea level or flooded by the river is a major and real concern for the settlers in the Lord Howe settlement.[26]

Given the threat of climate vulnerability and overpopulation facing the existing settlements of Fishing Village and Lord Howe, the next section evaluates the willingness of migrants to relocate to another potential relocation site within the proximity of Honiara.

B. Willingness to Relocate from Existing Settlement

An equal percentage of respondents showed interest in relocating from their current location in Honiara. Further, there is no discernible difference between genders with almost 50% of female respondents also agreeing to relocate. Regarding the expectations that the respondents had with regard to the relocation sites, most of them stated that they would want that the relocation sites be on registered land, be disaster-proof from flooding and landslides, accommodate essential services like road access and water, have enough space to grow food and be conducive for them to set up their business plans. They also want the relocation site to be close to Honiara, the centre of all economic activities. The findings of the study reveal that the relocation site and its proximity to economic activities is considered a key element in selecting the potential site for relocation of the population from these two settlements. In other words,

[24] Interview with Interviewee Number 5. 'Interviewed by Lovelyn Otoiasi' (15 March 2023).

[25] Ibid.

[26] See Fig. 3

the respondents preferred a relocation site in proximity to Honiara City due to the access to better education, health services, and employment. There is a high generation of cash and a market within urban centres such as Honiara. People from these two settlements showed a willingness to move, but they emphasised that the Government must ensure that they are relocated to areas near Honiara. As of now, the time period within which the government will relocate them is uncertain, and the Relocation Guidelines 2022 that were recently adopted are yet to be implemented by the authority responsible for the same. Furthermore, land availability within Honiara City or on its outskirts is constrained.

In contrast, those who disagreed with moving to a new location responded that priority should be given to those populations back home who are victims of the impacts of climate change. Others felt that moving far from the coast would not be compatible with their existing livelihood as they were fishermen and moving far from the ocean would not be feasible. The study shows that the older generation of settlers in these two communities are content to continue to dwell there, whereas the younger generation, or those who have recently migrated to Honiara, are keen to leave these two areas. Another reason for wanting to leave would be to accomplish personal aspirations of becoming a business owner or owning a property bought for their own family. In this scenario, the next part of the paper discusses the policy and legal initiatives undertaken to manage the migration and relocation in the Solomon Islands.

3 Policy and Legislative Framework Governing Migration and Relocation in the Solomon Islands

Being part of the Pacific Islands, the Solomon Islands is vulnerable to both sudden and slow climate-driven change.[27] On the international front, the Solomon Islands is a Party to the United Nations Framework Convention on Climate Change (UNFCCC),[28] it's the Kyoto Protocol and the Paris Agreement.[29] The Solomon Islands has also adopted the Sendai Framework for Disaster Risk Reduction.[30] Within the Pacific regional level, the Solomon Islands has endorsed the Pacific Plan, the Pacific Islands Framework for Action on Climate Change (PIFACC), and the Regional Framework on Disaster Risk Reduction and Disaster Management.[31] In pursuance of its international commitments, Solomon Islands has been evolving detailed policy and legal mechanisms to address the challenges associated with climate change and migration.

Before discussing specific policy initiatives, it is crucial to refer to the legal safeguards within in the Constitution that provide for the protection of fundamental rights, equity, and freedom of the entire citizenry. According to section 3 of the Constitution, every person born in the Solomon Islands 'has a right to the fundamental liberties and rights of the individual.'[32] These fundamental rights include the right to life, liberty, and security of the person, and protection of private property from deprivation. Freedom of movement within the Solomon Islands is notably protected in

[27] Michelle et al. [14, 2095–2100], Kumar et al. [23], Monson and Foukona [7, 1–2].

[28] Solomon Islands National Climate Change Policy: 2012–2017, 11; See United Nations Framework Convention on Climate Change (adopted 29 May 1992, entered into force on 21 March 1994) 1771 UNTS 107 (hereinafter UNFCCC).

[29] UNFCCC; Adoption of the Paris Agreement, Report No. FCCC/CP/2015/L.9/Rev.1 http://unfccc.int/resource/docs/2015/cop21/eng/l09r01.pdf. Accessed 10 February 2023 (hereinafter Paris Agreement).

[30] Sendai Framework for Disaster Risk Reduction 2015–2030 (United Nations 2015); National Disaster Management Plan (Solomon Islands Government 2018) 4; Solomon Islands National Climate Change Policy 2012 (n30), 11.

[31] Disaster Risk Reduction and Management in the Pacific (Asian Development Bank 2013).

[32] The Constitution of Solomon Islands 1978, Sect. 3 http://www.paclii.org/sb/legis/consol_act/c1978167.pdf. Accessed on 12 June 2023; Foukona et al. [24].

section 14.[33]Though the constitutional provisions provide for the State's commitment to uphold the protection of the rights of every person, it does not provide for social and economic rights or the rights to better housing and good health to be attained as a civil commitment.[34]

The following section analyses Solomon Islands law and policy relating to climate change-induced migration and relocation.

A. National Disaster Management Plan (NDMP)

The Solomon Islands' national disaster risk management strategy encompasses two primary objectives: 'disaster management for disaster preparedness, response, and recovery and disaster risk reduction for reducing hazard risks and potential impacts for disaster events'.[35]

In general, this NDMP strategy recognises the need to identify and comprehend hazards, especially climate change, as the foundation of disaster risk management. The Plan envisages a core role for the National Disaster Council, which was established under the National Disaster Council Act (NDC Act).[36] The National Disaster Council as the institutional arrangement will coordinate with the Provincial Disaster Committee to deal with concerns relating to disasters in the Solomon Islands.[37] Other significant aspects of the plan include its focus on human rights including individual rights, equity, justice, gender, and minority cohorts of society. Section 17 paragraphs (M) and (N)[38] of the NDMP ensures the protection of the fundamental rights of the most vulnerable members of the community. For instance, the plan captures the role of ward and village committees to ensure full participation by all members of the community in preparing plans to respond during disasters and after recovery periods. More is captured on disaster preparedness and displacement, which in this context of the plan, provides for affected populations such as management of evacuation centres and building resilient communities in Solomon Islands rather than relocation of climate-affected communities in Solomon Islands. Notably, the NDMP addresses disaster risk reduction and displacement of affected communities in the short term after a disaster occurs but does not make provision for planned relocation.

B. Policy on Climate Change 2012–2017 [CCP]

In pursuance of the country's commitment at the international level, Solomon Islands has taken steps to address climate change concerns. The National Climate Change Policy 2012–2017 recognises climate change as a sustainable development issue and develops solutions to meet the risk and vulnerability assessments of Solomon Islands' most vulnerable communities.[39] The guiding principles identified in the Climate Change Policy include stakeholder participation and collaboration, gender equity and involvement of youth, children, and people with special needs, and respect for the culture and rights of indigenous people.[40] It also specifically notes that climate change adaptation and disaster risk reduction are closely interrelated and shall be aligned.[41] The government's commitment to vulnerability adaptation and disaster risk reduction (DRR) strategies and measures is discussed in section 8.3 (a) of the National Climate Change Policy. The inclusion of community relocation in section 8.3 (m) highlights the urgent necessity for formulating the relocation guidelines.[42]

[33] See the Constitution of Solomon Islands (n34), Sect. 14.

[34] Foukona et al. [24].

[35] National Disaster Management Plan 2018 (n30), part 1, Sect. 1.

[36] National Disaster Council Act [NDC Act] 1989, Sect. 3, para [1].

[37] National Disaster Management Plan 2018 (n30), Sect. 3.

[38] Ibid., Sect. 17.

[39] Solomon Islands National Climate Change Policy: 2012–2017 (n30), Sect. 5.

[40] Ibid., Sect. 7.

[41] Ibid..

[42] Ibid., Sect. 8.

The policy places more emphasis on adaptation to climate change impacts, that is, building the capacity of communities to take adaptation-appropriate actions as stipulated in the National Adaptation Programme of Action 2008 (NAPA). An example of this is increasing food production in affected communities by adopting new techniques or using rainwater tanks to store water for domestic use on atoll islands. The policy considers relocating communities as a last resort action in adaptation, in the Solomon Islands. Perhaps with the ongoing revision of this climate change policy, there is leverage for issues related to relocation that will complement the relocation guidelines that will be explained further in the next section (b) of this chapter. Furthermore, the Climate Change Policy recognises the need to assess and develop a plan of action for urban adaptation.[43] The Honiara Urban Resilience and Climate Action Plan 2016 sets out a comprehensive framework of action activities. This citywide plan includes upgrading of informal settlements, flood proof housing and buildings, coastal protection, livelihood-based protection, ecosystem services and water, rivers, and disaster risk reduction and management. The communities of the Fishing Village and the Lord Howe settlement have been identified as priorities in the action plan with the instalment of early warning systems on site and the building of a sea wall protection which are also part of the plan. Realising the threat of climate change and consequent migration, Solomon Islands has developed a specific policy initiative focusing exclusively on planned relocation.

C. **Solomon Islands Resettlement Programme**

The Ministry of Lands, Housing, and Survey set up this initiative from 2014 to 2016, but the progress of the programme was not satisfactory; therefore, the budget committee abolished it in 2017.[44] The purpose of this programme was to assist the national government's resettlement programme in the aftermath of the 2014 flash floods, which caused massive economic losses and displaced many households in Honiara City. The goal of this initiative was for the Ministry of Lands, Housing, and Survey to expedite land acquisition via power of domain or purchase of private or customary land owners, prepare subdivisions of lots on the newly relocated location, and issue an allocation of plots of land to winning bidders. Under the programme, interested parties such as Ministries, NGOs, or private households could submit bids for the plot of land advertised by the National Land Board. This project was implemented in 2014, and the criteria to allocate land was based on those household owners at Koa Hill and the Lord Howe settlement whose homes were destroyed during the flood. Not all managed to obtain a land title for a fixed-term estate (FTE) due to the delay in payment of land premium as the value of land was too expensive for some.[45] Similar Programme was carried on in later years. However, it was more focused on activities, such as consultations and meetings with landowners, which were primarily logistical and administrative, with no actual land acquisition for resettlement.[46] The multiple obstacles in dealing with customary land and ownership hampered the smooth functioning of the programme. For instance, in customary land tenure, land boundaries were not demarcated and did not have proper official records.[47] Hence, the process of determining who owns the land took a long time. In the rural society of Solomon Islands, land symbolises the identity of the forefathers and it is considered as a wealth to sustain their future generations.[48] Land, including

[43] Ibid., Sect. 8.3.1, para [i].

[44] Interview with Ministry of Lands, 'Interviewed by Lovelyn Otoiasi' (4 September 2023).

[45] Ibid.

[46] Interview with Ministry of Lands, 'Interviewed by Lovelyn Otoiasi' (4 September 2023).

[47] Commonwealth of Australia, 'Making Land Work: Reconciling Customary Land and Development in the Pacific' June 2008. www.ausaid.gov.au/publications. Accessed 18 June 2022; Corrin [25], Burt [26].

[48] Corrin [25, 221–222], Monson and Foukona [7, 292–293].

all natural resources, must be protected for the sake of those who have passed and for the future generations.

Formal registration of land in the Solomon Islands is implemented under the Land and Titles Act [LTA], which regulates the tenure, acquisition, and registration of interest in land and allows indigenous Solomon Islanders to acquire a perpetual estate, which can be granted by the National Land Board.[49] The National Land Board has the power to allocate interest on land, determine land allocation, and consider and approve the level of rent, etc.[50] The Act provides for customary land acquisition and compulsory land acquisition for government interest or development.[51] These two divisions are as follows.

Division 1 allows the Commissioner or the provincial government to purchase or lease customary land. As per this Division, customary land may be sold or leased to the Commissioner or any Provincial Assembly, for mining operations, or the installation of telecommunication towers on unregistered land. Division 2 deals with compulsory land acquisition for public purposes.[52] Both these divisions are mandatory, whether customary landowners want it or not and the government was required to compensate customary landowners. However, the complications associated with these land acquisition procedures thwarted the Solomon Islands Resettlement Programme.

Learning from these drawbacks and challenges, Solomon Islands has taken the initiative to adopt all-inclusive and comprehensive guidelines on Planned Relocation known as the 2022 Planned Relocation Guidelines.[53] The following section evaluates the core features of the 2022 Planned Relocation Guidelines.

D. 2022 Planned Relocation Guidelines

The Solomon Islands Planned Relocation Guidelines of 2022 was developed under the project titled 'Solomon Islands: Developing Planned Relocation Guidelines in the Context of Slow and Sudden Onset Disasters' and was funded by the International Organisation for Migration (IOM).[54] It was developed through primary research, including baseline investigations conducted in several affected communities across the country and a secondary desktop review study on pertinent guideline toolkits[55] by different researchers contributing to this document.[56] Various government agencies, non-governmental organisations (NGOs), and stakeholders were consulted on the document through meetings and workshops. It contains important concepts and principles on orderly migration that benefits both migrants and society.[57] The SI Planned Relocation Guidelines is a living document, which means it is subject to changes and options as needed while being implemented.[58]

The main concepts embraced by the SI Planned Relocation Guidelines start with the need for inclusive and participatory planning and decision-making to be used in all four stages of the process before relocation, development of the relocation plan, during relocation, and after relocation. This calls for the inclusion of all stakeholders—including migrants, host communities, and others—in the planning and decision-making processes. This process is intended to make sure that all impacted communities and stakeholders are fully informed on the methods and pertinent mechanisms employed in the relocation process. To make sure that the relocation process does not impair the rights to access basic capital to both relocated and host

[49] Land and Titles Act [Chapter 133], part IIA, Sect. 8c [1].

[50] Ibid.

[51] Ibid., Sects. 60–85 of the Land and Titles Act [LTA] [133].

[52] Ibid., Sect. 71.

[53] Planned Relocation Guidelines 2022 (n7), Sects. 1–24.

[54] Ibid.

[55] A Toolbox: Planning Relocations to Protect People from Disasters and Environmental Change (International Organisation for Migration and UNHCR 2017).

[56] Planned Relocation Guidelines 2022 (n7), Sect. 1.

[57] Ibid.

[58] Ibid., Sect. 1.1.

communities' populations, it is also necessary to assess the livelihood security of nearby migrant and host communities. The third principle makes sure that the relocation process is coordinated and planned according to the necessary mechanisms that strictly adhere to the target timeline for all relevant stakeholders. This is to avoid the practice of 'Solomon time'[59] which would ordinarily require a long time to complete everything. This requirement includes provisions to ensure an integrated approach by many sectors and agencies to participate in the various stages of the relocation process to eliminate duplication and limit resource and time waste. For example, at the time of implementing the first phase, before relocation, it is mandatory for the Ministry of Environment, Climate Change, Disaster Management, and Meteorology to conduct a vulnerability assessment, while the Ministry of Lands and Housing identifies sites using sets of Standard Operating Procedures. Later in the phase of 'developing a relocation plan', it recommends the involvement of ideas from other disciplines as well. Hence, the process does not only involve a mono-disciplinary field of experts but also a collaboration from different sectors.

Aside from the aforementioned principles, any relocation procedure undertaken in the Solomon Islands must respect local customs, culture, and knowledge. The Solomon Islands' way of life and progress are profoundly rooted in their customs, beliefs, and traditional knowledge with respect to how they live and what rules or values were passed down from generation to generation. For instance, before relocation, it is critical that the culture of the relocated people be not undermined by the host communities, and relocated populations may wish to bring the remains of their loved ones to the relocated place. Before the relocation phase, therefore, the Standard Operating Procedures should incorporate the action plan on how this would be handled and implemented.

Should the necessity arise, the relocation plan guideline may be modified and changed. Thus, monitoring and assessment are offered because modifications and problems could occur during the actual implementation. In this case, a new development may be initiated through institutional structure and arrangement, such as a revision to the current climate change policy, a proposal for new climate change legislation, or through changes to the policy and planned SOP. The articulated guidelines will enable the safe, well-coordinated, and planned migration of vulnerable individuals to suitable locations in the Solomon Islands.[60]

The main features of the guidelines are divided into four (4) phases: (i) before relocation,[61] (ii) developing a relocation plan,[62] (iii) during relocation,[63] and (iv) following relocation.[64] These four sections were presented in a well-structured format, stating the main activities, responsibilities, and descriptions of specific tasks to be carried out under each phase.

The initial phase shall be conducted before the commencement of the relocation process. It involves the identification of communities to be relocated, the planning of relocation, and the identification of a potential relocation site. For these tasks to be completed concurrently during this phase, extra time will be needed.[65] For instance, the integrated vulnerability and adaptation assessments are to be deployed at this phase to determine all potentially affected communities and collect baseline data to inform planning, monitor change, and measure the impacts of adaptation action areas.[66] This Integrated

[59] Solomon time: whenever a project process in the Solomon Islands is typically delayed due to the practice or conventions of working with delayed or slow development, as targets are typically not met by the deadlines set by foreign agencies or external stakeholders.

[60] Planned Relocation Guidelines 2022 (n7), Sect. 6, paras [3–5].

[61] Ibid., Sect. 5.

[62] Ibid., Sect. 6.

[63] Ibid., Sect. 7.

[64] Ibid., Sect. 8.

[65] Ibid., Sect. 5.

[66] Ibid.

Vulnerability and Adaptation Assessment is expected to give information on the ability of communities to adjust to climate change risk as well as the duration of various hazards on the impacted population in either rural or urban settlements in the country.

Another important mechanism at this planning stage is the planning of the financial budget and the cost that will be incurred in the relocation process. It is important to explore potential financial sources whether through the existing national or provincial budget on disaster risk reduction, climate adaptation, or development planning; or community contribution in the form of cash, labour, and materials might also be considered as a source[67] and, further, external sources could be tapped to finance the relocation process.

The decision to be relocated is decided by the community through their Village Disaster Risk Committee (DRC)[68] assisted by faith-based and non-government organisations and a formal request is submitted to the National Climate Risk Resilience Committee (N-CRR). Another approach to this task is for the N-CRR to initiate the process based on the vulnerability and adaptation assessment and the community's decision that they would not be able to adapt to the risk of climate change impact. Through this approach, both village DR, Provincial Climate Risk Resilience Committees (P-CRR)[69] and the village DR committee shall together decide on whether a planned relocation is necessary. Simultaneously, the host communities and other communities in proximity shall establish a village DR committee, which shall accommodate the tribal leaders, church leaders, and the existing community organisation committee members. The sole responsibility of this committee shall be to facilitate matters to be regulated under the Standard Operating Procedures (SOP) that would be considered in issues related to the involvement of women, youths, and marginalised groups, the minimum standard for free, prior and informed consent, assessment of the potential impact of relocation on indigenous knowledge, practices, and artefacts.

Finally, the process of identification of relocation sites shall be deployed by the Ministry of Lands and the Ministry of Environment. Notably, the decision-making and planning process must include all individuals and households that will relocate. A proposed SOP shall be established to facilitate the process of identifying the relocation site through an active community-led process of negotiating new land arrangements, such as the type of land ownership, exploring land acquisition mechanisms, and considering marine and land resource access for the relocated and host communities.[70] The SOP on-site identification might as well consider learning from successful cases on land acquisition for relocation in the Solomon Islands, such as the acquisition of Choiseul Bay township land or the Bina Harbour land acquisition case. Taking the example of the Auluta-Bina Harbour land acquisition, a pilot project funded by AusAid through the Solomon Islands Institutional Strengthening Land Assistance Project, popularly known as the SIISLAP that was implemented by the Ministry of Lands, Housing and Survey. The Project had set out preconditions as per the Customary Records Act such as tribal genealogy or family tree justifying custodianship; agreed tribal boundaries; and tribal land authority through a long participatory process and in consultation with the communities and land resource owners. Despite challenges on technicalities of the system used to articulate and record genealogies, the project learned, through community wide consultation, to develop a functional framework on recording and registering customary land and to promote mediation and reconciliation to resolve disputes, rather than ending up in courts of law.[71] In the

[67] Ibid.

[68] Ibid., Sect. 1.3. Village DR Committee refers to Village Disaster Risk Committees established under the auspices of the National DM Plan.

[69] P-CRR Committee refers to Provincial Climate Risk Resilience Committees established under the auspices of the National DM Plan.

[70] Planned Relocation Guideline 2022 (n7), Sect. 5.

[71] Ibid. (n49) and (n50).

prior phase of relocation, there might be some community members who are not willing to be relocated. In such an instance, an SOP shall be developed to ensure that the land rights claims to their original site are not affected, just as those who relocated from alienated sites shall not claim their fixed-term titles. For customary land ownership types, households shall reclaim their rights to relocation; those who hold fixed-term land titles to a previous site shall not go back to their previous site but will retain the ability to return to those sites periodically.[72]

The next phase of the relocation process deals with the development of a relocation plan, strategies to be implemented during relocation, and evaluation measures following relocation. A relocation plan must be created in compliance with the 'Solomon Islands laws and policies, including but not limited to human rights law, development policies, and environmental and climate change policies'.[73] Any legislation or policies regarding provisions for water sanitation, education, and infrastructure development will be incorporated into the relocation plan. The specification of a relocation plan shall include measures to ensure smooth progress on the phases of relocation of the concerned individuals and establishment of housing, utility services (water, electricity, transportation infrastructure, education, health facilities, and access to markets), a budget to finance relocation, and measures to ensure the protection of the marginalised people in situations of vulnerability such as the elderly and people with disabilities, measures to mitigate or prevent risks, a full proposed site plan of the relocated site, and details of monitoring tools to evaluate possible complaints and feedback.[74]

Provisions of ongoing monitoring and evaluation are mandatory to ensure adequate social-economic welfare of the settlers.[75] Section 7 stipulates specific strategies to ensure the sustained development of the new site and the facilitation of new settlers into the economic and social life in their newly relocated environment.[76] Comprehensive monitoring and evaluation of the entire process, from initial planning to actual relocation, is detailed in section 8 of the relocation guidelines. Continuous social and environmental monitoring will be the responsibility of the ward and village DR committees or other suitable local authorities, with assistance from the P-CRR at intervals, to be decided with the relocated community. The N-CRR committee will take the lead and provide ongoing assistance for the climate-proofing of education, economic prospects, and resilience to climate change at the new relocation site.[77]

The planned relocation guidelines encapsulate a human rights approach concept to protect displaced persons rights. In the absence of Standard Operating Procedures, the implementation of the proposed relocation guidelines will remain pending. In this scenario, the following points are suggested based on the field study and assessment of the legal and policy measures initiated by the Solomon Islands.

4 Going Forward

The existing guidelines, particularly the NDMP 2018, address a specific situation that a person is in but do not provide answers for relocation.[78] The goals of the Solomon Islands Climate Change Policy will also need to be revised in order to be in line with the Ministry of Lands, Housing, and Survey's Solomon Islands Resettlement programme and the Planned Relocation Guidelines 2022.

The Constitution of the Solomon Islands recognises indigenous customary rights and protects customary ownership.[79] The Perpetual

[72] Ibid. (72), Sect. 5.5.

[73] Ibid., Sect. 6, para [1].

[74] Ibid., Sect. 6, paras [3–9].

[75] Ibid., Sects. 7.2–7.3.

[76] Ibid.

[77] Ibid., Sects. 8.1–8.3.

[78] Scott [27], Foukona et al. [24, 17].

[79] Ibid. (n32), 8–9.

Estate (PE)[80] ensures the rights of indigenous Solomon Islanders' to occupy and possess the land. The LTA that deals with land tenure and property rights as explained in section 3.1 has limitations when it comes to dealing with land acquisition or registration of customary land. Another pivotal aspect related to the LTA and planned relocation or resettlement of climate refugees is the interpretation of the existing provisions of the LTA.

It is important that the methods for customary land acquisition in Division 1 of the LTA be thoroughly reviewed as having one person dealing with compulsory acquisition land can result in officers involving in in corruption and bribery. Instead of an acquisition officer, a specific department under the Ministry of Lands should deal with land purchases that could be included in the standard operating procedures so that the land acquisition process can be facilitated by an appointed department or an existing unit within the current structure in the Ministry of Lands. This department shall conduct the process of land acquisition of the perpetuate estate[81] from landowners to registration of fixed-term estate.

Furthermore, the terms 'relocation settlement' and 'resettlement land' should be explicitly added to a section in the LTA to make provisions for the relocated site as per the Planned Relocation Guidelines. The language should be interpreted to accommodate the needs of climate change-induced migration or relocated communities rather than individuals in the context of land tenure ownership. This means that when a community is relocated on customary land, the site must be registered to community trustees on the PE or FTE as a whole lot or by a parcel number. This is not the case when the community is transferred to a place on registered land owned by the commissioners of lands. The proposed amendments to the LTA must be made to incorporate in the envisaged Standard Operating Procedures, which call for planning before relocating the communities in areas in danger from the effects of climate change.

A. Revise National Territorial Spatial Plan to Address Rural–Urban Migration

Migration is widely recognised as a climate change adaptation strategy.[82] In other words, rather than spending and directing government budgets towards disaster response and preparedness measures, transferring vulnerable communities that are always at risk of climate change-induced harm, would be more prudent.[83] The issue of regional disparity between rural and urban regions in terms of economic development and social infrastructure service delivery has become a controversial challenge to any society. For the Solomon Islands, as a least developing nation with a non-homogenous culture, ethnicity and vast, scattered islands, its physical characteristics and cultural diversity have been an obstacle to government planning towards effective goals for development. This has inflicted challenges to service delivery, communication, coordination of development programmes and prioritisation of development policies nationwide.[84] Rural–urban migration results from a spatial imbalance in development and service delivery in any geographical location as well as threats to livelihood due to environmental changes such as climate change. The country's National Development Strategy 2016–2030 is comprehensive and aims to enhance the social and economic circumstances of all Solomon Islanders.[85] The national development strategies over the past several decades had embarked on decentralisation and had stressed on ensuring balanced regional development.

[80] PE stands for 'Perpetuate Estate' for rights to use and occupy registered title under the Land and Titles Act. The two forms of land ownership and registered land are captured in the Land and Titles Act.

[81] Planned Relocation Guideline 2022 (n7), Sect. 5.5.

[82] Interview with Ministry of Environment Climate change division 'Interviewed by Lovelyn Otoiasi' (23 March 2023).

[83] Ibid.

[84] Prasad and Kausimae [28].

[85] National Development Strategy 2016–2035, 10–11.

The need to have development growth centres around the country for economic activities so that they are not concentrated only in the main city but also extend to rural areas as well, was also emphasised.[86] However, despite having stated these intentions for development plans for economic growth centres, there is no clear criteria mentioned in the national plans nor a blueprint on how to set up these designated centres, their location or the economic feasibility of implementing them. As a result, creating a national spatial plan that outlines clear guidelines for planning and laying out a road map for climate change relocation programmes and land purchases can close the gap in mass migration to urban areas. To balance spatial disparities in service delivery, employment opportunities, and development, the government must realign the revised National Development Strategy concerning other sectoral ministerial plans and policies such as the climate change policy, Planned Relocation Guidelines, provincial plans, industrial development policies, and sectoral agriculture plans. Noting that the Planned Relocation Guidelines 2022 emphasised sectoral integration into the process of planned relocation, the revision of the national development strategy and intended National Planning law, should make provisions for the relocation of vulnerable communities in the country's national comprehensive plan. A classic example of this is the designation of geospatial locations on blueprints that will indicate development opportunities in various sectors and the drawing of thresholds through integrated Vulnerability and Adaptation Assessment and Environmental Impact Assessment to determine the feasible areas for relocation and setting up of national economic centres. At first, the government will identify relocation sites based on the standard operating procedures set under the SI Planned Relocation Guidelines 2022 that align with other economic development strategies towards livelihood and job creation opportunities. Spatial planning would enable relevant stakeholders to assist the capacity of the communities to increase resilience and livelihoods that are threatened by environmental changes, thus discouraging climate-induced migration flooding into the central capital of Honiara. The government has already used this workable mechanism to counter, and although this exercise appears hard, it can potentially create a road map of the country's success in combating environmental-induced migration into urban centres.

In addition to this, it is necessary to re-establish the Site Development Fund, a distinct consolidated fund under the Ministry of Finance and Treasury, to enable completely serviced land for resettlement. This means that all utility services—water, power, road access, and schools—for relocated planned areas should be available on resettlement-registered land. This fund could be generated through the revenue on sale of land to provide well served sub-division on alienated land in the Solomon Islands.[87]

B. Creating New Climate Change Legislation

Without a defined legal framework on climate change-induced migration and relocation, unresolved challenges and issues connected to human rights and land ownership protection, in the Solomon Islands will remain in dispute. As previously stated, the national government has produced policies; however, a national Climate Change Act is required to supplement these government policies. In fact the climate change division believes that a national climate change law is critical.[88] The legal framework will provide for establishing an administrative board and strengthen the climate change unit to deal with climate change adaptation and mitigation measures, as well as any other issues related to climate change as well as the government's obligations under international treaties. Another

[86] Ibid., 31–32.

[87] Ministry of Lands, Housing and Survey 'Funding Public Infrastructure in New Land Sub-division' (Ministry of Lands, Housing and Survey 2016).

[88] Interview with Solomon Islands Climate Change Division Officer, 'Interviewed by Lovelyn Otoiasi' (23 March 2023).

significant aspect that would be captured in the proposed Climate Change legislation is the creation of trust funds that would serve the planning of prior, during, and after relocation of risk communities.[89] Fiji, for example, has adopted the Climate Change Act 2021. This Act came about three years after their Planned Relocation Guidelines were developed. Section 4(d) of the Act explicitly stated a provision for the establishment of an institutional and governance arrangement for the implementation of the Act.[90] Notably, the Fiji Climate Change Act 2021 makes provisions to address various matters subject to climate change mitigation and adaptation measures and mechanisms for implementation of the Act.[91] Section 78 of Part 12 of the said Act also makes provisions for a trust fund,[92] which will be used to facilitate the relocation and displacement of at risk communities. This is to be facilitated through a framework to secure funds from internal and external sources.[93] As a result, the Fiji Climate Change Act 2021 implies a mandate for setting up an institutional framework and legal mechanism that assists relevant authorities in implementing the relocation of vulnerable communities, as well as addressing climate change-related concerns associated with this Act. This is also an example of how the proposed Act for the Solomon Islands will enable and enhance existing climate change policies, Planned Relocation Guidelines, and other additional policies outlined previously.

However, with inadequate technical knowledge to draft and enact a climate law, the likelihood of adopting a law seems bleak.

[89] Ibid.

[90] Republic of Fiji, Climate Change Act 2021 (Act No. 43 of 2021).

[91] Ibid., Sect. 4(a) to (r).

[92] Republic of Fiji, Climate Relocation of Communities Trust Fund Act 2019 [21], Sect. 4.

[93] Republic of Fiji, Climate Change Act 2021(n92), Sect. 4(q) and Sect. 78.

5 Conclusion

To summarise, this paper demonstrates that environmental challenges, chances for work, and a better life in Honiara, are factors in the rural–urban migration and relocation. Sea level rise and food security severely threaten the atolls and artificial island communities of Solomon Islands. As a result, the government must act quickly to adopt a comprehensive approach and execute robust Standard Operating Procedures to supplement the implementation of the Solomon Islands' Planned Relocation Guidelines 2022. Three options are proposed in this study:

(i) Protect the rights of migrants through land rights protection; use the Land and Titles Act to facilitate the removal and resettlement of vulnerable communities through appropriate land acquisition and registration measures.
(ii) Align the National Development Strategy with other sectoral policies to handle the country's rapid urbanisation growth; a comprehensive approach is needed to address the expanding rural–urban migration by integrating these policies.
(iii) Create new climate change legislation—without a legal framework and revisions to the current LTA, the relocation of vulnerable communities, individual households, and the rights of migrants will not be taken seriously.

In addition, research on other thematic aspects of financial implication to justify relocation as a priority adaptation action rather than a last resort by the government would help influence and redirect government capital investment to urgently hear the call of the voices of the vanishing communities on the outer island atolls and artificial islands across the country.

References

1. Tabe T (2019) Climate change migration and displacement: learning from past relocation in the pacific. Social Sci, 1–18

2. Manou D, Mihr A (2017) Climate change migration and human rights. In: Manou D et al (eds), Climate change, migration and human rights: law and policy perspective. Routledge, pp 3–8
3. Klepp S (2017) Climate change and migration. Oxford Research Encyclopaedia of Climate Science, pp 1–30
4. David KJ, Jennifer MO (2020) The impact of climate change on migration: synthesis of recent empirical insights. Clim Change 518:281–300
5. Adoho F, Wodon Q (2014) Do changes in weather patterns and environment lead to migration? In: Wodon Q et al (eds) Climate change and migration: evidence from the Middle East and North Africa. The World Bank, pp 145–162
6. Population & Housing Census (2009). National Census (Solomon Island National Statistic office, pp 1–2. https://www.statistics.gov.sb/statistics/demographic-statistics/census. Accessed 12 June 2023
7. Monson R, Foukona J (2014) Climate related displacement and options for resettlement in the Solomon Islands. In: Leckie S (ed) Land solution for climate displacement. Routledge, New York, pp 291–316
8. Solomon Island National Statistic Office (2009) Population and housing census report on migration and urbanisation. https://www.statistics.gov.sb/statistics/demographic-statistics/census. Accessed 12 June 2023
9. Solomon Islands National Statistics (2019) Population and housing census report, pp 26–33. https://www.statistics.gov.sb/images/SolomonFiles/Survey_Statistics/Census_2019/Solomon-Islands-2019-Population-and-Housing-Census_National-Report-Vol-1-1.pdf. Accessed 8 October 2023
10. Campbell J, Warrick O (2014) Climate change and migration issues in the pacific. United Nations Economic and Social Commission for Asia and the Pacific, pp 6–7
11. Warn E, Adamo SB (2014) The impact of climate change: migration and cities in South America. World Meteorological Organization, p 2
12. Campbell JR (2014) Climate change migration in the pacific. The Contem Pacific 26(1):1–28
13. Oakes R (2019) Culture climate change and mobility decisions in pacific small island developing states. Population Environ 40:480–503
14. Michelle M et al (2022) Small islands. In: Pörtner HO et al (eds) Climate change 2022: impacts, adaptation and vulnerability contribution of working group II to the sixth assessment report of the intergovernmental panel on climate change. Cambridge University Press, pp 2043–2121
15. Ackerly BA et al (2017) Climate and community: the human rights, livelihood and migration impacts on climate change. In: Dimitra M et al (eds) Climate change, migration, and human rights: law and policy perspective. Routledge, pp 189–202
16. Brown O (2008) Migration and climate change. Migration Research Series 31 International Organization for Migration IOM, pp 16–29
17. Piggott-Mckellar AE et al (2019) Moving people in a changing climate: lessons from two case studies in Fiji. Social Sci 8(133):1–17
18. Campbell J, Bedford R (2014) Migration and climate change in Oceania. In: Piguet E, Laczko F (eds) People on the move in a changing climate: the regional impact of environmental change on migration. Springer, pp 180–181
19. Ranaa MP, Ilina IN (2021) Climate change and migration impacts on cities: lessons from Bangladesh'. Environ Challen 5:100242, 1–9
20. Laczko F, Piguet E (2014) Regional perspectives on migration, the environment and climate change. Global Migr Issues 1:1–2
21. Gray C et al (2020) The changing climate–migration relationship in China, 1089–2011. Clim Change 160:103–122
22. Foukona J (2022) Solomon Islands: flooding, displacement and durable solutions: the April 2014 Flood in Honiara. In: Scott M, Salamanca A (eds). Routledge
23. Kumar L et al (2020) Climate change and the pacific islands. In: Kumar L (eds) Climate change impacts in the pacific. Springer, pp 1–31
24. Foukona J, Bernard V, Scott M (2020) Solomon Islands national law and policy report; displacement in the context of disaster and climate change. Raoul Wallenberg Institute, pp 8–9
25. Corrin J (2011) Customary land in Solomon Islands: a victim of legal pluralism. Land Law Judicial Govern, 221–249
26. Burt B (1994) Land in Kwara'ae and development in Solomon Islands. Wiley on Behalf of Oceania Publications 64(4):317–335
27. Scott M (2020) The role of national law and policy in addressing displacement in the context of disaster and climate change in Asia and the Pacific. In: Scott M, Salamanca A (eds) Climate change, disaster internal displacement in Asia and the Pacific: a human rights approach. Routledge, pp 43–44
28. Prasad BC, Kausimae P (2012) Social policies in Solomon Islands and Vanuatu. Commonwealth Secretariat, p 6

Lovelyn Otoiasi holds a Master's in Urban Planning and Regional Development from the University of Seoul, South Korea, and a BA in Land Use Planning and Geography from the University of the South Pacific, Fiji. Before joining the Faculty of Agriculture, Fisheries, and Forestry, she worked with the Ministry of Lands, Housing, and Survey from 2012 to 2018. She was also a member of the task force for The "Greater Honiara Urban Development Strategy and Action Plan" (GHUDSAP), -a joint initiative of the Government of the Solomon Islands/Asian Development Bank to prepare an Urban Development Strategy and Action Plan (UDSAP) for the Greater Honiara Area (GHA) under the Ministry of Lands, Housing, and Survey (MOLHS). She also coordinated the development of Provincial Centres and Local Planning Schemes of Auki, Gizo, and Taro Township projects from 2016 to 2018. She has collaborated with and was a corresponding author of many research papers. She is an urban planning and resource management lecturer under the Environment Department at the Faculty of Agriculture, Fisheries, and Forestry.

Impact of Climate Change and Accessing Services in Papua New Guinea

Dora Kuir-Ayius

Abstract

With rising sea levels, changes in rainfall, increased frequency of El Niño-type conditions, and increased intensity of cyclones, the Pacific Islands region is exposed to climate change. A large proportion of the people are marginalised and are among the most vulnerable and least prepared to cope with its impacts. There are many competing definitions of the concept of vulnerability, but in the context of the Pacific Islands countries (PICs), it is referred to as the lack of power, influence, and control of those affected by climate change. The lack of power is reflected where access to resources is minimised. This can be viewed in various ways, including a lack of access to basic services such as enabling infrastructure, transport, hospitals, and others. In addition, the lack of access to income-generating opportunities further complicates the challenges imposed by the impact of climate change. In most of the PICs, including Papua New Guinea (PNG), Kiribati, Vanuatu, Marshal Islands, and Cook Islands, the chances of access to basic needs such as food, water, and land for survival are threatened. Approaches to dealing with climate change, displacement of people, and issues of traditional land ownership and relocation must be taken seriously. The paper presents some challenges of accessing services mainly caused by lack of effective management by the government of Papua New Guinea in the midst of climate change.

D. Kuir-Ayius (✉)
School of Humanities and Social Sciences,
The University of Papua New Guinea, Port Moresby,
Papua New Guinea
e-mail: dkuirayi@upng.ac.pg

1 Introduction

Papua New Guinea (PNG) is located within the Blue Pacific[1] continent countries in the Asia–Pacific region. It occupies the eastern half of

[1] Secretariat of the Pacific Community, 'Compendium of case studies on climate and disaster resilient development in the Pacific' (SPC 2015), 93.
The concept of the 'Blue Pacific' refers to the vision of harnessing the collective strength and potential of the Pacific Island nations and territories to address common challenges and pursue sustainable development in the region. It emphasises the significance of the Pacific Ocean as a shared resource and the need for cooperation among Pacific Island countries to protect and manage its vast expanse. The term 'Blue Pacific' originated from the Pacific Islands Forum, a regional inter-governmental organisation that promotes cooperation and dialogue among its Member States. It was first introduced in the context of discussions about the impacts of climate change and the need for a unified approach to address environmental and ocean-related issues affecting the Pacific Ocean.

S. Jolly et al. (eds.), *Climate-Related Human Mobility in Asia and the Pacific*,
Sustainable Development Goals Series, https://doi.org/10.1007/978-981-97-3234-0_11

the island of New Guinea, as well as numerous smaller islands and archipelagos. Apart from the 600 small islands and atolls, it has four big islands: Manus, New Britain, New Ireland, and Bougainville. PNG has diverse ecosystems, rich biodiversity, and a unique cultural heritage. It is home to hundreds of indigenous tribes, each with its distinct customs, rituals, and languages. Traditional practices, art, music, and dance continue to play a significant role in the country's cultural identity. Agriculture, mining, and natural resources drive the country's economy.[2] PNG is rich in mineral resources, with products such as coffee, cocoa, palm oil, and tropical fruits being significant exports. However, despite its abundant natural resources, the country faces poverty, limited infrastructure, and social inequality.[3]

PNG is located in one of the regions most exposed to natural hazards and climate change.[4] The low-lying nature of the islands makes them particularly exposed to even small increases in sea level. Some consequences of sea-level rise include saltwater intrusion, land loss, displacement, relocation, and food insecurity.[5] A large proportion of women, children, and persons with disabilities are marginalised and are among the most vulnerable and least prepared to cope with the impacts of climate change. There are many competing definitions of the concept of vulnerability, but in the context of the Pacific Islands countries, it is referred to as the lack of power, influence, and control of those affected by climate change.[6] The lack of power is reflected where access to resources is minimised. This can be viewed in various ways, including a lack of access to basic services such as enabling infrastructure, transport, hospitals, and others. In addition, the lack of access to income-generating opportunities further complicates the challenges imposed by the impact of climate change. Thus, PNG faces an uphill task in ensuring primary access to services threatened by climate change and extreme weather events.

This paper apart from this introduction is composed of five sections including the second section which contextualises the impact of climate change in PNG. The third section describes some characteristics of climate-related human mobility in the country. The fourth section focuses on challenges people face in access to land and its spillover impacts for climate or natural disaster-induced migrants. The fifth covers lack of planning and management of induced migrant resettlements and finally the conclusion.

2 Contextualising the Impact of Climate Change in PNG

Australian geographer Heathcote (1979) defined natural disaster as 'extreme geographical events greatly exceeding normal human expectations in terms of magnitude and frequency and causing significant material damage to man and his works'.[7] However, more recent works[8] attempt to understand the impact of these hazards especially with an appreciation on the social context inclusive of socio-economic, political and other factors, including dimensions of exposure, vulnerability, and capacities. For instance, the Council of Europe defines disaster stating that 'natural disasters are frequent and global occurrences that range from earthquakes, volcanic eruptions and tsunamis to crop failure, famine, drought and floods'.[9] This gives an understanding on the definition and how these disasters affect the country's physical as well as economic

[2] United Nations Development Programme, '2014 National Human Development Report: Papua New Guinea' (UNDP 2014), 99.

[3] World Bank Group, 'Climate Risk Country Profile Papua New Guinea' (World Bank 2021), 25.

[4] Lauer [1].

[5] Cox et al. [2].

[6] Secretariat of the Pacific Community (n1) 72.

[7] Blong [3].

[8] United Nations General Assembly, 'Report of the open-ended intergovernmental expert working group on indications and terminology relating to disaster risk reduction' (UNGA 2016), 41.

[9] Council of Europe, 'Pandemics and Natural Disasters as Reflected in History Teaching' (Council of Europe 2022), 95.

and social features. It is important to explore the impact of climate change in PNG in two parts; first as disaster, second the consequence of these impacts on the environment, as well as the socio-economic impacts on people who are affected.

A. **Disasters and their Impacts**

PNG continues to be ranked highest in the World Disaster Index.[10] It ranked 9th in 2021, and in 2017 as per the World Risk Index, it was ranked as the world's 12th, respectively. These results place PNG as one of the most at-risk countries globally. The risk factor is also reflected through ranking in the 2024 INFORM Risk Index, which assesses country-level vulnerability and coping capacity related to climate change.[11] PNG is categorised under very high climate risk, its ranking being determined by high levels of exposure to flood, drought, and cyclone.[12] The country's lack of coping capacity is ranked the 12th lowest coping capacity per the World Risk Report 2023. In other words, PNG is among the countries that are least prepared to cope with the aftermath of any disaster.

As a country, it has encountered various disasters including heat waves and drought, flooding and landslides, and cyclones and storm surge.[13] For instance, PNG has been facing increasing temperatures from 31–32 °C annually around November and has a prolonged period of excessively hot weather.[14] These heat waves have significant impacts on both natural ecosystems and human communities leading to heat exhaustion and heart stroke, especially in vulnerable populations including the elderly, children, and those with pre-existing health conditions. Additionally, increased temperatures can expand the range of disease-carrying vectors like mosquitoes, potentially leading to a higher incidence of diseases like malaria and dengue fever.[15]

PNG has experienced notable droughts over the years. The droughts in 1997 and 2015 resulted in widespread food insecurities.[16] The impact of the 1997 drought was severe, with destruction of crops, drying up of the local sources of drinking water, and intensified bush-fires from the coast to the crests of the main mountain ranges, destroying economic trees, subsistence food crops, and dwellings. Drought occurrences in the country in 2015 continue to impact the environment and its people. PNG currently 'faces an annual median probability of severe meterological drought of around 4%'.[17] The 2015 drought has also had severe and wide-ranging impacts on PNG affecting both the environment and human communities through reduced rainfall leading to water scarcity, live-stock losses, reduced agricultural productivity, and other negative impacts. Water scarcity contributes to hardships experienced by those especially in rural locations where people depend entirely on the environment for survival.

PNG, due to its geo-climate features, is prone to flooding.[18] Some of the key causes of flooding in the country include heavy rainfall and saturated soil leading to landslides. In 2012, a number of incidences of flooding and landslides occurred in several provinces, including Southern Highlands, Morobe, Western Highlands, Enga, Sandaun, National Capital District, Central, and Gulf.[19] Flooding resulted in extensive destruction of houses and casualties, including deaths across these provinces. Apart from these fatalities and injuries, there were displacement and economic losses, agricultural, and other impacts. PNG with its rugged

[10] World Vision, 'Papua New Guinea Climate Change Adaptation & Risk Mitigation' (World Vision 2023), 5.

[11] Frege [4].

[12] Thow et al. [5].

[13] Allen and Bourke [6].

[14] World Bank Group (n 3), 12.

[15] Imai et al. [7].

[16] Ibid.

[17] World Bank Group (n3), 10.

[18] Clark [8].

[19] International Federation of Red Cross and Red Crescent Societies 'Information Bulletin Papua New Guinea Landslide'(IFRC 2012), 12.

nature and widespread locations has on many occasions been unable to fully recover from the damages caused.

In addition, in the Pacific region, cyclones are a significant natural hazard.[20] When cyclones make landfall, they can bring about various impacts, including storm surges. The impacts of cyclones and storm surges can cause significant damage to buildings, infrastructure, and vegetation, which can lead to economic losses and displacement of communities.[21] In November 2008, Cyclone Guba petered down from Vanuatu from the north of Australia, swept through the Oro Province of PNG, causing swelling rivers, unstable soil as well as rough seas, consequently resulting in the destruction of homes, food gardens, infrastructure including roads and bridges.[22]

In short, disasters contribute to the displacement of communities and destruction of homes and infrastructure which has resulted in economic losses for the affected. Droughts, flooding, and landslides as well as cyclones and storm surges negatively affect agricultural fields and other community resources. This can have a significant impact on food security and livelihoods in affected areas. The spillover impacts of natural hazards in the next section demonstrate the impacts of climate change on people in PNG.

3 Human Mobility in the Carteret and Manam Islands

Since mobility can be a form of adaptation to climate change and other disasters; it is expected of populations who live in severely affected locations to move.[23] Similar to other parts of the world, the Pacific Islands too are expecting large-scale climate-induced mobility among the severely affected communities. PNG, despite being the only Pacific Island country that is not surrounded by the ocean, does have sinking atolls and coastlines being affected by climate change and people living within the affected areas are bound to migrate.

The relocation of people as a result of climate-related impacts and disasters causes friction when these people are relocated onto other lands. The basic cause of the friction originates from the concerns around migrants settling on lands that are traditionally owned by people in the destination. For example, the Carteret islanders in Tinputz are confronted with challenges including struggles such as social disarticulation.[24] This is because around 80% of the people in rural areas survive on subsistence agriculture and need their land for food cultivation. Moreover, in PNG, traditionally, clans or tribes own land communally. Most of this land is patrilineally owned, although there are a few locations where women also own land.[25] PNG has Land Acts that govern the land management framework but land ownership in PNG is also complicated by people's oral histories and social identities and their connections to the physical land features.[26] Land is tied to locality—whether terrestrial or marine—[and] is the basis for membership and nationality for most Melanesians. Research in general has summarised the significance of land and therefore highlighted the importance of land ownership: 'Land is our life and our physical life-food and sustenance. Land is our social life; it is marriage; it is status; it is security; it is policies; in fact, it is our only world'.[27] As Banks argues, resources are deeply connected to the creation

[20] Anzellini et al. [9].

[21] International Federation of Red Cross and Red Crescent Societies, 'Final Report Papua New Guinea Cyclone Guba' (IFRC 2009), 10.

[22] Auamoromoro and Luma [10].

[23] Campbell [11].

[24] Edwards [12].

[25] Caritas Aotearoa New Zealand, 'Caring for our Common Home: Caritas State of the Environment Report for Oceania' (Caritas 2015), 45.

[26] Independent State of Papua New Guinea, 'Land Act' (1996).

[27] Banks [13].

and continuation of kinship relations and identities of people.[28] Land ownership is often difficult to identify because people have stories and diverse connections with landscape and social life. This can cause complications in separating people from the physical features they associate their identity with, and in turn, this can affect the delivery of infrastructure and services when land ownership is contested.

The complicated land ownership in PNG has also contributed to numerous conflicts. Jorgensen cites the following reasons for these conflicts.[29]

- multiple claimants to particular pieces of land;
- complex land histories giving rise to claims from conflicting clans;
- the State's recognition of indigenous land rights giving rise to pressure by these landowners on the State.[30]

Numerous candidates contesting ownership over an area of land have caused difficulties in identifying the rightful owners of that land. This has been the case with the climate-induced migrants of Carteret and the Manam islands. Their chances in securing land for cultivation are slim.[31] Several clans often claim ownership over one stretch of land, which is usually complicated by the absence of written records, the lack of specific surveyed measurements, and the use of imprecise natural features such as a mountain range or valleys as boundaries.[32]

It is important to explore the issue of relocation in PNG in the context of the Manam islanders who are victims of volcanic eruption. This approach is taken to understand the manner in which displaced populations are resettled in the country.

In this sub-section, two cases concerning planned relocation in the Carteret and Manam Islands are presented to demonstrate how human mobility is coordinated. Induced migration in these cases refers to those who are forced to relocate. These are people who have been relocated to new locations.

A. **Tinputz Atolls**

The Carteret Islands, also known as Tinputz Atolls, are made up of six small main islands and are among the atolls that are less urbanised—Carteret Islands have gained significant attention as it is often cited as one of the first places to be affected by climate change-induced sea-level rise.[33] The islands are located in the Autonomous Region of Bougainville (ARoB), and none of these islands are higher than two metres above sea level.[34] Rising sea levels create susceptibility among those who live along the coast, leading to increased coastal erosion, flooding, saltwater intrusion, and inundation. Storm surges have led to the intrusion of saltwater into freshwater sources, impacting drinking water supplies and agricultural land. Coastal inundations caused by cyclones have led to severe coastal erosion, damage to infrastructure, and the loss of coastal habitats.[35] As per Caritas reports from PNG, many coastal villages and atolls have lost land due to sea-level rise. An example from the Carteret Islands confirmed the impacts as the sea is 'basically eroding our shorelines (and) we have lost 60–70 m of land'.[36]

Agriculture and fishing are the primary sources of livelihood for the Carteret islanders. Climate change affects their ability to sustain themselves through these activities due to changes in rainfall patterns, increased storm frequency and intensity, and the degradation of marine ecosystems. The scenario becomes

[28] Ibid.
[29] Jorgensen [14].
[30] Ibid., 601.
[31] Ibid.
[32] Banks (n 27), 25.

[33] Kimiafa [15].
[34] Connell [16].
[35] Caritas Aotearoa New Zealand (n 25), 31.
[36] Ibid.

complicated because it is one of the most densely populated island groups, that lack available land for agriculture and gardening due to sea-level rise and changes in rain patterns. This has led to food security challenges, increased economic vulnerability, and consequent displacement of communities.[37] The rising sea level has affected the livelihood of the islanders to an extent that most of the islanders cannot sustain themselves on their island homes. The destruction of the natural environment extends to soil erosion and sedimentation of water bodies. The existing infrastructures such as roads and communications cables are destroyed. These impacts threaten homes, and access to essential services, resulting in land loss and community displacement and relocation.[38]

As per the results from a case study[39] on Carteret Islands, the resettlement agenda has been around for several decades mainly due to climate change and food insecurities. Several families were relocated along the Kieta-Buka Highway but many returned to the islands due to land disputes and failure of migrants integrating with the receiving community.[40] After the Bougainville Civil War, frustrated islanders saw unfulfilled promises from the Bougainville Government as a delay in their relocation and they initiated their own arrangements to relocate.

The current relocation of climate-induced migrants for resettlement on Tinputz is the initiative of the Carteret islanders through the Tulele Peisa, the local NGO endorsed by the islanders and is led by Ms. Ursula Rakova, a social worker from this island, who is the prominent voice for the people of the Carteret Islands.[41] This NGO has secured 81 hectares of land donated by the Catholic Church in Tinputz.[42] The relocation of the people in Tinputz took place in 2009. As per the Post Courier, a national newspaper, various partners, mainly NGOs including the Lutheran Pentecostal Church of Germany, City Pharmacy Group of Companies in PNG, Catholic Coalition on Climate Change of the US, Caritas New Zealand, and Sisters of Mercy, Brisbane, Australia, assisted in various ways to resettle some of the islanders. The relocation plan initiated by the Carteret islanders through Tulele Peisa began in 2007, and a proposal for the relocation of half of the population was advocated for.

Results from a case study on the relocation of the Carteret islanders revealed that ABG had conducted feasibility studies to relocate the IDPs on Karoola, Buka Island. The ABG was just about to secure land when the islanders successfully secured land in Tinputz. In spite of the government's efforts, Tulele Peisa went ahead with their collaboration with the Catholic Church on Bougainville to settle islanders on Tinputz.[43] There appears to be a lack of information to explain the reasoning behind this arrangement.

There also seems to be a lack of the Government of PNG's (GoPNG) involvement in this relocation process. The Tinputz initiative was organised by Tulele Peisa, the local NGO of the Carteret Islands in partnership with other partners. Since there was little or no involvement of the government, multifaceted challenges have occurred. To prevent compounded impacts such as settlers and landowners of those who relocated, getting into conflicts currently experienced in Madang and Tinputz, action must be taken to rectify the current situation as well as setting a course for future relocation centres around the country. A call for an approach to cater for displaced climate-induced migration has been around for a while now but without much attention from the government. In 2015, Kaigabu Kamnanaya, an employee of the National Disaster Centre, PNG made a request, but nothing has been developed thus far.[44]

[37] Ibid.

[38] United Nations Development Programme (n 2), 57.

[39] Edwards (n 24), 55.

[40] Ibid.

[41] Nangoi [17].

[42] Ibid.

[43] Edwards [12], 55.

[44] Ingipa [18].

The experience from Tinputz Atolls highlights the complex challenges faced by the relocated population, particularly in relation to land use and communal land ownership. As per Edward's study, the relocation process involves socio-economic and cultural aspects.[45] Relocated Carteret islanders encountered challenges including landlessness, joblessness, homelessness, marginalisation, food insecurity, loss of common lands and resources, increased health risks, and social disarticulation.[46] The settlers are confronted with multiple challenges including access to land for gardening and fishing. The original landowners of the Tinputz area often deny access to land use for gardening to the migrants. In a recent conversation,[47] Ursula Rakova, the chief executive of Tulele Peisa, admitted that 'land and the host community were seen as the enemies, and the islanders were very much keeping to themselves for a while and not mixing with mainland people'. These were the times when there was a lack of understanding from both parties: the settlers and the landowners. Despite the challenges, there have been some positive stories on relationships between the two groups of people in Tinputz. Tulele Peisa with another local NGO[48] 'facilitated post-relocation counselling'. Supporting both groups with counselling as well as training on farming skills to the settlers have enabled these people to work the land allocated to them. This training was important as the settlers whilst living on the island did not make gardens for survival due to land insecurity. Also, the government of PNG provides those on the island, with food every quarter to supplement their local diet of coconuts and fish[49] There seems to be a lack of available information on the success rate on continuing peace among the settlers and the landowners. Therefore, research is required to further understand the relationship of the two groups of people in Tinputz. Currently, there seems to be a lack of emphasis on the negative aspects.

The case of Manam is also important as it points to the impact of a natural disaster in the form of volcanic eruptions forcing people to migrate. The experience of Manam can contribute to the understanding of relocation of induced migrants as it is one of the two major relocation experiences in PNG. Therefore, it is important to view closely the experience of the Manam islanders as induced migrants.

B. Manam Islanders

The case of Manam islanders in Madang, PNG, further illustrates complications in relocating people to another location. Manam Island is located along the Pacific Ring of Fire, whose people have been continuously displaced due to the volcanic eruption.[50] The Manam islanders have been officially declared as Internally Displaced Persons (IDPs) by the Office of the United Nations High Commission for Refugees.[51] Displacement of the islanders had initially been temporary as they would shift for short periods, from 2–3 months during the volcano's eruption. Earlier, islanders sought refuge in the mainland with their 'hereditary exchange of resources such as nuts and fish on the mainland'.[52] The islanders exchanged foodstuff, including pigs prior to their displacement. They capitalised on these traditional relationship ties to relate to the mainlanders. However, this relationship has been challenged, now that the islanders are without these resources and cannot reciprocate as the eruption has destroyed their villages as well as their resources.

The 2004–2005 volcanic eruptions witnessed a lot of Manam islanders migrating to the mainland. A majority of the displaced people live in makeshift houses built of traditional materials. After several eruptions, over 9000 IDPs were

[45] Edwards [12], 60.
[46] Ibid.
[47] Rakova and Salote Soqo [19].
[48] Ibid.
[49] Ibid.
[50] Naser [20].
[51] Ibid.
[52] Connell and Lutkehaus [21].

relocated to the mainland near Bogia into care centres in 'Potsdam, Asuramba, and Mangem, as well as several informal centres'.[53] Displaced young men built temporary shelters which have deteriorated but people still live in them.

After six months of settlement in the care centres, there were land use issues as reported by the UNHCR.[54] The landowners whose land the care centres were established on did not accept the settlers residing on their land. The displaced people have experienced challenges in meeting their basic needs such as having access to safe drinking water, privacy, and lack of access to food. The local village near the Asurumba care centre asked their local MP to evict the Manam islanders.[55]

As per an IOM report[56] despite 15 years of resettlement, the settlers continue to face challenges including land tenure, agriculture, education, health, shelter, and others. Although in the case of Manam relocation was induced by volcanic eruptions, settlers are now confronted with sea-level rise due to their settlement on the sea shore where erosion is inevitable. Settlers do not have access to proper shelters, education, and land for gardening. Connell and Lutkehaus' work demonstrated seven challenges namely: alteration of livelihood, little available land, absence of piped water, lack of funds for school fees, land ownership issues, weak traditional leadership, and increase in disputes with local host landowners.[57]

The latest experience on Manam in 2021 illustrated clearly the lack of preparedness to deal with natural disasters such as volcanic eruptions in PNG. Despite the occurrences of volcanic eruptions, over three decades serious lack of services continues. As per a situation update, the settlements continued to lack basic services including health and proper housing with the recent eruption. The Madang Provincial Government through the Manam Restorative Authority appears to have limited capacity to provide necessary services for the victims of volcanic eruptions.[58]

As noted, challenges in land ownership and relocation of induced migration including climate-induced migrants and other disasters such as volcanic eruptions can contribute to complications or problems in the relocation sites. Although the example on Manam Island is not directly related to climate change, climate change will be an issue in the near future. As per the World Bank PNG Country Profile, 'sea-level rise also threatens the integrity of Papua New Guinea's coastal resources and biodiversity' including land.[59] It is crucial to highlight that both Tinputz and Manam settlements experience similar challenges whilst living in their respective settlements. It is also important to unpack some challenges surrounding access to basic services in the destination of the induced migrants.

4 Challenges in Access to Land and Its Impact for Climate-Induced Migrants

Having access to land and basic services is paramount to any individual, family, or a community, as this provides comfort and convenience for survival in any context including for those affected by climate change and other natural disasters such as volcanic eruptions. In the midst of climate change and natural disasters in PNG, those affected face serious challenges. Most communities lack access to land and to basic services including health and sanitation facilities and Tinputz and Manam resettlements as induced migrant communities are no exception. This section has a twofold aim; first it will contain discussions around the lack of access

[53] Naser [20], 25.

[54] Connell [16], 20.

[55] Naser [20], 79; See also Another Manam? The Forced Migration of the Population of Manam Island, Papua New Guinea, due to Volcanic Eruptions 2004–2005, (International Organisation of Migration 2016).

[56] Ibid.

[57] Connell and Lutkehaus [22].

[58] Naser [20], 82.

[59] World Bank Group (n 3), 17.

to land and incidental effects thereof, while the second will focus on the spillover impact of the former.

A. Challenges in Access to Land

The experiences of these two communities (Manam and Tinputz) illustrate the challenges encountered by the induced settlers. It is evident as per the studies of Connell and Lutkehaus and Edwards, that resettlements of Tinputz and Manam islanders on the coast of Bogia, Madang, lack basic services.[60] As it turns out, both communities are confronted with some common issues including landlessness as the major issue, challenges in adjusting to a new livelihood, food insecurity, lack of funds to send children to school, lack or absence of piped water, and other challenges.

Landlessness in the context of PNG especially in rural areas such as in Tinputz and Manam is a serious issue. Being landless refers to those who do not own land and miss out on being self-sufficient, as land is the source of livelihood. In the observation of Connell and Lutkehaus 'disputes increased with the local 'host' landowners, as islanders struggled to adapt to the mainland'.[61] It is also clear that the 'host villages' or original landowners did not welcome the settlers. As the population of Manam grew, so did the pressure for their needs. Densely populated settlements compounded with 'inadequate access to land and lack of resources created problems and conflicts'.[62] These challenges have been continuous and have caused ongoing challenges for the original landowners as well as the settlers with little or no intervention from the government and its partners.

Tinptuz also experienced similar issues with lack of access to sufficient land especially the one similar to their original home. As Edwards points out, 'default low-lying atolls have little relief, wide-open vistas, and lots of sand, a hard environment to replicate.[63] Low-lying islands have a high ratio of shoreline to land area, and it can be observed that all atoll dwellers live within a few metres of the sea. Acquiring a similar stretch of coastline on the mainland with ready access to the shore is unlikely. Coastal sites are prime locations and local landowners are reluctant to give up a narrow, shoreline strip. This in itself may pose challenges to the settlers who are not used to engaging in activities on the mainland of Bougainville. For instance, the Carteret islanders will have to adjust themselves to the new environment when engaging in their daily tasks.

Being relocated to a new environment can alter the livelihood of the settlers. These settlers will have to change the way they think and engage in activities such as gardening, fishing, and other usual tasks they were engaged in at their previous destination. Islanders who are used to mainly fishing on their spacious shoreline strip would probably have a beachfront that is half the size of the beach on their original destination. Displacement from disaster can lead to disruption in 'people's livelihoods, housing conditions, health, education, security, social life, and environment'.[64] People feel neglected in their destination as they are placed in a helpless situation. In a country like PNG where almost 80% of the people heavily rely on the environment for survival, this is disastrous. Impacted communities and their people would suffer especially if there is little or no support from the government or other stakeholders.

Security in the context of climate change and natural disasters, as in the case of the Carteret Islands and Manam can be seen in the light of the law and order issues that erupt from relocation and its multifaceted challenges.

Lack of funds for school fees is another issue the settlers have to deal with. Settlers therefore struggle to send their children to school. The Manam Islanders previously while living on their island could sell copra to make money,

[60] Connell and Lutkehaus [22], 20.

[61] Connell and Lutkehaus [22], 20.

[62] Ibid.

[63] Edwards [12], 74.

[64] Anzellini et al. [9], 13.

while currently in the settlement they have nothing. As a result, a lot of school-age children stay at home, 'Although education is still highly valued among the Manam islanders, the high cost of school fees has meant that often it is only the children from elite families' who get to attend school.[65] And this contributes to the increase in uneducated or illiterate youth within the community in the care centres as well as the temporary shelters within the villages.

Many have echoed the issue of food insecurity being encountered by many displaced settlers. Many families within the settlements suffer from lack of food supply. As per the results from the International Organisation for Migration, little or no land for gardening contributes to many of the challenges faced by the settlers.[66] As demonstrated by Teddy Basse, a settler in a Manam Settlement—'vegetables and other sources of protein are in short supply. Many children suffer from potbellies, while pregnant and lactating mothers eat the same food as everybody else, thereby lacking the necessary nutrients, the assessment said. Insufficient access to land and overuse of the soil have exacerbated food shortages, and families routinely skip a meal. We survive on what little we can grow on the land around us'. There is no doubt that settlers encounter compounded issues, relating to land ownership in their new destination.

Families in both Tinputz and Manam are struggling to provide sufficient food supply for their homes. Issues on access to sufficient land contribute to families not having enough food. The change of environment also plays a part in the lack of food supply. For example, Manam islanders struggle to cope with working on land on the settlement as they complain about the land being infertile and their food crops not yielding as much as their gardens back on Manam, a volcanic and fertile island.

The original landowners on the other hand have occasionally shown their frustrations through disagreements, by getting into conflict with the settlers. There is insufficient information to present their experiences in sharing land with the settlers. Perhaps research is required to explore the process and procedures followed by the government of PNG to acquire land for the settlers. It would only be fair to the original landowners if the government of PNG and other stakeholders heard their side of the story on how their land was acquired in the resettlement process. Apart from that, if there are benefits for the landowners both in the short and long-term to at least give these people some peace of mind as well as preventing occurrences of conflict over land ownership issues. As pointed out by Edwards,[67] any 'resettlement project needs culturally appropriate compensation for both hosts and those displaced, and greater commitment from national and provincial government, especially in relation to long-term planning'. Lack of culturally appropriate negotiations with the original landowners on the initial settlement of disaster-induced Manam islanders continues to be an issue today. A recent source illustrated[68] aggrieved landowners of Bunabun village expressing concerns regarding the growing rift between them and the settlers causing lawlessness in the community. Mr. Joe Masep, a landowner called for the 'government and other agencies to step in to settle the issue of resettling the islanders'. It is important to include the original landowners and the settlers from the beginning to the end of the decision-making process to allow for understanding among all partners involved.

The issue of land ownership is one of the most complicated in a Melanesian context and has contributed to numerous conflicts. This, as Jorgensen argues, is due to three main reasons: numerous claimants to particular pieces of land, complex land histories giving rise to claims from conflicting clans, and the State's recognition of indigenous land rights giving

[65] Naser [20], 15.
[66] Ibid.
[67] Ibid., 74.
[68] Wemuru [23].

rise to pressure by these landowners on the State.[69]Having no land in a climate-induced environment together with experiences of other disasters can exacerbate land ownership issues as in the case of Manam and Tinputz. In PNG, numerous candidates contesting ownership over an area of land has caused difficulties in identifying the rightful owners of that land. Several clans can often claim ownership over one stretch of land which is usually complicated by the absence of written records, the lack of specific surveyed measurement, and the use of imprecise natural features such as a mountain ranges or valleys as boundaries.[70]

In the background where land rights have assumed the status of a stumbling block in the way of relocating the climate change-induced people, it should be noted that land rights issues in PNG cannot be eliminated, but they can be managed in a way that the landowners benefit from the use of their land. For instance, the State can pay rent to the landowners. Cultural approaches in dealing with climate change and displacement of people from one location to another as well as issues of traditional land ownership and relocation must be taken seriously. In the relocation process, relevant stakeholders including non-governmental organisations, churches, and the government, must coordinate with landowners on relocation destination or land for climate-induced migrants to prevent unforeseen challenges such as violence over land use.

B. Other Impacts

Other impacts refer to some spillover effects of climate change and other disasters on induced migrants. These impacts are multifaceted as they are the results of various factors. The other impacts of climate change and natural disaster-induced migrants are many and include: lack of basic essential services such as health, water and sanitation, electricity, lack of income and others.

[69] Jorgensen [14], 600.

[70] Banks [24].

It is important to unpack some encounters of induced migrants regarding the availability and access to service provisions through the lens of these migrants in Tinputz and Manam.

The health of the displaced population is usually at risk especially when they are confronted with a lack of proper water and sanitation facilities, an essential part of a person. The practice is that many including both within induced migrant settlements as well as original landowner villages, still drink from well water in PNG. Despite the availability of tanks and piped water, this practice continues in many isolated communities in the country. This practice is unhealthy and contributes to health issues faced by induced migrants in the settlements. Many faces 'common health issues such as diarrhoea and water-related diseases'.[71] The services available for these settlers remain the same as these results reflect the experiences of Manam settlers in 2016, seven years after the result of people drinking contaminated well water within a generally unhealthy environment was reported.[72] Today the difficulty in access to and shortage of medical supplies continues as there is 'still (lack of) basic services with medicine running out'.[73] This could mean that not much effort has been invested into service provision to improve the health of the settlers, and that water and sanitation facilities within the induced migrants settlements in PNG is neglected.

Apart from issues relating to water and sanitation, malaria was another challenge. Change of location caused an increase in cases of malaria among the Manam.[74] The Carteret islanders on the other hand also experienced an increase in cases of malaria while they were still on the island. Results from Edwards revealed that the fact that these people suffered from the 'inadequate diets and the risk of increased prevalence

[71] World Bank Group and others, 'Vulnerability, Risk Reduction, and Adaptation to Climate Change: Papua New Guinea' (World Bank Group 2011), 12.

[72] Ibid.

[73] Wemuru [23].

[74] Imai et al. [7], 2.

of certain diseases means that, prior to relocation, the physical health of islanders may be poor'.[75] Cases of malaria were common among the Carteret islanders. This was due to the constant contact with the wet soils on Han Island which served as an incubator for the breeding of mosquitoes resulting in incidences of malaria.[76] This is the impact of climate change on the island and the lack of available health services at their original destination to address health issues.

In addition, seeking medical assistance in the care centres is difficult as there is lack of money to fund for transport and required medical services, as revealed by Manam islanders.[77] There is little or no effective medical services in the induced migrant resettlements. Access to basic services is a significant aspect of any community and settlements of climate or natural disaster-induced locations are no exception. Services such as built infrastructure including roads, classrooms, hospitals, power poles and electricity supply, water supply, and any other services must be made available and accessible to those affected. Experiences of the Carteret and Manam settlers have illustrated several challenges in service delivery including water sanitation and hygiene. There are implications that the government has neglected its responsibility to provide essential services in induced migrant settlements in PNG.

5 Lack of Planning and Management of Induced Migrant Resettlements

Planning and management of any community is important and the induced migrant resettlements are no exception. As such, planning for a resettlement of climate or other natural disaster-impacted people is significant. However, there is a lack of effective management in the overall operations of services, including health facilities, especially at the district level of government. The effects of this have trickled down to service provision within the communities and also include the induced migrant settlements. Campos-Outcalt, Kewa, and Thomason focused on decentralisation and administration of health services at the sub-district level and revealed some staggering results.[78] They have demonstrated difficulties such as a lack of finances, transport to do health runs, and limited professional support. Work by Howes and others confirms the financial difficulties encountered in delivering health services in PNG.[79] This trend is also evident in the climate or natural disaster-induced settlements as in the case of Carteret and Manam.

The Carteret islanders experience multifaceted challenges including[80] land insecurity, homelessness, lack of basic services, and others. Challenges associated with re-establishing their livelihoods, especially when they were dependent on fishing which is being disrupted by climate change, also abound. Government support in providing alternative livelihood options is crucial for the economic sustainability of these communities. Although the government has played an important role in supplying food rations on a quarterly basis to those who still live on the islands; its role in the relocation process is currently unclear. It is also believed that both the Bougainville and PNG Governments know about our proposal but have not helped at all. They have not spent a *toea* (cent) to help us with the relocation.[81] In other words, the State has done little to support the Carteret islanders.

Lack of government support in these resettlements has manifested in several ways. Among other challenges, lack of infrastructure and basic services including education and basic healthcare services are quite visible. After almost two decades post the initial resettlement of Manam

[75] Edwards [12], 74.

[76] Ibid.

[77] Connell and Lutkehaus [25].

[78] Campos-Outcalt et al. [26].

[79] Howes et al. [27].

[80] Edwards [12].

[81] Komai [28].

islanders, a lack of basic services still exists. A recent source[82] illustrated that 'there is also a need for a school and other services'. The experience of both resettlements is evident through the challenges encountered due to a lack of infrastructure.[83] Manam resettlement encountered overcrowding in the community due to lack of available land. The government struggles to provide essential infrastructure such as water supply and sanitation in the resettlements. This can be elaborated to illustrate the lack of concern or oversight on the part of the PNG government and its stakeholders. The lack of basic services can significantly impact the well-being of the relocated communities.

Those responsible at the provincial and district levels were ill-prepared to take up responsibilities to manage the funds provided to deliver services to the local communities. Several other studies on the impact of decentralising health services in PNG have shown similar results.[84] There are more negative impacts than positive ones regarding decentralisation on the overall delivery of health services. These results demonstrate a lack of support that can be provided for climate-induced migrants in a new location with little or no available health services. The question is how can the spillover impacts of climate change be effectively addressed when there is a lack of effective management in the government system? Although the government of PNG is working in partnership with international partners through NGOs and other renowned actors including Red Cross, Asian Development Bank, and others, it needs to look for alternatives to turn the situation around to take on challenges created by climate change.

The role of the government is unclear in terms of resettlement of the induced migrants. And also, it has shown little interest in supporting the displaced people despite its efforts on the general agenda on climate change. Despite initial arrangements in relocating the victims of the eruptions, there is little or no information available for further understanding the role of the State, 'uncertainty, lack of political will, and the absence of a government policy on internally displaced persons have hindered attempts to provide definitive and durable solutions'.[85] Therefore, most importantly the government of PNG has an outstanding task to develop a relocation policy for its climate affected population. For PNG as a country that is located in a high-risk zone that is exposed to climate change and natural disaster, having a relocation policy is important. The resettlement of induced migrants is an important agenda, especially in the midst of climate change. Therefore, it is important that the resettlement process is well thought out and carefully planned. Planning is an important part of any project and it is significant for any government to consider various factors involved in the initial organising of resettlement of induced migrants.

In addition, the government of PNG has done little to obtain land for people forced to move from the Carteret Islands to the now-called Tipputz Atolls on the mainland of Bougainville. Most land disputes among the islanders are due to customary land ownership systems which have been examined in the earlier section. These challenges appear to be the result of a lack of vision and planning by the government of PNG. The government of PNG through its Enhanced Nationally Determined Contribution 2020 outlined nine priority areas for adaptation including coastal flooding, inland flooding, food security, cities and climate change, climate-induced migration, damage to coral reefs, malarial and vector-borne diseases, water and sanitation, and landslides. However, this is not evident in the

[82] Wemuru [23].

[83] 'Another Manam? The Forced Migration of the Population of Manam Island, Papua New Guinea, due to Volcanic Eruptions 2004–2005 (International Organisation for Migration 2016). https://reliefweb.int/report/papua-new-guinea/another-manam-forced-migration-population-manam-island-papua-new-guinea-due-to-volcanic-eruptions-2004-2005. Accessed on 25 Jan 2024.

[84] Thomason et al. [29].

[85] Naser [20] 74.

locations where the impact of climate change has posed challenges, in the case of planned relocation of climate-induced migrants.

6 Conclusion

PNG has been experiencing severe climate change issues through landslides, floods, cyclones, droughts, earthquakes, and the problems caused by sea-level rise. As it turns out, the country is currently confronted with the relocation of climate change–affected populations and access to basic services. The cases of the Manam and Carteret islanders in PNG have demonstrated problems encountered by induced migrants in their destinations. Both relocation sites experience land ownership issues that affect the overall well-being of the relocates. In such a situation, the government must intervene to rectify the spillover impacts of unplanned relocation efforts.

Social and economic challenges of climate change include loss of livelihoods, increased poverty, food insecurity, and displacement and eventual relocation of coastal communities that exacerbate existing social and economic challenges and can lead to social unrest. Furthermore, the country's limited resources and infrastructure make it difficult to adapt effectively to these changes. Addressing the impacts of climate change and improving access to services in PNG, require a multifaceted approach. It involves international cooperation, adaptation measures, sustainable development practices, and investments in infrastructure, health care, education, and disaster management. Efforts should also empower local communities, promote climate resilience, and ensure equitable access to services, particularly for vulnerable populations. PNG, like many other countries, needs to follow adaptation and mitigation strategies to address these impacts. This includes implementing measures to improve coastal protection, investing in resilient infrastructure, promoting sustainable agriculture practices, conserving biodiversity and ecosystems, and enhancing health care and social safety nets.

Additionally, international cooperation and support are crucial to assisting PNG in building resilience and mitigating the effects of climate change. The PNG government also needs to formulate a planned relocation policy for climate-induced migrants. PNG requires a comprehensive approach that integrates climate change adaptation and mitigation strategies with efforts to improve access to basic needs and services. This includes investing in resilient infrastructure, sustainable agricultural practices, water resource management, healthcare facilities, and education. Furthermore, enhancing community resilience, promoting sustainable livelihoods, and empowering vulnerable populations are crucial for ensuring equitable access to basic needs and services in the face of climate change.

References

1. Lauer M (2016) Governing uncertainty: resilience, dwelling, and flexible resource management in Oceania. Conserv Soc 14(1):34–47
2. Cox et al (2023) Climate change as a driver of conflict in Papua New Guinea. US Institute of Peace, p 68
3. Blong R (1986) Natural hazards in the Papua New Guinea highlands. Mountain Res Develop 6(3):233–246
4. Frege I et al (2023) WorldRiskReport 2023, focus: diversity. Bündnis Entwicklung Hilft, p 75
5. Thow A et al (2022) INFORM climate change: quantifying the impacts of climate and socio-economic trends on the risk of future humanitarian rises and disasters. Office of the European Union, p 23
6. Allen B, Bourke M (2009) The 1997–98 drought in Papua New Guinea: failure of policy or triumph of the citizenry? In May RJ (ed) Policy making and implementation: studies from Papua New Guinea. ANU Press, pp 325–344
7. Imai C et al (2023) Associations between malaria and local and global climate variability in five regions in Papua New Guinea. Tropical Med Heal 44(23):1–9
8. Clark D (2020) Papua New Guinea. In: Barnes P (ed) A pacific disaster prevention review. ASPI, pp 12–16
9. Anzellini V et al (2022) Disaster displacement: Papua New Guinea country briefing. IDMC and ADB, p 23
10. Auamoromoro M, Luma D (2023) Floods destroy gardens, displace 700 People (22 April 2020). The National 22 Apr. 2020. https://reliefweb.int/report/papua-new-guinea/flooding-finally-subsides. Accessed on 14 August 2023.

11. Campbell J (2014) Climate-change migration in the pacific. The Contemp Pacific 26(1):1–28
12. Edwards J (2013) The logistics of climate-induced resettlement: lessons from the Carteret Islands, Papua New Guinea. Refugee Surv Quart 32(3):52–78
13. Banks G (2008) Understanding 'resource' conflicts in Papua New Guinea. Asia Pacific Viewpoint 49(1):23–34
14. Jorgensen (1997) Who and what is a landowner? mythology and marking the ground in a Papua New Guinea mining project. Anthropol Forum 7(4):599–627
15. Kimiafa K (2017) A perspective from the Bougainville Islands, atolls, 5 May. The National. https://www.thenational.com.pg/a-perspective-bougainville-islands-atolls/. Accessed 10 July 2023
16. Connell J (2015) Vulnerable islands: climate change, tectonic change, and changing livelihoods in the Western Pacific. The Contemp Pacific 27(1):1–36
17. Nangoi D (2017) Article on Tulele Peisa disputed, 24 May. https://postcourier.com.pg/article-tulele-peisa-disputed/. Accessed on 5 May 2023
18. Ingipa J (2019) People told to take disaster seriously, 28 June. The National. https://www.thenational.com.pg/people-told-to-take-disaster-seriously/. Accessed on 13 June 2023
19. Rakova U, Soqo S (2021) Strengthening our ancestral community across the seas, 6 December. Cultural Survival. https://www.culturalsurvival.org/publications/cultural-survival-quarterly/strengthening-our-ancestral-community-across-seas. Accessed on 24 January 2024
20. Naser M (2016) Assessing the evidence: migration, environment and climate change in Papua New Guinea. IOM, p 95
21. Connell J, Lutkehaus N (2017) Environmental refugees? a tale of two resettlement projects in coastal Papua New Guinea' 48(1):85
22. Connell J, Lutkehaus N (2017) Escaping Zaria's fire? the volcano resettlement problem of Manam Island, Papua New Guinea. Asia Pacific Viewpoint 58(1):14–26
23. Wemuru S (2021) Growing rift between Manam Islander, locals, 17 February. The National. https://www.thenational.com.pg/growing-rift-between-manam-islanders-locals/. Accessed on 24 January 2024
24. Banks G (2006) Mining, social change and corporate social responsibility: drawing lines in the Papua New Guinea mud. In Firth S (ed) Globalisation, governance and the pacific islands. ANU Press, pp 259–274
25. Connell J, Lutkehaus N (2016) Another Manam? the forced migration of the population of Manam Island, Papua New Guinea, due to volcanic eruptions 2004–2005. IOM, p 87
26. Campos-Outcalt D, Kewa K, Thomason J (1995) Decentralization of health services in Western Highlands Province, Papua New Guinea: an attempt to administer health service at the subdistrict level. Social Sci Med 40(8):1091–1098
27. Howes S et al (2014) A lost decade? service delivery and reforms in Papua New Guinea 2002–2012. The National Research Institute, p 205
28. Komai M (2022) Proposed loss and damage finance facility agreed to at COP27 is 'too late' to compensate for more than 15years of suffering the Carteret Island, 24 November. https://pasifika.news/2022/11/carteret-island-needs-k14-million-to-move-350-families-by-2027/. Accessed on 24 January 2024
29. Thomason J, Mulou N, Bass C (1994) User charges for rural health services in Papua New Guinea. Social Sci Med 39(8):1105–1115

Dora Kuir-Ayius is the Deputy Dean (Academic and Student Affairs) for the School of Humanities and Social Science, University of Papua New Guinea. She is also a senior Social Work lecturer and a commissioner with PNG's Constitutional and Law Reform Commission. She attained her Ph.D. in Development Studies from Massey University, New Zealand, Masters in Social Work from La Trobe University, Australia, Postgrad Diploma in Education, and Bachelor of Arts in Social Work from The University of PNG. Her Ph.D. thesis titled, 'Building Community Resilience in Mine-Impacted Communities: A Study on Delivery of Health Services in Papua New Guinea' focused on building resilience in mining communities. Dora has over 30 years of teaching experience in PNG, with ten years in primary and secondary schools and 20 years in university. She has researched and analysed data on various issues, including gender-based violence, biodiversity conservation, socio-economic impact on road usage, child protection, and family sexual violence. She is also involved in research on the effects of climate change on families. She specialises in building resilience by developing community capital to achieve sustainable communities.

State Responsibilities and International Obligations in Responding to Climate Mobilities: What Should International Assistance Look Like?

Liam Moore

Abstract

Climate change-related events are already pushing people towards mobility, and it is likely this dynamic will only escalate as the effects of the climate crisis are more readily felt. Despite knowing this, before 2018, there were very few examples of state-level policies to manage climate mobilities. Unpacking the two path-leading policies designed by Fiji and Vanuatu in this space, we can see a significant recognition of the rights that individuals faced with mobility have and the obligations that states have to take to protect these rights and prevent harm. Although these policies are still in their infancy, we can already see significant barriers to implementation developing. In addition, considering the unequal distribution of responsibility for the drivers of climate change, there is a compelling legal and moral argument that the international community has a duty to step forward and assist states like these in discharging their responsibilities and managing climate mobilities effectively. In this chapter, I argue that the early experiences of Fiji and Vanuatu show us that the international community has a responsibility to help and I attempt to showcase what this responsibility should look like. It is my assertion that the international support must be locally led and tailored to local needs and context. Those who know how to best implement policies are already on the ground, so instead of parachuting in foreign experts, local actors should be supported and assisted with technical and financial support that builds their capacity.

1 Introduction

Climate change-related events are already pushing people towards mobility, and it is likely that this dynamic will only escalate as the effects of the climate crisis are more readily felt. This has been acknowledged in the international community for almost three decades.[1] Most of this movement will occur within states rather than across borders to neighbouring states. Despite knowing this for some time, it is only recently that states have begun to develop detailed domestic-level policies for specifically managing climate mobilities within their borders. Fiji and Vanuatu were among the first states to

L. Moore (✉)
International Politics and Policy,
James Cook University, Townsville, Australia
e-mail: liam.moore@jcu.edu.au

[1] 'The Changing Atmosphere: Implications for Global Security' (Secretariat of the World Meteorological Organization 1989) 20.

S. Jolly et al. (eds.), *Climate-Related Human Mobility in Asia and the Pacific*,
Sustainable Development Goals Series, https://doi.org/10.1007/978-981-97-3234-0_12

adopt these types of policies in 2018, and the Solomon Islands has since followed them in 2022. While these policies offer concrete pathways towards better management of climate mobilities by state-level actors, their implementation has been hamstrung by issues including financial limitations. Therefore, the question I explore in this chapter is how the international community can better fulfil its obligations to assist states like these to protect people within their borders from harm associated with climate mobilities. Under international law, it is clear that states have the primary responsibility to assist when people face harm related to current or future mobilities within their borders.[2] However, the international community also has an obligation to assist states in overcoming barriers to implementing climate mobilities management and protection plans. This is especially the case when the states in question have not significantly contributed to drivers of climate change.[3] I argue that this assistance should be designed to allow states to implement their policies in line with existing rights-based approaches by filling capacity gaps through financial and technical contributions. To do this, the assistance needs to be delivered in a way that is locally led and focused on the contextual needs of the recipients, not the donors.

I make this argument in four sections. The first section looks at the obligation of the international community to assist states in protecting the rights of climate-mobile individuals. Second, I look at how Fiji and Vanuatu have approached the management of climate mobilities and the barriers they have faced in implementing their policies. Thirdly, I pivot towards the role the international community can and should play in supporting states to fulfil their protection responsibilities around climate mobilities. I look at what should be done under existing international obligations and how this could be improved to better address the specificities of climate mobilities. Then, I argue that for this type of international assistance to be successful, it must be locally led and focused on the needs of the community and not the donors. Recognising agency and improving processes so that development and assistance partners work with, instead of acting for, these states and communities is key to ensuring those affected have the best chance at finding durable solutions to their situations of mobility. Finally, I offer policy prescriptions on how to develop and improve climate mobilities-related assistance. Throughout the chapter, I consciously use the term climate mobilities over alternatives like migration or displacement. This is because it better captures the diversity of the 'multiple forms, directions, and multiplicities of human movement in the context of climate change',[4] including immobility, relocation, and circular mobility.[5]

To piece together these narratives, I adopt a process tracing methodology. Within this, I supplement desk-based research with data from eight semi-structured elite-level interviews conducted with key policymakers through 2022 and 2023 in locations across Suva, Fiji, as well as remotely.[6] The interview participants were initially identified through desk research, and the pool was expanded through methods of snowballing. The University of Wollongong Ethics Committee granted the ethics approval, and all participants have agreed to be identified at various levels in this research.[7]

---

[2]Ferris [1], McAdam and Ferris [2], Bellamy [3], UN High Commissioner for Human Rights [4]; UN General Assembly, Resolution 46/182 1991 [46/182 (paragraphs 3–5)].

[3]Annual Report of the United Nations High Commissioner for Human Rights and Reports of the Office of the High Commissioner and the Secretary-General: Report of the Office of the United Nations High Commissioner for Human Rights on the relationship between climate change and human rights 2009 [A/HRC/10/61] paras 84–7; 'United Nations Framework Convention on Climate Change' (United Nations 1992) FCCC/INFORMAL/84 Articles 3–4.

[4]Boas et al. [5].

[5]Boas et al. [6].

[6]Gusterson [7].

[7]Ethics was granted on 13/September/2022 under Ethics number 2022/230 from the University of Wollongong's Human Ethics Committee.

2 The Protection and Assistance Obligations of States and the International Community Around Climate Mobilities

Two fundamental aspects of international law form the foundations of this chapter. The first is that individuals have basic rights.[8] Beyond basic rights, international human rights treaties and their guidance note that specific groups of people are owed additional protections based on their needs and circumstances.[9] For instance, people forced to move can claim rights and protections due to displacement.[10] This is applicable in the case of cross-border displacement due to 'persecution' in the form of refugee protection, adjacent types of protection for conflict displacement, and other forms of unrest.[11] However, these protections are not limited to those who cross an international border; in fact, those forced to move within their borders—whether due to conflict, environmental hazards, disasters, or a combination thereof—can also claim specific protections related to their circumstances.[12]

This is important since most climate-related movements will be internal—within countries. Therefore, it is the state that holds the primary responsibility for protection.[13] States' protection responsibilities have been explicitly identified as including disasters and environmental hazards.[14] Further, these hazards need not be experienced, but merely foreseeable for these responsibilities to apply.[15] So, while there are no specific global treaties or laws on protection in the context of climate mobilities, as McAdam has argued, we already have clear frameworks from international human rights and humanitarian law that could guide our actions in this space.[16] We are generally missing the specifics of applying these rights and obligations in the context of climate mobilities. In this section, I offer a broad idea of the direction this should take, focusing on the international community's responsibilities and financial assistance.

[8] 'Universal Declaration of Human Rights' (United Nations 1948) A/RES/217(III). https://documents-dds-ny.un.org/doc/RESOLUTION/GEN/NR0/043/88/PDF/NR004388.pdf?OpenElement. Accessed 17 July 2023; 'International Covenant on Civil and Political Rights' (UN General Assembly 1966) 14668. https://treaties.un.org/doc/publication/unts/volume%20999/volume-999-i-14668-english.pdf. Accessed 17 July 2023; Human Rights Committee, 'General Comment No. 6: The Right to Life (Art 6)' (United Nations Human Rights Committee 1982) para 1; Pictet [8], Shue [9].

[9] 'Convention on the Rights of the Child' (UN General Assembly 1989) 44/25. https://www.ohchr.org/sites/default/files/crc.pdf. Accessed 17 July 2023; 'Convention on the Rights of Persons with Disabilities' (UN General Assembly 2007) A/RES/61/106; 'Convention on the Elimination of All Forms of Discrimination against Women' (UN General Assembly 1979). https://www.ohchr.org/sites/default/files/cedaw.pdf. Accessed 17 July 2023; 'United Nations Declaration on the Rights of Indigenous Peoples' (UN General Assembly 2007) A/61/L.67 and Add.1. https://social.desa.un.org/sites/default/files/migrated/19/2018/11/UNDRIP_E_web.pdf. Accessed 17 July 2023; 'United Nations Principles for Older Persons' (UN General Assembly 1991). https://social.un.org/ageing-working-group/documents/fourth/AWAZUNprinciplesforolderpersons.pdf. Accessed 17 July 2023; UN High Commissioner for Refugees [10], African Union [11]; Kälin [12]; Colloquium on the International Protection of Refugees in Centra America, Mexico, and Panama, Cartagena Declaration on Refugees 1984; McAdam [13], Pérez [14].

[10] UN High Commissioner for Refugees (n 9); Kälin (n 9); African Union (n 9); Colloquium on the International Protection of Refugees in Central America, Mexico, and Panama, Cartagena Declaration on Refugees 1984 (n 9).

[11] McAdam (n 9) 6.

[12] Kälin (n 9); Oloka-Onyango [15].

[13] UN General Assembly Resolution 46/182 (n 2) paras 3–4; UN High Commissioner for Human Rights (n 2) 226; Ferris (n 2) 66; Thakur [16]; Bellamy (n 2) 619; McAdam (n 9) 6.

[14] UN General Assembly Resolution 46/182 (n 2) paras 3–4; *Case of Öneryildiz v Turkey* [2004] European Court of Human Rights 48939/99 [71–2]; Kälin (n 9); International Law Commission [17].

[15] Human Rights Committee, 'Views Adopted by the Committee under Article 5 (4) of the Optional Protocol, Concerning Communication No. 2728/2016' (United Nations Human Rights Committee 2020) CCPR/C/127/D/2728/2016 para 9.4; *Case of Budayeva and Others v Russia* [2008] European Court of Human Rights 15339/02, 21166/02, 20058/02, 11673/02 and 15343/02 [128–9, 135]; Foster and McAdam [18].

[16] McAdam [19].

A. The Responsibilities of the International Community

While states may hold the primary responsibility to uphold the rights of those within their borders, they are not the sole duty bearers. The specific complexities and dynamics of climate change and climate mobilities add weight to the argument that the international community has a significant role in supporting states and stepping in when they are unable to adequately fulfil their responsibilities. Three broad points can be raised to support this proposition.

Firstly, almost all states have committed to some level of emissions reduction at various UN Framework Convention on Climate Change (UNFCCC) and Conference of the Party (COP) meetings in recent years.[17] Mitigation measures like these are the foundational aspects of reducing risk for climate-vulnerable communities. The process was first crystallised at the 1992 UN Conference on Environment and Development in Rio, which produced the UNFCCC and the Rio Declaration. Principle 16 of the Rio Declaration states that polluting states should bear the costs associated with pollution.[18] Article 4 of the UNFCCC states that developed countries should provide financial assistance to developing countries to help them limit their emissions. Article 4 of the UNFCCC also imposes individual obligations on all state parties to mitigate climate change by limiting anthropogenic greenhouse gas emissions.[19]

Mitigation actions are a collective obligation for the international community to act in concert to reduce emissions and limit the drivers of climate change where possible. However, not every state has the same level of obligation. Article 3 of the UNFCCC creates the idea of common but differentiated responsibility, where some states will need to shoulder more of the burden and responsibility than others.[20] These collective but differentiated obligations were further spelled out by the 1997 Kyoto Protocol,[21] and the 2015 Paris Agreement.[22] Article 9 of the Paris Agreement states that developed countries 'shall provide financial resources to assist developing country Parties with respect to both mitigation and adaptation', whilst also continuing 'to take the lead in mobilising climate finance from a wide variety of sources, instruments and channels'. However, the agreement only required developed countries to communicate projected levels of available finances and keep developing countries informed on the support on offer.[23] There were no binding commitments to ensure developed states committed significant assistance measures—financial or otherwise.

Paragraph 14 (f) of the Cancun Adaptation Framework was the first UNFCCC document to explicitly link issues of climate change and 'displacement, migration, and planned relocation', although it also stops short of assigning obligations and instead 'invites' states to enhance action whilst 'taking into account their common but differentiated responsibilities and respective capabilities'.[24] While discussions of climate mobilities in the forum of the UNFCCC have not yet consolidated around concrete outlines

[17] 'Paris Agreement' (UN Framework Convention on Climate Change 2015), article 4. https://unfccc.int/files/essential_background/convention/application/pdf/english_paris_agreement.pdf. Accessed 15 July 2023.

[18] UN General Assembly, Report of the United Nations Conference on Environment and Development (Rio Declaration on Environment and Development 1992, para 16.

[19] 'United Nations Framework Convention on Climate Change' (n 3), Article 4.

[20] Lee [20].

[21] Meguro [21], Hermwille et al. [22]; Lee (n 21) 77; United Nations, 'Kyoto Protocol to the United Nations Framework Convention on Climate Change' (UN Framework Convention on Climate Change 1998). https://unfccc.int/resource/docs/convkp/kpeng.pdf. Accessed 12 September 2022.

[22] 'Paris Agreement' (n 17); Meguro (n 21) 943.

[23] 'Paris Agreement' (n 17) article 9.

[24] UN Framework Convention on Climate Change, 'Report of the Conference of the Parties on Its Sixteenth Session, Held in Cancun from 29 November to 10 December 2010' (UN Framework Convention on Climate Change 2011) UNFCCC/CP/2010/7/Add.1 para 14(f). https://unfccc.int/resource/docs/2010/cop16/eng/07a01.pdf. Accessed 15 July 2023.

of state obligations, existing understandings of the protection obligations of both states and the international community in the context of climate change and human rights are already well established.

Secondly, the basis for international obligations extends far beyond the remit of the UNFCCC. In a report to the UN General Assembly, the Office of the High Commissioner for Human Rights noted that while states have the primary obligation to protect and uphold human rights where they are impacted by climate change,[25] this exists alongside an obligation for all members of the international community to cooperate to address these issues effectively.[26] Drawing on the precautionary principle, they confirmed these obligations could extend across time to encompass questions of intergenerational equity as well.[27] Therefore, there are two layers of obligations here. While states hold the responsibility of primary protection, the international community must act to assist states and avoid harm wherever it may occur across time and space. Bringing this into conversation within the climate context, the Inter-Agency Standing Committee's Operational Guidelines on the Protection of Persons in Situations of Natural Disasters suggest that a disaster can become an issue of international concern if the capacity of the authorities to fulfil their responsibilities 'is/are insufficient'.[28] In this case, it is up to the international community to step in and assist.

[25] Annual Report of the United Nations High Commissioner for Human Rights and Reports of the Office of the High Commissioner and the Secretary-General: Report of the Office of the United Nations High Commissioner for Human Rights on the relationship between climate change and human rights (n 3) paragraphs 72–4.

[26] Ibid., 84–7.

[27] Ibid., 89–91.

[28] Inter-Agency Standing Committee, 'IASC Operational Guidelines on the Protection of Persons in Situations of Natural Disasters' (Brookings—Bern Project on Internal Displacement 2011) 6. https://www.ohchr.org/Documents/Issues/IDPersons/OperationalGuidelines_IDP.pdf. Accessed 28 August 2020.

Thirdly, adding to the transnational nature of the problem, when people are forced into mobility due to climate-related reasons, their decisions to move are often driven by the localised effects of a broader international crisis. Climate change is a global process that members of the wider international community are responsible for. It follows that those most responsible should be, in some way, accountable for the emissions that have led to these climatic changes and have a particular obligation to assist those states that have contributed the negligible emissions, to protect their populations from environmental hazards.[29] States with the least capacity and who have contributed the least to the root problem should not have to face these issues alone; cooperation from high-emitting states is essential.[30] Cooperation, in this case, extends beyond helping with immediate disaster response. Contributions must be made towards long-term recovery, adaptation, and risk reduction projects, as less-developed states that will be most affected by climate-related hazards often lack the financial and technical capacity to fully implement adaptation policies on their own—particularly at the scale and speed required.[31] So, while the mobilities may be internal, and states have the primary responsibility for protecting those within their jurisdiction, the transnational nature of climate change and the varied capacities and differentiated responsibilities of states means that the international community does have an obligation to assist states in discharging their protection obligations. The international community, particularly higher-emitting states, has a legal and normative obligation to assist least-emitting states to fulfil their protection obligations towards their citizens, prevent harm, and provide compensation when harm is unavoidable. One form these obligations manifest in is the need to offer financial assistance to developing states to help them adapt to the realities of climate change.

[29] Cohen and Bradley [23].

[30] Farbotko et al. [24].

[31] Charan et al. [25].

B. International Financing Mechanisms

Having states acknowledge and fulfil these obligations has proven extremely difficult, however. Vanuatu and the Alliance of Small Island States campaigned vigorously over a long period for a global fund and insurance pool based on the Polluter Pays principle to be established during negotiations to develop the UNFCCC.[32] However, it was not until 2007 that the term 'loss and damage' was referenced in a UNFCCC document.[33] From this point, it was another five years until the Warsaw International Mechanism on Loss and Damage (WIM) was established. The WIM has a mandate to enhance 'action and support, including finance, technology, and capacity-building, to address loss and damage associated with the adverse effects of climate change'.[34]

The WIM was given a durable legal basis in Article 8 of the 2015 Paris Agreement.[35] However, while provisions for the WIM to be strengthened in the future were included, paragraph 52 of the decision accompanying the Paris Agreement specifically stipulates that Article 8 'does not involve or provide a basis for any liability and compensation'.[36] As a result, there is a great degree of ambiguity in both the mandate and how loss and damage was framed in discussions. The concerns of developed states resulted in the terms 'liability' and 'compensation' being excluded from the document as well.[37] However, despite this, developing states saw the establishment of a process that could address loss and damage from climate change as a significant breakthrough.[38]

After 30 years of campaigning, COP27 in 2022 saw another breakthrough with an international agreement on establishing a global fund to pay for climate-related loss and damage. However, the Conference had been preceded by a long-drawn struggle to get loss and damage on the agenda since the United States and other wealthy large emitters had attempted to block negotiations. It was only after marathon negotiations into the wee hours on the first day of the Conference that developing countries and small island states succeeded in getting loss and damage onto the agenda.[39] Therefore, gaining a formal agreement to develop a loss and damage fund is an extremely significant success.[40] It was effectively admitted that prior efforts had failed to prevent climate change, and that adaptation now has to sit alongside mitigation on the agenda.[41] The caveat here is that significant questions remain about the form the fund will take, the amount of funds required, who will pay into the fund, whether contributions will be voluntary or mandatory, who is entitled to claim compensation from the fund, and what even counts as loss and damage in this context.[42] Additionally, while an agreement was finally reached on beginning the process to establish this fund, China's initial response was that they would not be contributing to the fund. Key petrostates also pushed back on the idea of being compelled to contribute.[43]

As Harris argues, the idea of the fund is 'ripe with possibility', but there is a real danger that

[32] 'United Nations Framework Convention on Climate Change' (n 3) article 4.8; Wewerinke-Singh and Salili [26].

[33] 'Bali Action Plan' (UN Framework Convention on Climate Change 2007) Conference of the Parties UNFCCC/CP/2007/6/Add.,1 article 1(c) (iii). https://www.preventionweb.net/files/8376_BaliE.pdf. Accessed 15 July 2023.

[34] 'Report of the Conference of the Parties on Its Eighteenth Session, Held in Doha from 26 November to 8 December 2012' (UN Framework Convention on Climate Change 2013) UNFCCC/CP/2012/8/Add.1 article 5(c). https://unfccc.int/sites/default/files/resource/docs/2012/cop18/eng/08a01.pdf. Accessed 15 July 2023; Wewerinke-Singh and Salili (n 32) 4.

[35] 'Paris Agreement' (n 17) article 8.

[36] Wewerinke-Singh and Salili (n 32) 4.

[37] Wewerinke-Singh and Salili (n 32) 4.

[38] Ibid.

[39] Peel [27], Donald [28].

[40] McDonald [29], Lillo et al. [30].

[41] Harris [31].

[42] Ibid.; McDonald (n 39).

[43] Lillo and others (n 39).

larger states will stonewall or slow roll the process to avoid financial liabilities.[44] As in the Paris Agreement, the final text of the agreement on loss and damage again avoided using terms like liability or compensation—in this case largely due to campaigning from the United States. Additionally, by 2023, there is only a commitment to establish a transitional committee that will make recommendations for consideration at COP28 in 2023. While recent progress is positive, there is still the chance this could turn into another 'placebo fund' that never eventuates at the promised or required scale.[45]

Vanuatu's submission to the Executive Committee of the WIM (Warsaw International Mechanism) on loss and damage finance adds specificity to how they believe the fund should function. A range of activities that should be covered by loss and damage are outlined in the proposal, including the costs of relocating coastal communities; reconstruction after extreme weather; a range of social programmes including those around gender equality and livelihood transformation and protection; subsides for insurance premiums; emergency financial and contingency planning support for governments; and technological cooperation and transfer.[46] To strengthen their claims, Vanuatu has successfully petitioned the International Court of Justice (ICJ) to issue an advisory opinion to clarify the nature and scope of obligations industrialised nations have for climate change and resulting loss and damage.

This campaign took an important step forward on 29 March 2023, when the UN General Assembly adopted a resolution by consensus to request an advisory opinion from the ICJ on states' obligations regarding climate change. The resolution was co-sponsored by more than 120 countries; however, the US was notable for its absence as a co-sponsor and the reservations it expressed during discussions. Despite this, the motion was passed with unanimous state support—the first time this has happened for a General Assembly request for an advisory opinion.[47] While advisory rulings are non-binding, the findings will shape how states understand their international obligations around climate change.[48]

The two recent developments, of COP27's loss and damage agreement and the successful request for an ICJ advisory ruling, show that there is a growing understanding and acknowledgment that those states that have the ability and have contributed the most to driving climate change, are obligated to assist and compensate those smaller states who are bearing the worst of a problem they did not create. However, precisely what these obligations should look like is still in question. The ICJ's opinion is awaited, and there is a lack of clarity around how the loss and damage fund will be structured. Therefore, I offer a proposal of how this assistance should look. This is based on the lessons learnt from two of the first states to develop policies on climate-related mobilities—Fiji and Vanuatu. Beyond the lessons these cases offer for other states developing their own climate mobilities policies, I argue that they also show how the international community needs to step forward. Assistance must proceed in a way that supports states in protecting the rights and futures of those at risk of mobilities—it must be locally led and driven by the context-specific needs of communities rather than the interests of donors. With this objective, the following section evaluates the policy initiatives in Fiji and Vanuatu.

[44] Harris (n 40).

[45] Najam [32].

[46] Wewerinke-Singh and Salili (n 32) 2.

[47] UN Press, 'General Assembly Adopts Resolution Requesting International Court of Justice Provide Advisory Opinion on States' Obligations Concerning Climate (**GA/12497** 29 March 2023 /03/20023). https://press.un.org/en/2023/ga12497.doc.htm. Accessed 10 April 2023; Lakhani [33], Peel and Nay [34].

[48] Peel and Nay (n 46).

3 Pioneering Approaches to Climate Mobility Management: Fiji and Vanuatu

A. Fiji

Fiji is comprised of more than 330 islands and 500 islets, with around 110 of the islands permanently inhabited. Rather than a small island nation, Fiji and other Pacific Island states have re-framed themselves as large ocean nations in the last decade.[49] More than 98% of Fiji's 1.3 million square kilometre territory is comprised of ocean.[50] Even on land, the majority of Fiji's approximately 850,000 citizens live in coastal areas, with 76% living within 5 km of the coast and 27% within 1 km.[51] These characteristics and the broader location of Fiji leave the state extremely exposed to the effects of climate change. Since 1992, the ocean around Fiji's islands has been rising at an average of around 5.5 mm per year, almost twice the global average.[52] This trend may be accelerating. In 2019, the Intergovernmental Panel on Climate Change stated that sea levels in the Western Pacific Ocean were rising at around three times greater than the global mean.[53] This shows that climate change poses a clear material threat to Fiji and its people.

The 2014 relocation of Vunidogoloa is now widely accepted as the world's first state-sponsored climate change-related community relocation project.[54] This was the first of six relocations the Fijian Government has been involved in, with another three due in 2023.[55] Legislative and policy development has supported these practices. Recent developments have included the 2018 Planned Relocation Guidelines,[56] the 2019 Displacement Guidelines in the Context of Climate Change and Disasters,[57] the 2019 Climate Relocation and Displaced Peoples Trust Fund for Communities and Infrastructure,[58] and the 2021 Climate Change Act, which writes these ambitious climate-related policies into law.[59] However, despite an abundance of will to implement positive protective practices, the projects have been hamstrung by problems. Interviews with government officials have shown that the most significant of these issues is often finding adequate financing to carry out projects at the scale required and in the ways outlined in these key policy documents.

One example of the complexity of the problems faced in implementing these policies is the proposed relocation of the village of Nabukadra. This was one of the initial 42 candidates identified by the Government for relocation. Heavily impacted by rising sea levels and the Tropical Cyclone Winston in 2016, plans were undertaken to relocate the community to a new hillside site. However, a protracted and fractured decision-making process meant that even by 2019, only a few houses had moved. A lack of available financing from the government or donors was a key barrier to relocation here.

[49] Chan [35], Morgan [36].

[50] Fiji has a land area of 18,270 km^2 and an Exclusive Economic Zone of 1,289,978 km^2; Hume et al. [37].

[51] Andrew et al. [38].

[52] Martin et al. [39].

[53] Church et al. [40].

[54] Tronquet [41], McNamara and des Combes [42].

[55] Interview with Lebaiatelaite Gaunavinaka, Fiji Climate Change Division, 'Interviewed by Liam Moore' (13 January 2023). The six relocations include the initial relocation of Vunidogoloa, alongside Vunisavisavi, Denimanu, Nagasauva, Narikoso, and Tukuraki.

[56] Ashna Kumar, 'PM Launches Our Adaptation Plan at COP24' *Fiji Sun* (Suva, Fiji, 15 December 2018). https://fijisun.com.fj/2018/12/15/pm-launches-our-adaptation-plan-at-cop24/. Accessed 24 April 2019.

[57] Sarina Theys, 'COP25 and Pacific Island States: "We Are Keen to Lead, Not to Be Led"' (*International Affairs Blog*, 21 October 2020). https://medium.com/international-affairs-blog/cop25-and-pacific-island-states-we-are-keen-to-lead-not-to-be-led-ced74ece0f4e. Accessed 29 October 2021.

[58] Fiji Ministry of Communications, 'World's First Ever Relocation Trust Fund for People Displaced by Climate Change Launched by Fijian Prime Minister' (Ministry of Communications 2019). https://www.fiji.gov.fj/Media-Centre/News/WORLD%E2%80%99S-FIRST-%E2%80%93EVER-RELOCATION-TRUST-FUND-FOR-PEOP. Accessed 29 October 2021.

[59] Republic of Fiji, Climate Change Act 2021.

The households that did move were restricted to only rebuilding slightly further up the hillside, where land could be cleared by hand. Without the investment and technical expertise needed to remove and level land further up the hillside, the community was limited to rebuilding within the boundaries of their existing village.[60]

Since 2016, issues of funding and capacity have persisted. In 2023, two of the three relocations slated to occur have faced significant financing issues. For the relocation of the community of Cogea, the Fijian Government did not have the funding to implement the project. Instead, the project will be implemented by the Fijian Council of Social Services, a civil society organisation, and the funding is coming from the Christian charity, Bread for the World.[61] Interviews held with officials at the Fiji Climate Change Division revealed that while the Government would have preferred to conduct the relocation themselves, they are grateful for the support, as the project is unlikely to have been funded in the immediate future.[62] It was also stated that funding had not been secured for the second state-led planned relocation scheduled for 2023, and that the project is currently stalled. The third relocation was allocated funding, but was then held up for almost three years due to difficulties in finding suitable land and securing its rights. As of May 2023, however, a site has been secured and funding is in place.[63] Officials estimate these projects to cost around FJ$1 to 4 million (USD$ 440,000 to 1.75 million) each to implement,[64] with projects taking between 18 and 24 months to complete in best-case scenarios.[65]

Compounding these issues is the knowledge that over 40 other communities are on a list to be relocated, however, currently the government team that assesses and plans for relocations can only manage one project at a time as they are a three-person team and simply do not have the bandwidth to manage more than this.[66] Government officials did confirm, however, that in 2023, in partnership with the Platform on Disaster Displacement, the Climate Change Division had hired a new staff member in the role of embedded climate finance advisor. This is part of the Project to Avert, Minimise, and Address Displacement Related to the Effects of Climate Change and is funded through the Norwegian Agency for Development Cooperation. The advisor's role is focused on donor coordination and increasing the profile of the trust fund to increase access to more sources of funding.[67] It is hoped this can help address some funding shortfalls and increase capacity to implement solutions in the future.

Fijian officials confirmed that a key part of making Fiji a more attractive prospect for donors and financiers has been the move to formalise climate-related policies and guidelines.[68] Perhaps the most important recent development was the approval of the Standard Operating Procedures (SOPs) associated with the Relocation Guidelines in March 2023. The SOPs offer extremely detailed operational guidelines on how the principles from the relocation guidelines should be applied during all stages of implementation—including identification, planning, relocation, and long-term

[60] Nichols [43].

[61] Vakasukawaqa [44]; Interview with Filimone Ralogaivau, Fiji Climate Change Division, 'Interviewed by Liam Moore' (3 September 2023).

[62] Interview with Filimone Ralogaivau, Fiji Climate Change Division (n 60).

[63] Interview with Lebaiatelaite Gaunavinaka, Fiji Climate Change Division (n 54).

[64] Interview with Christine Fung, 'Interviewed by Liam Moore' (30 January 2023); Interview with Lebaiatelaite Gaunavinaka, Fiji Climate Change Division (n 54). Christine Fung estimated between FJ$ 2 and 4 million, with Lebaiatelaite Gaunavinaka saying the government was currently budgeting for around FJ$1 million for a standard relocation.

[65] Interview with Christine Fung (n 63).

[66] Interview with Lebaiatelaite Gaunavinaka, Fiji Climate Change Division (n 54).

[67] Interview with Filimone Ralogaivau, Fiji Climate Change Division (n 60); Interview with Lebaiatelaite Gaunavinaka, Fiji Climate Change Division, 'Personal Communication' (6 July 2023).

[68] Interview with Filimone Ralogaivau, Fiji Climate Change Division (n 60).

recovery.[69] Those in the Government see the SOPs as key to attracting donations to the trust fund.[70]

Financial concerns were a key driver behind Fiji's establishment of the Climate Relocation and Displaced Peoples Trust Fund in 2019. Described as a mini loss and damage fund for Fiji,[71] it is financed partially through their Environment and Climate Adaptation Levy and partly through voluntary donations.[72] The fund's primary goal is to provide the necessary financial resources to relocate communities; however, these resources can also be re-purposed to provide emergency assistance in the case of acute-onset displacement events.[73] New Zealand was the first state to interact with the fund, pledging $US2 million in 2020 as part of a broader $150 million package of climate change assistance.[74] Interviews have shown that this pledge was conditional, with a fraction of this funding released to support the development of Fiji Standard Operating Procedures for the Planned Relocation Guidelines.[75] The rest of the funding is conditional on these procedures being approved by Fiji's Parliament. Additionally, no other state has committed to contributing to the fund, leading some government officials to express scepticism over whether the fund will be able to solve the financial barriers to policy implementation on its own. Other officials, however, have noted that regional actors like Australia are likely to donate once these procedures are approved.[76] One interviewee stated that other traditional regional donors were following the lead of New Zealand. Once the new Parliament formally endorses the SOPs, it would increase the likelihood of more donations.[77]

As officials said in interviews, however, outcomes are generally better when the funding has fewer strings attached. This is one of the drivers behind the SOPs and the Trust Fund approach, as it is designed to reassure donors that Fiji is not looking for a 'blank check', but has 'a very advanced structure for using funding for loss and damage'.[78] Establishing good knowledge and trust between donors and those implementing the policies allows for a greater degree of agency and more context-specific solutions to be developed. This is key for any frameworks developed around international obligations on loss and damage and climate financing—states must be able to use the funds in ways that apply to their circumstances and will provide the most context-appropriate solutions. As Fiji shows, having competent governance and political will is sometimes not enough—financial and technical capacity is also vital to effectively implement policies at the required scale.

B. Vanuatu

Fiji's Pacific Island neighbour Vanuatu is another state both living with the reality of climate change and taking world-leading steps to prepare for and manage climate mobilities. While Fiji has received international recognition for its policies and relocation practices, Vanuatu published its key climate mobilities policy some months before Fiji, in 2018.[79] Vanuatu's policy action on climate mobilities, in addition to their long campaign to compel higher polluting

[69] Climate Change Division [45].

[70] Interview with Filimone Ralogaivau, Fiji Climate Change Division (n 60).

[71] Ibid.

[72] Climate Relocation of Communities Trust Fund Act 2019 [21] Clause 12.

[73] Lund [46].

[74] Goering [47].

[75] Interview with Filimone Ralogaivau, Fiji Climate Change Division (n 60); Interview with Lebaiatelaite Gaunavinaka, Fiji Climate Change Division (n 54); Interview with Daniel Lund, Special Advisor to Government of Fiji, 'Interviewed by Liam Moore' (12 August 2022); Interview with Christine Fung (n 63).

[76] Interview with Filimone Ralogaivau, Fiji Climate Change Division (n 60).

[77] Interview with Interviewee 1, 'Interviewed by Liam Moore' (13 March 2023).

[78] Interview with Daniel Lund, Special Advisor to Government of Fiji (n 74).

[79] Vanuatu National Disaster Management Office [48], Ferris [49]; Wewerinke-Singh and Van Geelen [50].

states to pay for the loss and damage caused by climate change, leave them well placed to be a global leader on climate change and climate mobilities.

Vanuatu is consistently ranked the most hazard-prone state in the world by the World Risk Index.[80] In practice, this means that around half of the population of Vanuatu's 80 islands is impacted by extreme climate-related or geo-hazard events each year.[81] Constantly responding to common and recurring hazards stretches resources and limits the amount of risk reduction, preparation, and long-term focused work that can be carried out—authorities rarely have time to plan for the future because they are constantly responding to hazards.[82] As Minister of Foreign Affairs, International Cooperation, and External Trade, Ralph Regenvanu said at the COP24, the effects of climate change on top of existing hazards have pushed Vanuatu into a 'constant state of emergency'.[83]

One key example of the risks Vanuatu faces from major events was the impact of the Tropical Cyclone Pam in 2015. At the time, it was the most potent recorded cyclone to hit Vanuatu. Over 150,000 people are estimated to have been affected—over half the population—and eighty per cent of the national housing stock was damaged or destroyed.[84] This was a key trigger for developing the National Policy on Climate Change and Disaster-Induced Displacement (NPCCDID).[85] One expert noted specifically that the combined influence of Cyclone Pam and a volcanic eruption on the island of Ambrym, crystallised the need for a formal displacement policy and pushed policymakers into accelerating the development process.[86]

Developed with assistance from the International Organisation for Migration (IOM), the NPCCDID represents 'one of the world's most progressive policies on climate-driven displacement'.[87] The Minister of Climate Change Adaptation at the time, Ham Lini Vanuaroroa, said that due to the NPCCDID, Vanuatu could mitigate challenges and 'prepare, plan, and respond to the short term and long term needs of displacement'.[88] The IOM's Chief of Mission, Dr Lesikimakuata Korovavala noted during a 2017 consultation that 'Vanuatu is among the first countries in the world to prepare such a policy and is venturing into new territory with this initiative'.[89] Broadly, the policy:

> Aims to help guide emergency and development planners to work together with the Government of Vanuatu to address the needs of all communities affected by displacement, including people at risk of displacement, displaced people, internal migrants, people living in informal settlements, and host communities.[90]

The policy plans to achieve this by working towards minimising the drivers of displacement, minimising the negative impacts of displacement when it does occur, striving for durable solutions for displaced populations, and ensuring displaced persons can make voluntary informed choices about their futures and participate in the planning for solutions; facilitate well-managed and safe migration; promote access to affordable, disaster-resilient housing; and integrate human mobilities concerns into key policy areas around climate, disaster, environment, health, education, land, housing, infrastructure, planning, food, and livelihoods security.[91]

---

[80] McDonnell [51].

[81] Ibid.

[82] Wewerinke-Singh and Van Geelen (n 79) 27.

[83] Wewerinke-Singh and Salili (n 33) 3.

[84] Bolo et al. [52].

[85] Vanuatu National Disaster Management Office (n 79).

[86] Vinke et al. [53]; Bolo and others (n 84) 8; Interview with Christopher Bartlett, Special Advisor to Government of Vanuatu, 'Interviewed by Liam Moore' (20 December 2022).

[87] Wewerinke-Singh and Van Geelen (n 79) 17.

[88] International Organisation for Migration, 'Vanuatu Launches National Policy on Climate Change and Disaster-Induced Displacement' (2018) Press Release. https://www.iom.int/news/vanuatu-launches-national-policy-climate-change-and-disaster-induced-displacement. Accessed 12 July 2022.

[89] Wewerinke-Singh and Van Geelen (n 78) 17.

[90] Vanuatu National Disaster Management Office (n 78) 7.

[91] Ibid., 18.

While the policy itself is well regarded, both internally and externally, implementation has lagged.[92] Interviewees stated that a lack of knowledge and awareness across government sectors and a lack of ownership from the implementing agency have meant that the implementation of this policy has slipped off the radar domestically.[93] Additionally, officers from the National Disaster Management Office have been quoted as noting the lack of adequate staffing, training, and expertise within Government ranks. They expressed their concern over the difficulties they faced in implementing these policies.[94] An academic summarised the situation by noting that while Vanuatu's legal and policy framework is 'very strong… there is limited capacity to implement laws and policies'.[95] Several others from government agencies and non-government organisations in Vanuatu have expressed similar concerns, with one government employee saying, 'what is written on the policy is very good, but we lack the resources to implement it'.[96] The NPCCDID itself notes that without 'adequate mobilisation of financial and technical resources', implementation will be constrained.[97]

There are two levels to this problem, according to Jane McAdam. On the one hand, states like Vanuatu, which contract out the development of their policies, are participating in quite a wasteful process. The tender process tends to result in a lot of duplication, is extremely expensive, and takes additional time. Secondly, bringing in foreigners to develop policies often means they approach things from an outside and top-down perspective. Trying to write the perfect policy without sufficient consideration can result in policies that are too abstract to implement, 'where things become so high-level, they are almost meaningless… [and they are] completely disconnected with what is needed on the ground'.[98] These issues are widely recognised among those experienced in the space. Cosmin Corendea, the initial drafter of the Fijian Planned Relocation Guidelines, said that the goal must be for states 'to have a policy which is not sitting in a drawer or is just a nice policy on your website…. [but is] taken by the people on the ground and implemented'.[99]

Even within state-level institutions, however, there needs to be knowledge and ownership from those who will be implementing the policies. An interviewee working in the Vanuatu Department of Strategic Policy Planning and Aid Coordination, based out of the Prime Minister's Office, noted that there were significant implementation issues with the NPCCDID. During development, the policy was under the remit of the Ministry of Climate Change and Disaster Risk Reduction, however, it was decided that the Ministry of Internal Affairs would instead manage the implementation. Due to the lack of coordination here, as well as a lack of physical infrastructure, human resources, and finances, 'the implementation hasn't begun', the 'policy hasn't been functioning', and it is currently 'still sleeping'. Despite the attention this flagship policy has received at a regional and international level, the official stated that 'we haven't done much work on it, in terms of its implementation'.[100]

However, in spite of the difficulties in its implementation, the NPCCDID has been key to Vanuatu's campaign to attract funding and financial support. A government official described it as a useful tool 'to attract more funding in

[92] Vinke and others (n 85) 95; Yamamoto [54]; Bolo and others (n 83) 8; McDonnell (n 79) 286.

[93] Interview with Christine Fung (n 63); Interview with Zoe Ayong, National Disaster Management Office of Vanuatu, 'Interviewed by Liam Moore' (30 January 2023).

[94] Wewerinke-Singh and Van Geelen (n 78) 27.

[95] Vinke and others (n 85) 95.

[96] Ibid.

[97] Vanuatu National Disaster Management Office (n 78) 47.

[98] Interview with Jane McAdam, Academic and Consultant, 'Interviewed by Liam Moore' (28 September 2022).

[99] Interview with Cosmin Corendea, Academic and Consultant, 'Interviewed by Liam Moore' (27 October 2022).

[100] Interview with Zoe Ayong, National Disaster Management Office of Vanuatu (n 92).

terms of adaptation and mitigation'.[101] Much of Vanuatu's recent efforts in the climate finance space have been directed at garnering support for a loss and damage mechanism to compensate and support smaller developing states to adapt to and recover from the effects of climate change. Calls for an insurance mechanism, or a 'collective loss-sharing scheme to compensate victims of projected sea-level rise' have long been a key pillar of ni-Vanuatu foreign policy and the broader climate diplomacy of small island states.[102] It was Vanuatu in 1991 that tabled the proposal on behalf of the Association of Small Island States (AOSIS) to develop a climate funding mechanism.[103] This highlights how recent successes around loss and damage have not come out of the blue but were the culmination of decades-long advocacy from states like Vanuatu. The long-term focus from Vanuatu has been so strong that even when signing onto the Paris Agreement in 2015, a note was made that this in no way constituted a renunciation of any of Vanuatu's rights to claim compensation for loss and damage caused by larger polluting states and entities.[104]

Vanuatu recently put a specific price tag on the costs of climate adaptation and loss and damage domestically. In their Nationally Determined Contribution (NDC) to the UNFCCC, authorities outlined that they expected costs around loss and damage alone to run to over USD$177.5 million by 2030, with the total cost of achieving their revised NDC at over USD$1.2 billion. It is noted that Vanuatu's 'limited national resources and technical capacities for scaling up climate action' will need to be bolstered through financial, technological, and capacity-building support from global funds, bilateral and multilateral agencies, and development partners.[105]

In Vanuatu's case, while domestic implementation has lagged, it has been extremely clear to the international community about the support they believe should be provided to climate-affected states. Increased support would likely improve implementation. However, as this experience shows, the policies being implemented still need to be locally owned and specific to a state's particular context. Without ownership and the ability to fit within existing institutions and practices, ensuring that implementation will occur is harder. This is why assistance on its own may not suffice—whether technical or financial, it must be locally led and focused on the needs of affected communities. With this in mind, the following section examines the role the international community should play in helping states overcome barriers to implementing climate mobilities and adaptation policies.

4 The Role of the International Community in Overcoming the Implementation Barriers

The COP27 showed that there is a widespread desire among the majority of the international community to establish some type of financial assistance and compensation mechanism. However, the declaration on the pathway

[101] Ibid.

[102] Government of Nauru, 'Submission of Nauru on Behalf of the Alliance of Small Island States: Views and Information on Elements to Be Included in the Recommendations on Loss and Damage in Accordance with Decision 1/CP.16' (UN Framework Convention on Climate Change 2012) 4. https://unfccc.int/sites/default/files/aosis_submission_on_loss_and_damage_submission_2_october_2012.pdf. Accessed 16 July 2023.

[103] Government of Vanuatu, 'Elements for a Framework Convention on Climate Change: Proposed by Vanuatu on Behalf of States Members of the United Nations and of the Specialized Agencies That Are Members of The Alliance of Small Island States' (UN Framework Convention on Climate Change 1991) para 19. https://aosischair.sharepoint.com/sites/aosiscontentpublishing/Published%20Documents/Forms/AllItems.aspx?id=%2Fsites%2Faosiscontentpublishing%2FPublished%20Documents%2F1991%2E6%2E4%20%2D%20Submission%20%2D%20Elements%20for%20a%20Framew%5FWebsite%20User%2Epdf&parent=%2Fsites%2Faosiscontentpublishing%2FPublished%20Documents&p=true&ga=1. Accessed 16 July 2023.

[104] Wewerinke-Singh and Salili (n 34) 5.

[105] Government of Vanuatu [55].

towards a loss and damage fund lacked specificity on how to implement this in practice.[106] Significant questions remain about the form the fund will take, the amount of funds required, who will pay into the fund, whether contributions will be voluntary or mandatory, who is entitled to claim compensation from the fund, and what even counts as loss and damage in this context.[107]

As Harris argues, the idea of the fund is 'ripe with possibility', but there is a real danger that larger states will stonewall or slow roll the process to avoid financial liabilities.[108] The final text of the COP27 agreement was careful to avoid using terms like liability or compensation—largely due to the campaigning of the United States.[109] The commitment to address loss and damage was furthered by an agreement on day one of COP28 in Dubai to operationalise the fund.[110] However, concerns remain. For one, against the wishes of the Alliance of Small Island States, the G77 and China group, and other developing countries, the fund will be administered by the World Bank and not a new independent entity. Secondly, financial contributions to the fund remain voluntary, there are no long-term commitments, and the pledges made are woefully inadequate considering the scale of the issue.[111] While recent progress is positive, there remains the chance this could turn into another 'placebo fund' that never eventuates at the promised or required scale.[112] To avoid these dangers, policymakers, practitioners, academics, and other stakeholders in the development and humanitarian spheres need to ensure that states trying to implement climate mobilities policies are supported and assisted in the right ways. In most cases, responses can be strengthened by ensuring local needs and local leadership guide them.

A. The Timing and Accessibility of Assistance

To improve how the international community works to assist states in preparing for and managing climate mobilities, the focus needs to be moved forward so support is available before a crisis response is needed. It is crucial for climate financing to be made available as early as possible so that anticipatory action can be taken to minimise or prevent harm from occurring. However, traditionally, humanitarian responses follow a pathway of immediate triage, assessing damage and needs, and only then searching for funding. There is a need to flip this for climate-related mitigation and adaptation—i.e. plans and actions must be focused not on impacts, but on risks.[113]

B. Locally Led Assistance

In general, we can say that local knowledge of climate-related risks, vulnerabilities, and capacities will help construct better solutions. I note that defining 'the local' is often fraught and should not be essentialised as in opposition to the international or romanticised as the solution to fixing the deficiencies of international efforts.[114] However, borrowing from the post-development literature available, we can see that local and pluralist solutions that value connections to the place and the knowledge that comes with this, are crucial to developing approaches that provide better outcomes for

[106] 'Funding Arrangements for Responding to Loss and Damage Associated with the Adverse Effects of Climate Change, Including a Focus on Addressing Loss and Damage' (UN Framework Convention on Climate Change 2022) Decision-/COP.27-/CMA.4. https://unfccc.int/sites/default/files/resource/cma4_auv_8f.pdf. Accessed 16 July 2023.

[107] Harris (n 40); McDonald (n 39).

[108] Harris (n 40).

[109] Najam (n 44).

[110] 'Operationalization of the new funding arrangements, including a fund, for responding to loss and damage referred to in paragraphs 2–3 of decisions 2/CP.27 and 2/CMA.4' ((UN Framework Convention on Climate Change 2023) Draft decision-/CP.28-/CMA.5. https://unfccc.int/event/cop-28?item=8%20g. Accessed 21 January 2024.

[111] McDonland [56], Lakhani [57], Moore [58].

[112] Najam (n 44).

[113] Zommers [59].

[114] Roche et al. [60].

people, communities, and the comprehensive non-human environment.[115]

This is where the international community must step in to help overcome the barriers to support the implementation of solutions like those developed in Fiji and Vanuatu. I note that these are barriers, not limits—the difference is that limits imply thresholds, while barriers can be overcome.[116] In the context of climate mobilities, even where thresholds are reached for staying in a specific location, the barriers to finding appropriate, rights-based, locally led, durable solutions to the situations can be overcome. As Ayers and Forsyth argue, there is a professed obligation inherent in climate adaptation funding, that the developed Global North states that benefited from higher emissions must support those in the Global South to manage best the challenges posed by climate change.[117]

C. Availability of Adequate Financial Resources

In reality, adaptation funding is minimal and current models have not been able to contribute significantly to the overall annual funding required for the range of mitigation, adaptation, and loss and damage costs incurred by climate-exposed and developing states.[118] Despite recent pledges of dedicated loss and damage funding from states like Scotland and Denmark, they fail to scratch the surface of the tens or hundreds of billions of dollars of estimated global costs of climate-related losses and damages.[119] In the 25 years that climate finance has been on the international agenda, there has been limited progress towards justice. Instead, we have seen a focus on voluntary actions, large states avoiding commitments to fault, liability, or responsibility, and a growing emphasis on private, market-based strategies.[120] These shifts and the relative failures of previous collective international funding models mean that states may be forced away from relying on these large funds and towards more state-based solutions that rely on regional or bilateral agreements to fund adaptation projects to manage climate mobilities. As one representative of the Government of Vanuatu noted, they were now applying a 'no stone unturned' policy to finding funding and support for climate adaptation and mobilities programmes.[121] When told about how the ni-Vanuatu approach had been described, a Fijian official from the Climate Change Division noted that they were applying a similar type of logic to finding funding for their trust fund and relocation projects.[122]

Even when funding does exist, barriers persist. One of the critical problems identified has been the discrepancies between local needs and the priorities of donors and funders. For example, aid can often be given based on strategic foreign policy concerns,[123] or what donors see as priorities significantly differ from what the communities want to prioritise.[124] As the policies of Fiji and Vanuatu highlight, community consent, engagement, and active participation are vital to achieving durable solutions in cases of potential climate mobilities. This must also apply in cases where polices and projects are externally funded. Despite where the money comes from, following best practices here would ensure that projects are still locally led and tailored to the community's needs and what they have requested.

D. Beyond Financial Assistance

In most cases, support must extend well beyond simple financial assistance. Technical support and expertise, knowledge and information, and skilled staffing resources can all assist states

[115] Ibid., 143.

[116] Moser and Ekstrom [61], Piggott-McKellar et al. [62].

[117] Ayers and Forsyth [63]; Piggott-McKellar and others (n 116) 384.

[118] Chong [64].

[119] Harvey [65].

[120] Khan et al. [66].

[121] Interview with Christopher Bartlett, Special Advisor to Government of Vanuatu (n 86).

[122] Interview with Filimone Ralogaivau, Fiji Climate Change Division (n 60).

[123] Robinson and Dornan [67].

[124] Betzold [68].

like Fiji and Vanuatu. The timing of assistance is key here, as inadequate resources can limit the planning and implementation of adaptive programmes, leading to barriers becoming limits and options for in situ adaptation becoming more limited.[125] The complex application processes for international funding and assistance can hamstring implementation, making accessing funding difficult and time-consuming.[126] Opposite these acute time pressures, there are barriers around long-term timeframes. Donors can run a project for several years, but when it concludes and funding ends, the situation can revert to how things were before, and adaptive measures can be undone.[127]

Financial limitations are one set of barriers Fiji hopes to surmount with its Trust Fund and SOPs—it is expected that encouraging contributions towards this project will ensure donors and financial contributors are assured of where their money is going and that Fijian actors will be able to access funding promptly and use it as required for each local context.[128] In contrast, Vanuatu has had minimal success in implementing its flagship policy. While some of this is down to institutional issues, it can also be attributed to a lack of available resources—including financial resources, human resources, and existing infrastructure.[129]

5 Lessons from the Pacific: What the International community Needs to do to Improve Assistance

To improve outcomes now and into the future, I suggest that international support and assistance should strive to be locally led and focused on the needs of local communities instead of donors. We can learn from the experiences of Fiji and Vanuatu to develop a more practical approach to international assistance in the context of climate mobilities—one that ensures rights are upheld, harms are minimised, and durable solutions to mobilities are found for those who do have to move. This approach would build upon that of the early adopters from the Pacific—borrowing the good, improving points of contention, and avoiding known problem areas.

While Fiji's Planned Relocation Guidelines were hailed as a world-leading project at their publication, those tasked with turning these ideas into practice faced many difficulties. While the authors of the policy conducted consultations, these meetings and the document that resulted from them have been criticised. The result was that when it came to developing the Standard Operating Procedures for the Guidelines, Fijian officials had to be brought on board to translate the broad ideas into something that could be implemented in the Fijian context.[130] It is hoped that the knowledge of local political structures, customs, and communities will help ensure that this policy can be effectively implemented to assist those in need.

A. Improving Drafting Processes

Watching the example of the Solomon Islands will be an interesting comparison here, as both the Solomon Islands and Vanuatu have partnered with the International Organisation for Migration (IOM) and contracted external Australian legal academics to develop their policies. As we have seen with Vanuatu, the lack of local ownership of a foreign-designed policy and the limited resources to implement it can result in an idea that looks good on paper but fails in practice.[131] This issue was also present in the Fijian case, with an interviewee confirming that foreign-authored documents had to be

[125] Moser and Ekstrom [69].

[126] Robinson and Dornan (n 121) 1112–3; Weir et al. [70].

[127] Betzold (n 124) 486.

[128] Interview with Filimone Ralogaivau, Fiji Climate Change Division (n 60).

[129] Interview with Zoe Ayong, National Disaster Management Office of Vanuatu (n 92).

[130] Interview with Interviewee 1 (n 77); Interview with Christine Fung (n 63).

[131] Interview with Zoe Ayong, National Disaster Management Office of Vanuatu (n 92).

significantly revised by local experts after they were handed over to the Fijian Government.[132] What can be said, though, is that when the policies themselves advocate for local participation and tailored solutions for at-risk communities, the policy development and implementation should also be predominantly locally led.

What allowed the policies to be locally led and tailored to Fijian needs, in this case, was a particularly context-specific solution from the Fijian and New Zealand governments. While New Zealand was unwilling to release its entire committed contribution to Fiji's Trust Fund based on the guidelines alone, they said they would be willing to do so once the SOPs were enacted. To facilitate this, New Zealand signed off on having a smaller portion of these funds released to support the development of these SOPs and to fund the team tasked with developing them in consultation with local, civil, and government partners.[133]

B. Building Collaborative Partnerships

Additionally, the German Development Fund (GIZ) has been a key partner for over a decade in Fiji. This long-standing partnership has evolved into one with a level of mutual trust, where local authorities are largely trusted to implement projects in ways that are more likely to work in the Fijian or local context. It has not been smooth sailing, with an interviewee noting that significant difficulties were faced in the first relocation GIZ supported—however, this relationship has now evolved, and Fijian staff are now leading the agency's local offices and key projects with a great degree of autonomy. The assistance of GIZ has been invaluable in developing policy and implementing practices. However, this relationship took time to develop and has deepened over time. If key partners in Vanuatu, and other countries implementing similar policies, like the Solomon Islands, can adopt similar perspectives, it could strengthen working relationships, encourage local ownership of policies and practices, and ultimately improve implementation and outcomes for affected communities.

In contrast, though, at the policy level in Vanuatu, the IOM has been criticised for its tight control over what policies should look like. This has meant that there is a distinct lack of ownership by the Government and little appetite or will to implement the policies once they are handed over. Rather than focusing narrowly on universalising language on climate mobilities in policies across different countries and contexts, international organisations like the IOM have to find a way to allow these principles to be localised in a way that makes them implementable and gives the people who will oversee and practice the policies, ownership of them.

C. Recognising the Agency of Large Ocean States

Traditional boilerplate policies and humanitarian responses are ineffective here because they are not contextually specific or locally led, but they also fail to recognise the changing needs of island states in a warming climate. There needs to be recognition of the calls for assistance through campaigns for loss and damage, the ICJ advisory ruling, and the changing tenor of international discourse around these issues as signalled at the COP27.

A key part of how these approaches must change is how donors, development partners, and those providing assistance must critically consider and redress how they conceptualise and frame these states. As we have seen in the Pacific, there is a strong movement away from being termed as 'small island states'. Instead, these people prefer the term 'large ocean nations', as it better reflects their size and how they view themselves.[134] As Theys has argued, smallness in this sense is socially constructed—it is not solely dependent on material factors

132 Interview with Christine Fung (n 63).

133 Interview with Lebaiatelaite Gaunavinaka, Fiji Climate Change Division (n 54); Interview with Filimone Ralogaivau, Fiji Climate Change Division (n 60).

134 Chan (n 48); Morgan (n 48).

but also on narratives, perceptions, and practices.[135] Fiji and Vanuatu have shown that they are not small states but global leaders in the climate space. The partners who work with them to implement climate mobilities protection policies should view them as such.

6 Conclusion

As one Fijian official said in an interview, we know what the solutions are and 'it would be a shame if we can't do it'.[136] This highlights the unique position states like Fiji and Vanuatu are in right now. They have developed world-leading policies to address climate mobilities within their states. However, small states like these lack the outright resources to respond to these issues at the scale and speed required. In addition, their lack of contribution towards the drivers of climate change adds weight to the argument that the international community is obligated to assist them in implementing these policies and protecting the rights of those within their borders threatened by climate mobility-related harms.

This chapter has looked at the obligations the international community has to assist states in managing climate mobilities and what form this assistance should take. First, the existing international legal and normative protections of states and the international community were examined, with a specific focus on frameworks surrounding the international climate finance regime. Secondly, the two path-leading cases of Fiji and Vanuatu were explored. These cases looked not just at the innovative policies coming out of the small island states, but also how they have tried to implement these and the barriers they have faced. Next, the role of the international community in assisting states to overcome implementation barriers was outlined, focusing on the timing and accessibility of assistance, the need for these projects to be locally led, the importance of adequate available financing, and requirements beyond the financial—such as technical support and human expertise. Finally, the chapter suggests three ways the international community and states can better work together to improve outcomes for those affected by climate mobilities. I argue that improving policy drafting processes so they are locally led, building long-term collaborative partnerships that value local knowledge and expertise, and a deeper recognition of the leadership and agency of large ocean states would significantly reorientate how the international community engages with and assists states around climate mobilities.

Problematically, there is a likelihood that larger developed states will not agree to contribute the required amounts of aid or provide support in the ways required. However, as COP27 showed, the attitudes of the broader international community are shifting. There is momentum behind the calls for a loss and damage mechanism and compensation for smaller states bearing the brunt of climate change. This momentum must be harnessed and used to ensure that international assistance around climate change—and particularly climate mobilities management—is not just increased, but that the support and assistance are provided in the right manner.

[135] Theys [71].

[136] Interview with Daniel Lund, Special Advisor to Government of Fiji (n 74).

References

1. Ferris E (2011) The politics of protection: the limits of humanitarian action. Brookings Institution Press, 66, 203
2. McAdam J, Ferris E (2015) Planned relocations in the context of climate change: unpacking the legal and conceptual issues. Camb J Int Comp Law 4(137):158
3. Bellamy AJ (2008) The responsibility to protect and the problem of military intervention. Int Aff 84(615):619
4. UN High Commissioner for Human Rights (2012) The relationship between climate change and human rights. In: Leckie S, Simperingham E, Bakker J (eds) Climate change and displacement reader. Earthscan 226
5. Boas I et al (2019) Climate migration myths. Nat Clim Change 9:901–903
6. Boas I et al (2022) Climate mobilities: migration, im/mobilities and mobility regimes in a changing climate. J Ethn Migr Stud 1:2

7. Gusterson H (2008) Ethnographic research. In: Klotz A, Prakash D (eds) Qualitative methods in international relations: a pluralist guide. Palgrave MacMillan, 104
8. Pictet J (1979) The fundamental principles of the Red Cross (II). Int Rev Red Cross 19(184):185
9. Shue H (1996) Basic rights: subsistence, affluence, and U.S. Foreign Policy, 2nd edn. Princeton University Press, 189–194
10. UN High Commissioner for Refugees (2010) Convention and Protocol relating to the Status of Refugees. United Nations High Commissioner for Refugees
11. African Union (2009) African Union convention for the protection and assistance of internally displaced persons in Africa. African Union
12. Kälin W (2008) Guiding principles on internal displacement annotations. Stud Transnatl Leg Policy 189
13. McAdam J (2012) Climate change, forced migration, and international law. Oxford University Press 52
14. Pérez BF (2018) Beyond the shortcomings of international law: a proposal for the legal protection of climate migrants. In: Climate Refugees Routledge, 217
15. Oloka-Onyango J (2010) Movement-related rights in the context of internal displacement. In: Kälin W et al (eds) Incorporating the guiding principles on internal displacement into domestic law: issues and challenges. American Society of International Law
16. Thakur R (2006) The United Nations, peace and security: from collective security to the responsibility to protect, 2nd edn. Cambridge University Press, 311
17. International Law Commission (2016) Draft articles on the protection of persons in the event of disasters. International Law Commission
18. Foster M, McAdam J (2022) Analysis of "imminence" in international protection claims: Teitiota v New Zealand and Beyond. Int Comp Law Q 71:975
19. McAdam J (2011) Swimming against the tide: why a climate change displacement treaty is not the answer. Int J Refugee Law 23(2):26
20. Lee J (2020) The Paris Agreement and the prospects for socialization into the climate norm: lessons from the U.S. and China. Int Stud Rev 21:73, 76
21. Meguro M (2020) Litigating climate change through international law: obligations strategy and rights strategy. Leiden J Int Law 33(933):942
22. Hermwille L et al (2017) UNFCCC before and after Paris—what's necessary for an effective climate regime? Climate Policy 17:150, 162
23. Cohen R, Bradley M (2010) Disasters and displacement: gaps in protection. J Int Humanit Legal Stud 1:95, 117
24. Farbotko C et al (2018) Transformative mobilities in the Pacific: promoting adaptation and development in a changing climate. Asia Pac Policy Stud 5(393):395
25. Charan D, Kaur M, Singh P (2017) Customary land and climate change induced relocation—a case study of Vunidogoloa Village, Vanua Levu, Fiji. In: Leal Filho W (ed) Climate change adaptation in Pacific countries: fostering resilience and improving the quality of life. Springer International Publishing, 20
26. Wewerinke-Singh M, Salili DH (2019) 'Between negotiations and litigation: Vanuatu's perspective on loss and damage from climate change. Clim Policy 1:4
27. Peel J (2022) It's the Big Issue of COP27 climate summit: poor nations face a $1 trillion "loss and damage" Bill, but rich nations won't pay up. Conversat, 10 November 2022. http://theconversation.com/its-the-big-issue-of-cop27-climate-summit-poor-nations-face-a-1trillion-loss-and-damage-bill-but-rich-nations-wont-pay-up-194043. Accessed 15 July 2023
28. Donald R (2022) COP27: climate loss & damage talks now on agenda, but U.S. Resistance Feared. Mongabay Environ News 8 November 2022. https://news.mongabay.com/2022/11/cop27-climate-loss-damage-talks-now-on-agenda-but-u-s-resistance-feared/. Accessed 15 July 2023
29. McDonald M (2022) COP27: One Big breakthrough but ultimately an inadequate response to the climate crisis. Conversat, 20 November 2022. http://theconversation.com/cop27-one-big-breakthrough-but-ultimately-an-inadequate-response-to-the-climate-crisis-194056. Accessed 10 April 2023
30. Lillo A et al (2022) COP27 "loss and damage" fund: a historic decision amid discouraging results. Conversat, 27 November 2022. http://theconversation.com/cop27-loss-and-damage-fund-a-historic-decision-amid-discouraging-results-195186. Accessed 10 April 2023
31. Harris PG (2023) COP27: from preventing dangerous climate change to salving loss and damage. PLOS Clim 2. https://doi.org/10.1371/journal.pclm.0000150. Accessed 10 April 2023
32. Najam A (2022) COP27's "loss and damage" fund for developing countries could be a breakthrough – or another empty climate promise. Conversat, 21 November 2022. http://theconversation.com/cop27s-loss-and-damage-fund-for-developing-countries-could-be-a-breakthrough-or-another-empty-climate-promise-194992. Accessed 10 April 2023
33. Lakhani N (2023) United Nations adopts landmark resolution on climate justice. Guard, 29 March 2023. https://www.theguardian.com/environment/2023/mar/29/united-nations-resolution-climate-emergency-vanuatu accessed 10 April 2023
34. Peel J, Nay Z (2023) The UN is asking the International Court of Justice for its opinion on states' climate obligations. What does this mean? Conversat, 3 April 2023. http://theconversation.com/the-un-is-asking-the-international-court-of-justice-for-its-opinion-on-states-climate-obligations-what-does-this-mean-202943. Accessed 10 April 2023

35. Chan N (2018) "Large ocean states": Sovereignty, small islands, and marine protected areas in global oceans governance. Glob Gov 24:537
36. Morgan W (2022) Large ocean states: Pacific Regionalism and climate security in a new era of geostrategic competition. East Asia 39:45
37. Hume A et al (2021) Towards an ocean-based large ocean states country classification. Mar Policy 134:104766
38. Andrew N et al (2019) Coastal proximity of populations in 22 Pacific Island countries and territories. PLoS ONE 14(e0223249):8
39. Martin PCM et al (2018) Responding to multiple climate-linked stressors in a remote Island context: the example of Yadua Island, Fiji. Clim Risk Manag 21(7):2
40. Church JA et al (2013) 2013: sea level change. In: Stocker TF et al (eds) Climate change 2013: the physical science basis. Contribution of Working Group I to the Fifth Assessment Report of the Intergovernmental Panel on Climate Change. Cambridge University Press, 1148 https://www.ipcc.ch/site/assets/uploads/2018/02/WG1AR5_Chapter13_FINAL.pdf. Accessed 8 May 2023
41. Tronquet C (2015) From Vunidogoloa to Kenani: an insight into successful relocation. State Environ Migr 2015(121):122
42. McNamara KE, des Combes HJ (2015) Planning for community relocations due to climate change in Fiji. International J Disaster Risk Sci 6:315, 316–7
43. Nichols A (2019) Climate change, natural hazards, and relocation: insights from Nabukadra and Navuniivi Villages in Fiji. Clim Change, 264–265
44. Vakasukawaqa A (2023) SOPs a step in right direction. The Fiji Times (Suva, 23 April 2023). https://www.fijitimes.com/sops-a-step-in-right-direction/. Accessed 28 April 2023
45. Climate Change Division (2023) Standard operating procedures for planned relocation in the Republic of Fiji. Office of the Prime Minister. https://fijiclimatechangeportal.gov.fj/wp-content/uploads/2023/04/Standard-Operating-Procedures-for-Planned-Relocation-in-the-Republic-of-Fiji-1.pdf. Accessed 28 April 2023
46. Lund D (2021) Navigating slow-onset risks through foresight and flexibility in Fiji: emerging recommendations for the planned relocation of climate-vulnerable communities. Curr Opin Environ Sustain 50(12):14
47. Goering L (2020) New Zealand makes first donation to Fiji climate relocation fund. Reuters, London, 27 February 2020. https://www.reuters.com/article/us-climate-change-fiji-newzealand-trfn-idUSKCN20L04K. Accessed 29 Oct 2021
48. Vanuatu National Disaster Management Office, 'National Policy on Climate Change and Disaster-Induced Displacement 2018' (Vanuatu National Disaster Management Office 2018)
49. Ferris E (2019) Climate change, migration, law, and global governance. N C J Int Law 44(425):437
50. Wewerinke-Singh M, Van Geelen T (2018) Protection of climate displaced persons under international law: a case study from Mataso Island, Vanuatu. Melb J Int Law 19:17
51. McDonnell S (2021) The Importance of attention to customary tenure solutions: slow onset risks and the limits of Vanuatu's climate change and resettlement policy. Curr Opin Environ Sustain 50(281):281
52. Bolo N et al (2021) Sudden-onset hazards and the risk of future displacement in Vanuatu. Internal Displacement Monitoring Centre, 11 https://www.internal-displacement.org/sites/default/files/publications/documents/21_0907_IDMCVanuatuRiskprofile.pdf. Accessed 14 July 2022
53. Vinke K et al (2020) Home lands: island and archipelagic states' policymaking for human mobility in the context of climate change. Deutsche Gesellschaft für Internationale Zusammenarbeit, 94
54. Yamamoto L (2020) Climate relocation and indigenous culture preservation in the Pacific Islands. Georg J Int Aff 21(150):151
55. Government of Vanuatu (2022) Vanuatu's revised and enhanced 1st nationally determined contribution 2021–2030. Government of Vanuatu. 36–7, 39. https://unfccc.int/sites/default/files/NDC/2022-08/Vanuatu%20NDC%20Revised%20and%20Enhanced.pdf. Accessed 28 April 2023
56. McDonald M (2023) COP28 climate summit just approved a 'loss and damage' fund. What does this mean? Conversat, 1 December 2023. https://theconversation.com/cop28-climate-summit-just-approved-a-loss-and-damage-fund-what-does-this-mean-218999 Accessed 21 Jan 2024
57. Lakhani N (2023) $700m pledged to loss and damage fund at COP28 covers less than 0.2% needed. Guard, 6 December 2023. https://www.theguardian.com/environment/2023/dec/06/700m-pledged-to-loss-and-damage-fund-cop28-covers-less-than-02-percent-needed. Accessed 21 Jan 2024
58. Moore B (2023) Beyond COP28—loss and damage: from ambition to reality. European Policy Centre, 15 December 2023. <https://www.epc.eu/en/Publications/Beyond-COP28-loss-and-damage-From-ambition-to-reality~56fd88. Accessed 21 Jan 2024
59. Zommers Z (2022) Addressing climate risks in Small Island States—opportunities for action. SDG Knowledge Hub https://sdg.iisd.org:443/commentary/guest-articles/addressing-climate-risks-in-small-island-states-opportunities-for-action/. Accessed 1 May 2023
60. Roche C et al (2020) The characteristics of locally led development in the Pacific. Polit GovAnce 8(136):136
61. Moser SC, Ekstrom JA (2010) A framework to diagnose barriers to climate change adaptation. Proc Natl Acad Sci 107(22026):22026–22027

62. Piggott-McKellar AE et al (2019) What are the barriers to successful community-based climate change adaptation? A review of grey literature. Local Environ 24(374):376
63. Ayers J, Forsyth T (2009) Community-based adaptation to climate change. Environ: Sci Policy Sustain Dev 51:22, 29
64. Chong J (2014) Ecosystem-based approaches to climate change adaptation: progress and challenges. Int Environ AgreemS: Polit, Law Econ 14:391, 395–6
65. Harvey F (2022) Denmark Offers "loss and damage" funding to poorer countries for climate breakdown. Guard, 21 September 2022. https://www.theguardian.com/environment/2022/sep/21/denmark-offers-loss-and-damage-for-climate-breakdown-as-protests-gather-pace. Accessed 22 Sept 2022
66. Khan M et al (2020) Twenty-five years of adaptation finance through a climate justice lens. Clim Change 161(251):265
67. Robinson S, Dornan M (2017) International financing for climate change adaptation in small island developing states. Reg Environ Change 17(1103):1105
68. Betzold C (2015) Adapting to climate change in small island developing states. Clim Change 133(481):486
69. Moser S, Ekstrom J (2010) A framework to diagnose barriers to climate change adaptation. Proc Natl Acad Sci 107(51):22026, 22029
70. Weir T, Dovey L, Orcherton D (2017) Social and cultural issues raised by climate change in Pacific Island countries: an overview. Reg Environ Change 17(1017):1022
71. Theys S (2022) Small states reconsidered: small is what we make of it. J Int Aff 74:81

Dr. Liam Moore is a Lecturer in International Politics and Policy at James Cook University, Australia, and holds a Ph.D. in International Relations from the University of Wollongong, Australia. His research identifies the conditions under which small states create change and how states in the Pacific respond to climate mobilities, displacement, and relocation in novel ways. Liam has a strong background in research on forced displacement and human mobility, particularly around climate-related mobilities, internal displacement, and mobilities in Pacific Island states. He also often works with IR normative theory, particularly around norms, contestation, uncertainty, and change.

Marginalisation of Adat Communities: Intersectionality of Land Grabbing, Human Rights, Climate Adaptation, and Human Mobility in Indonesia

James Reinaldo Rumpia

Abstract

This chapter seeks to examine issues concerning the Adat community, a community whose very existence is based on a strong link between indigenous knowledge and the environment. It will also examine the community's role in climate change mitigation and adaptation. The chapter addresses the problem of marginalisation of indigenous people through the intersectionality of land grabbing, fulfilment of human rights, climate adaptation, and human mobility. It uses an indigenous—local knowledge—human rights-based approach to understand the knowledge, traditions, and dynamics of the rights of indigenous people. The chapter uses the case of the Sikep and Rendu communities in Indonesia and shows that their lives as farmers doing traditional farming are very much a part of conservation efforts. And that their marginalisation because of the expansion of industries, specifically, the cement industry and various dam projects, has resulted in land grabbing and consequently, migration. In addition, the fragmentation of law and policy creates multiple vulnerabilities due to the absence of specific laws concerning the rights of indigenous people, limited legal recognition, bureaucratic-administrative nature of land registration, and the lack of meaningful participation by the indigenous people in environmental and climate policy.

1 Introduction

In recent years, the knowledge and experience of indigenous peoples and local communities have been recognised as necessary in understanding climate change issues; for example, the Intergovernmental Panel on Climate Change (IPCC) Assessment Report 2022 talks of indigenous knowledge and science.[1] In fact, the IPCC report mentions tenure rights for indigenous and local communities, close to 58 times, hammering home the relevance and importance of indigenous communities.[2] For Indonesia, the normative framework for recognition and commitment to involving indigenous peoples is contained in Law No. 16 of 2016 concerning the Ratification of the Paris Agreement to the United Nations Framework Convention on Climate Change (UNFCCC) and the Nationally

J. R. Rumpia (✉)
Center for Public Policy and Human Rights Studies, Faculty of Law, Lampung University, Bandar Lampung, Indonesia
e-mail: jamesreinaldor@gmail.com

[1] Moggridge et al. [1].

[2] IPCC [2]; See also Locke and Childress [3].

S. Jolly et al. (eds.), *Climate-Related Human Mobility in Asia and the Pacific*, Sustainable Development Goals Series, https://doi.org/10.1007/978-981-97-3234-0_13

Determined Contribution of 2022. Indonesia's Nationally Determined Contribution (NDC) 2022 mentions the role of indigenous peoples in addressing climate change, such as their use of non-timber products, social forestry, and conservation.[3]

So far though, the targets and achievements of various NDCs in mitigating the climate change problem are yet to show real progress. The United Nations Environment Programme notes that countries need to catch up on the realisation of NDCs. The gap in the NDCs is approximately three gigatons of CO_2 for unconditional NDCs and six gigatons of CO_2 for conditional NDCs for 2030, so further annual GHG emissions must be reduced by 45%.[4] The World Economic Forum stated that the problem of climate change mitigation and adaptation will still be a severe risk in the next ten years.[5] This risk comes from irreversible carbon dioxide concentrations on the planet and various other impacts such as ecosystem damage, a decline in ecosystem function, species extinction, and changes in hydrology.[6]

The complications and difficulties associated with solving climate problems require alternative solution windows/mitigation and adaptation alternatives, including from indigenous knowledge. Various conventions relating to the environment, climate, and indigenous peoples recognise and affirm the existence of indigenous knowledge. The Rio Declaration[7] and the Biological Diversity Convention[8] guarantee the implementation of the rights of indigenous peoples, affirming the Indigenous Peoples Conventions of 1989 and 2007.[9] The Paris Agreement emphasises adapting indigenous knowledge in climate policy and socio-economic action..[10] For the last few decades, studies and findings related to indigenous knowledge of the environment, known as Traditional Ecological Knowledge (TEK), have developed as a typical 'hot' topic among social and natural scientists.[11] Even though TEK is intertwined with science in a system of ideas and practices, it is distinguished from belief, use in a specific context, locality, traditional methods, cultural context, and sustainability for certain indigenous groups.[12] In the Indonesian context, research on TEK is adopted according to the contextuality and system of each indigenous group, for example, TEK as TEK in Indonesia (communities in Tasik Chini and communities in Lawu),[13] TEK as being associated with local wisdom (Tunjung and Bajau Dayak communities),[14] and TEK as a knowledge system (Dayak in Danau Sentarum).[15]

Human mobility for indigenous communities can result in forced displacement and land eviction if done without considering social and economic situations. In human mobility studies, for example, adaptation and survival sometimes are far more rational preferences than migration, because people feel at home and are tied to the community and want to live close to friends and relatives[16]; also socio-economic factors can determine who stays and who goes.[17] The issue of mobility then must be assessed, determined, and considered based on an appropriate approach to guarantee human rights. A study on rural migration for 218 households in Cambodia shows maladaptive events after moving

[3] Indonesia Government [4].

[4] United Nations Environment Programme [5].

[5] World Economic Forum [6].

[6] Solomon et al. [7], IPCC [8].

[7] Rio Declaration on Environment and Development 13 June 1992 31 ILM 874 Preamble and Article 22.

[8] Convention on Biological Diversity 5 June 1992, 31 ILM.818, Article 8 and Article 10.

[9] International Labour Organization (ILO) [9], UN General Assembly [10].

[10] Paris Agreement Article 7, para 5; See Decision 1/CP—21 (Adoption of Paris Agreement).

[11] Molnár and Babai [11].

[12] Berkes [12], Van Eijck and Roth [13], Ballard [14], Cajete [15].

[13] Seftyono [16], Sumarwati [17].

[14] Siahaya et al. [18], Ariando [19].

[15] Yuliani et al. [20].

[16] Buchori et al. [21].

[17] Zickgraf [22].

locations, leading to poverty traps.[18] Practically, migration limits must also be assessed and oriented based on four important questions starting from livelihood issues, adaptation objectives along with priority scale, mechanisms for determining how and under what conditions migration is carried out, and the consequences of implementing migration and distributed spending.[19] In the case of hydro-dam constructions, for example, complex challenges associated with developing water/hydro-modern infrastructure need to reflect or quantify hydro-social rupture with its social and ecological consequences, including migration as a potential response.[20]

This chapter reflects on insights from qualitative research focusing on indigenous peoples, known as the Adat community, in Indonesia through the intersection of indigenous knowledge, the environment, and climate change adaptation. Research is carried out through desk research, which analyses content from books, journal articles, proceedings, research reports, theses, dissertations, policy papers, law, policies, news, websites, Indonesian government reports, and reports.[21] The Sedulur Sikep Community and the Rendu Community were chosen for the case study because these two communities are unique in describing the inter-relation and phenomena of climate change, indigenous people, and human mobility. They also represent marginalised groups pushed back by economic development, extreme weather and the violation of human rights. They can also use indigenous knowledge to survive climate change and pressure from external forces such as cement factories and dam construction. Case studies of the Sedulur Sikep community in Central Java and the Rendu community in East Nusa Tenggara (NTT) are employed to support two arguments. First, the case study explores the practicality of indigenous knowledge through the daily lives and way of life of indigenous people. The complexity of science and indigenous knowledge elements in TEK will not be presented, considering that the research only uses literature studies.[22] For clarity of terms and scope, this chapter uses indigenous knowledge as a way of life in environmental preservation and sustainability, while maintaining identity in the process of adapting and developing.[23] Second, this case study reveals the extent to which the State guarantees the implementation of indigenous rights in the face of political and land recognition. Finally, human mobility is also in question, resulting from violations of the rights of indigenous people, problems of land tenure policies and land confiscation, and the impact of environmental change.

The chapter is divided into introduction, discussion, and conclusion. First, the introduction explains the reasons, arguments, and existence of indigenous knowledge as part of climate mitigation, adaptation solutions, and the dynamic of human mobility for indigenous people. Second, the problem of marginalisation of indigenous peoples is examined through the intersectionality of land grabbing, fulfilment of human rights, climate adaptation, and human mobility. The discussion then moves on to the importance of indigenous knowledge practices in the Sikep and Rendu communities, human rights violations, land grabbing, and national policies that create vulnerability for indigenous peoples. The conclusion reflects the discussion results and essential notes for indigenous knowledge and climate change research.

2 Way of Life and Conservation: Case Studies of the Sedulur Sikep and Rendu Community

A case study is presented of the Sedulur Sikep people and the Rendu community with respect to their indigenous knowledge of sustainable

[18] Jacobson [23].

[19] Sakdapolrack et al. [24].

[20] Miller et al. [25], See also Middleton [26].

[21] All of this secondary data is used to understand case studies in the two communities that are the focus of the chapter.

[22] Molnár and Babai (n 11).

[23] Nadasdy [27], Berkes [28]; Ballard (n 12).

environmental management. These two indigenous communities may not directly mention the climate narrative and its mitigation but could relate to protecting the ecosystem. Shiva mentions this condition as 'original participation', which emphasises the integral relationship of indigenous peoples with the world to represent the interests of living things.[24] This chapter focuses on the Sedulur Sikep Community in Baturejo Village, Sukolilo District, Pati Regency. For the Rendu Community, those in Nagekeo Regency and areas which rejected the Lambo Dam are selected.

A. Contextualising the Sedulur Sikep and Rendu Community

The Sedulur Sikep community usually lives in parallel residential areas that tend to be close to each other and for those who live in Pati Regency, Bombong and Kudus villages are far away. The distance is only 8 kms or around 40 min by motor vehicle from the district capital of Pati Regency.[25] The natural environment around the area is mountainous, with rich karst geology and rice fields. In the Sedulur Sikep community, the teachings of the Adamic religion are behavioural and provide guidelines for life, while dealing with the environment and humans. Shiraishi calls Adam's religion a journey (life), wherein wong sikep is also called sukma/life and symbolises the essence of all forms of life. Life can take various forms (tangibles), but all forms are divided into two: wong (humans) and clothing (food and clothing).[26] This belief reflects the Sedulur Sikep community in their daily actualisation as farmers.[27]

The Rendu community lives in Rendu village, Aesesa Selatan District, Nagekeo Regency, and NTT. The distribution of the Rendu community is across several villages, namely Rendu Butowe, Ulupulu, Labolewa, and Rendu Ola, the oldest.[28] Based on the area, the Rendu community can also be called part of the Nagekeo people.[29] They are given abundant accessible natural wealth, such as fertile soil and rich natural resources, including various food crops, such as tubers, rice, and corn.[30] From the perspective of geographical conditions and natural potential, the Rendu community has many natural resources that must be appropriately utilised.[31]

B. Sedulur Sikep and Rendu Communities and their Integral Association with the Land

The land has a real and deep connection embodying the knowledge, beliefs, and traditions of the Sedulur Sikep and Rendu communities. For the Sikep people, the land is a manifestation of Mother Earth, which must be cared for and maintained so that human life can be in harmony with nature.[32] The Rendu community appreciates and respects the land symbolically and ritually through ceremonies such as the life cycle, clearing forests, planting, harvesting, and other activities.[33] A mark of ensuring respect for the land is the Rendu community's belief that the inheritance from their ancestors must be maintained because the land is life.[34]

Following the Sedulur Sikep thought process, as a consequence of this reverence for the land, farmers too are considered deeply important in protecting nature because it is farmers who can transform the wind, water, land, and wood while the State does not 'create' natural resources.[35] This condition proves that indigenous communities carry out improvisation, learning, and hybridisation of technology and knowledge systems

[24] Shiva [29].

[25] Personal experience during the research on Sedulur Sikep Community year 2021.

[26] Shiraishi [30].

[27] Korver [31].

[28] AMAN [32].

[29] Hendro [33]

[30] AMAN, 'Gaung Aman Media Informasi Dan Komunikasi Masyarakat Adat' (2022) *April-Juni 2022* 20.

[31] AMAN (n 28).

[32] Safitri [34].

[33] Hendro (n 29).

[34] Aman, 'Kabar Nusantara' (n 28).

[35] Peluso [35].

without losing their deep links with the land by caring for the environment and creating resilience through indigenous knowledge.[36] Indigenous communities thus become the most sensitive and dynamic actors in dealing with environmental change, including the contextualisation of indigenous knowledge in climate resilience.

The Rendu community employs several traditional farming practices that maintain sustainability. Firstly, due to the relatively dry land, the people in Nageko adapt to planting drought-tolerant cassava on their farms. Cassava does not require too much water and is in fact tolerant to dry land and resistant to pests, mild diseases, and requires minimal maintenance.[37] Secondly, the people of NTT found and replanted traditional sorghum crops as a source of food security and sovereignty.[38] Thirdly, palm fruit, or Lo Koli, is a local tradition for the Rendu community and has various benefits. The sap water can be processed into brown sugar, the leaves are made into mats, and the fruit flesh is used in preparing sweets.[39]

All the benefits from the palm tree are optimised, using traditions that have been passed on through generations. The Rendu community often utilises palm leaf fronds as a material for various woven crafts, such as mats (*tee koli*), covers for huts in the garden (*ghubu keka*), containers for storing crops (*bola koli*), and containers for local seeds, such as corn, beans, and rice (*kaka koli*).[40] Palm leaves are also used to support the construction of houses for the Rendu community. The roof of the house is made of palm leaves, and the walls are made of other natural materials such as woven bamboo, coconut leaves, and pandanus, while the floors are made of bamboo or planks.[41] All these materials were taken from around the Rendu community's area.

In comparison, in Wajo Village, the community built a traditional residential complex based on considerations of spiritual, social and environmental values. The dominant aspect of the spatial layout of the Wajo customary village is the orientation and hierarchy of the top-down space: the top is the sacred zone, and the bottom is the public or profane zone.[42] The principle of top-down hierarchy is formed from tribal order and natural factors (topography). The dominant aspect of the identification principle is the datum, which is formed due to adherence to traditions, cultural symbols, and community kinship.[43]

The knowledge of the Sedulur Sikep community and the Rendu community shows that their daily lives as farmers are pivotal to the efforts to protect the environment and maintain ecological functions, whether they realise it directly or not. Throughout the world, indigenous and local knowledge systems have succeeded in the long term in managing ecosystem functions, providing inputs for scientific efforts and preventing further degradation.[44] For example, during springtime, the Sedulur Sikep community ensures that the trees remain sustainable by guarding them against the exploitation of rocks (which are believed to damage hydrology).[45] For the Sedulur Sikep community, the Kendeng mountains also have a deep connect with preserving animal and vegetation habitats and have evidence of human history, such as ancient caves and historic places.[46] In NTT, the Rendu community farmers encourage land preparation, crop/vegetation diversification through mixed cropping patterns, and raising livestock

[36] Ford et al. [36].

[37] Ngongo et al. [37].

[38] Miranda and Eghenter [38].

[39] AMAN (n 28).

[40] Ibid.

[41] Hendro (n 29).

[42] Lake et al. [39].

[43] Ibid.

[44] Victor Reyes-García, 'Indigenous and Local Knowledge Contributions to Social-Ecological Systems' Management in *The Barcelona School of Ecological Economics and Political Ecology: A Companion in Honour of Joan Martinez-Alier*. Cham: Springer International Publishing 75.

[45] Mojo et al. [40].

[46] Putri [41].

in a crop-livestock system.[47] The Rendu community also benefits in the context of food security with the cultivation of a local crop known as sorghum. The cultivation of sorghum is essential for diversifying tough crops that resist various climates and soils while having good nutritional content and being gluten-free.[48] The above reflection on indigenous knowledge and local farming is an introduction to the next section which dwells on external problems and situations that disrupt the actualisation of agricultural life for indigenous communities to the point that their other human rights are neglected.

3 Indigenous Peoples in the Forefront of Conservation in Threats of Human Rights Violations, Land Confiscation, and Human Mobility

This section discusses the external factors and situations that create pressure over the Sedulur Sikep and Rendu communities rendering them unable to apply their indigenous knowledge. The discussion is based on the Sedulur Sikep community's resistance against mining and the Rendu community's resistance to the national dam development project. Tracing the current situation of the cases and their impact on indigenous communities is a start for discussing the threats to human rights and land rights. The discussion in this section also includes the inter-relation of indigenous communities with climate change issues, especially implementing mitigation and adaptation amidst marginalisation efforts and human mobility.

A. Sikep Community Movement Against the Expansion of the Cement Company in Kendeng

The cement industry was started around 2006 in the Kendeng mountains and has potentially uprooted the culture and identity of the Sikep Sedulur community, including their mobility choices, as a consequence of maintaining life. The forest area in the Kendeng mountains borders directly with the site of the PT Semen Indonesia industry. There are 1142 ha of limited production forest and 1310 ha of permanent production forest in the Gunem District administrative area based on the Mantingan Forest Management Unit (KPH), Rembang Regency.[49] This statistic shows that the Sikep community's concerns about environmental damage are well-founded and deeply connected with residential areas and forests, which act as buffer zones and catchment areas in the Kendeng mountains. Below is a map of the project location and a photo of the cement factory taken from personal documentation (Figs. 1 and 2).

The destruction that arises from cement mining activities is apparent as one sees the condition of the land belonging to the residents. The story of Prin from the Sikep community and Dulah, residents of Tegaldowo Village, Gunem, whose fields were damaged because they were close to a rock mining area, reveals the extent of the issue at hand.[52] Prin's chilli garden was affected by the pollution caused by limestone mining activities carried out by PT Semen Indonesia, resulting in crop failure.[53] Meanwhile, on Dulah's land, the mine not only polluted the corn plantation but also, there was a time when chunks of rock from the explosion almost hit the owner.[54] The ill-effects of this pollution on the Sikep community get exacerbated when it aligns with the fact that the cement industry is one of the largest contributors to the greenhouse effect, accounting for 8 per cent of all global emissions.[55] The

[47] Ngongo et al. [42].

[48] Arif [43].

[49] BPS Kabupaten Rembang [44].

[52] Kresna [45].

[53] Ibid.

[54] Ibid.

[55] Ramsden [46]. The impact of greenhouse gas emissions on the climate can be traced to 467 pathways covering human health, air, food, economy, infrastructure and security. This finding can be seen in Mora et al. [47].

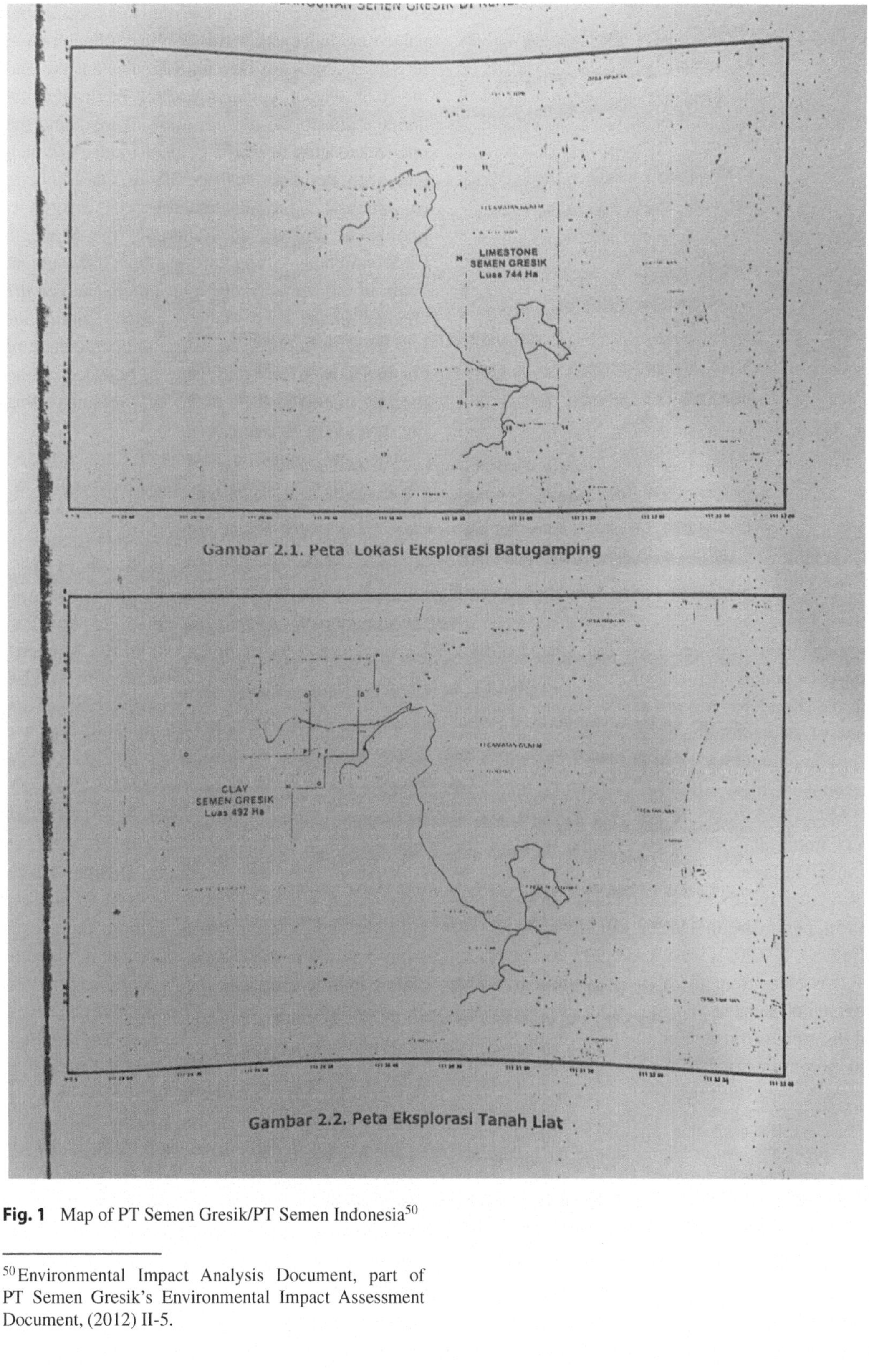

Fig. 1 Map of PT Semen Gresik/PT Semen Indonesia[50]

[50] Environmental Impact Analysis Document, part of PT Semen Gresik's Environmental Impact Assessment Document, (2012) II-5.

Fig. 2 Cement factory[51]

Strategic Environmental Study Team 'Policy for Sustainable Utilization and Management of the Kendeng Mountains' revealed that the loss of cement industry activities would also cause economic losses in the health sector, with a count of those affected: 64,544 people × $ 318,54 = 20.559.845,76 per year.[56] What is also very important with respect to the issue of the cement industry in Kendeng is that the cumulative damage to the Watu Putih Groundwater Basin per year up to 2020 is 38.753.447,040 (litres) × $ 0,00,011 = $ 2.466.903,26.[57]

It is clear from the above that rampant environmental damage and the increasing frequency of disasters is threatening the lives of the Sedulur Sikep community.[58] At the beginning of 2023, the Sedulur Sikep community visited the presidential palace to convey that the flooding phenomenon in Kudus, Pati, and Demak Regencies at the end of 2022 was a consequence of environmental damage due to the deforestation of trees and the unchecked growth of the cement industry, which had been permitted to operate in the Kendeng Mountains.[59] Through research related to morphological changes in Rembang, the Strategic Environmental Study from the Independent Team of the Ministry of Environment showed the changes in the karst of the Kendeng Mountains will have an impact, increasing the percentage of channel flow in the rainy season and decreasing the rate of mixed flow in the dry season so that the springs become dry.[60]

The environmental damage has affected water conditions directly and has even increased the frequency of disasters. Mining area expansion in the Kendeng Mountains is likely to continue through the expansion of existing companies such as PT Semen Indonesia and investments by other companies. In the Pati Regency area (on the other side of the Kendeng Mountains), a small-scale cement mine is also operating and is being followed by plans from P. T. Sahabat Mulia Sakti (a subsidiary of PT Indocement), which is part of the Heidelberg Cement Group (a company based in Germany), that wants to build a mine and factory with an area of around 2800 ha.[61]

What is apparent is that the Sedulur Sikep community was not directly evicted from their land but has experienced a systemic impact on their traditions due to environmental damage and cement mining activities which has affected crop fertility and livestock health and destroyed livelihoods in agriculture, effectively rendering the place almost impossible to inhabit.[62] For Sedulur Sikep, their belief is that only by farming in Kendeng can the history and values of their ancestors be preserved.[63] Furthermore,

---

[51] Personal Documentation from 2020 Field Research (not specifically created for this chapter but documentation when visiting the location).

[56] Tim Pelaksana KLHS, 'Kajian Lingkungan Hidup Strategis (KLHS) Kebijakan Pemanfaatan Dan Pengelolaan Pegunungan Kendeng Yang Berkelanjutan' (2017) III-68–69.

[57] Ibid.

[58] Putri [48].

[59] WALHI [49].

[60] Tim Pelaksana KLHS (n 56) III-60.

[61] Wagner [50].

[62] PRAKARSA [51].

[63] Sumarlan and Rumpia [52].

Sedulur Sikep interprets all the elements in Kendeng: water, plants, animals, and non-material elements such as the symbol of the giver of life, Mother Earth.[64] Rituals in the Kendeng mountains are carried out through sacred sites, including Watu Payung, by the Sedulur Sikep community by praying for specific purposes, as well as water sources and caves that irrigate Sikep houses and rice fields which receive special blessings and offerings.[65]

In the context of disaster and climate change, the Sedulur Sikep community is aware that the environmental problems affecting them are intertwined with food issues and that the floods that destroyed their paddy fields in November 2022–March 2023 correlate with a cement factory.[66] In the Sedulur Sikep community context, adaptation means replanting hardwood trees/perennials to withstand flooding.[67] This adaptation technique is also followed in all Sedulur Sikep's traditions and knowledge; for example, during the COVID-19 pandemic, the community encouraged the availability of village rice barns for food security.[68] This logic is also strengthened by the traditional farming methods of the Sikep people, starting from respecting rice fields and farming, to not using chemicals because farming must not damage the environment.[69]

B. Rendu Community Survives the National Strategic Dam Project

In the case of the Lambo Dam, the Rendu community has experienced conflict since 2000. The Nagekeo Local Government consultant team surveyed three villages in 2001 in Ngada District (before the division of the area) with planned locations in three villages, namely Rendu Butowe, Labolewa, and Ulupulu, located in three districts.[70] The Ministry of Public Works and Public Housing also stated that every dam project is oriented towards mitigating climate change through water reserves and hydropower availability.[71] Water reserves are intended for optimisation when the dry and rainy seasons change.[72] However, this mitigation does not consider the community's existence and participation in dam construction, and it has even ended in land confiscation.[73] The community feels that they are not being involved. What is even more critical is the usage regarding customary land, which is the target of the Lambo Dam project. In fact, security forces crushed the community's protest, which ultimately resulted in violence.[74] And even though it was delayed for several years, the Lambo Dam finally commenced its construction from 2023, through the National Strategic Programme initiated by the Government of President Joko Widodo and is expected to be completed in 2025.[75]

The impacts and threats to the Rendu community's rights post the completion of the Lambo Dam project are accompanied by various crucial problems, such as environmental damage and cultural shifts due to tourism and agricultural problems. First, the construction of the Lambo Dam disrupts the Rendu community's relationship with nature, such as traditional weaving. One of the women of the community, Rosina Wonga (Mama Ros) revealed that the protest against the Dam by the Rendu community is primarily because of the loss of water sources, cotton, and natural dyes, the loss of traditions, and the issue of there being ancestral graves in places that were part of the project

[64] Primi (n 46) 89.

[65] Ibid.

[66] Utami [53].

[67] Ibid.

[68] Sumarlan and Rumpia (n 63) 43–45.

[69] Kurniasari [54].

[70] de Rosary [55].

[71] Kementerian PUPR [56].

[72] Ibid.

[73] Land grabbing in the name of the environment, also known as green grabbing, has become a new term legitimised through new legal and market mechanisms. See Fairhead et al. [57].

[74] Konsorsium Pembaruan Agraria [58].

[75] News related to the development progress of the Lambo Dam Project can be accessed and searched on the web page. https://sda.pu.go.id/berita/view/penggalian_terowongan_pengelak_bendungan_mbay_lambo. Accessed on 24 September 2023.

area.[76] The Rendu community emphasised that they had accepted the dam construction and had suggested that the dam be built on empty land in other locations which did not take up land adjacent to their settlements and ancestral land.[77] The people of Rendu felt that being dispossessed of this land meant not only their eviction from their ancestral land and residence but also the loss of their productive land which was the mainstay of their lives.[78] The condition becomes increasingly worrying when there is no clarity regarding relocation after completing the Lambo Dam construction project.[79] The impact of the construction of the Lambo Dam is yet to be quantified in specific data, and how much of the dam project claims the Rendu community's land is yet to be calculated. However, according to statements made by persons like Mama Siti and Mama Mince's, the impact of building the dam will take away their customary land along with their villages, gardens, schools, traditional ritual sites, and ancestral graves.[80] The Indonesian Government stated that the project area of the Lambo Dam had reached 499.5 ha.[81] The project took away some of their land, the Rendu community continued to move against the dam operation.[82]

With regard to the NTT, the impact of the construction of the Lambo Dam as a National Strategic Project was considered alongside other projects such as tourism infrastructure and the Food Estate programme. The attachment to Presidential Regulation Number 109 of 2020 concerning the Third Amendment to Presidential Regulation Number 3 of 2016 concerning the Acceleration of Implementation of National Strategic Projects mentions12 types of projects starting from bridges and toll roads, ports, airports, economic zone sectors, housing, trains, dams and irrigation, clean water and sanitation, coastal embankments, energy, technology, and education. The Forum for the Environment (WALHI) states that the linkages between National Strategic Projects are often intertwined and extend to various developing fields; for example, National Strategic Projects (PSN) also extend to energy, food and electricity projects, Special Economic Zones (KEK), and National Tourism Strategic Area (KSPN).[83] The Ministry of Public Works and Public Housing stated that the reason for building the Lambo-Mbay Dam was a strategy to support the provision of raw materials and irrigation which would in turn support the Food Estate project and reduce flooding that occurred in NTT.[84] One of the most crucial needs of the Food Estate project is the land issue and agricultural modifications, which simultaneously give rise to an agrarian crisis and ultimately threaten farmers.[85]

In the tourism context, NTT, especially Labuan Bajo, is a super-priority destination set up by the Indonesian Government and the Ministry of Tourism and Creative Economy. This super-priority destination, known as the '10 New Balis' project, also focuses on the creative economy, regional planning, and infrastructure to support Labuan Bajo as an international tourist destination.[86] This priority scale is undoubtedly a signal and an illustration of the fact that the community will be swamped with tourism infrastructure. Hotel accommodation in NTT in the year 2022 totalled 498 units.[87] Even now, the Rendu community is projected as part of an eco-tourism scheme; for example, in Rendu

[76] Dalidjo [59].

[77] Kurniasih [60].

[78] Arianto and Al Imran [61].

[79] Ibid.

[80] Nurita [62].

[81] Kementerian PUPR [63].

[82] See de Rosary [64].

[83] TINJAUAN Lingkungan Hidup 2023, [49] 9.

[84] Kementerian PUPR [65].

[85] Konsorsium Pembaruan Agraria (n 74) 36–37.

[86] More details can be seen at: https://kemenparekraf.go.id/hasil-pencarian/infografik-5-destinasi-super-prioritas, dan, https://setkab.go.id/tujuh-arahan-presiden-dan-transformasi-labuan-bajo-sebagai-destinasi-pariwisata-super-prioritas/. Accessed on 29 September 2023.

[87] Badan Pusat Statistik Provinsi Nusa Tenggara Timur [66].

Tutubadha Village, tourist activities begin from the natural hot springs *are petu meze*, Mount Lambo, followed by the traditional village of Rendu Tutubadha, and the Etu ritual (harvest and farming rituals), trekking and birdwatching on Mount Lambo.[88] Lessons learnt from other places reveal that eco-tourism is not always profitable and often has the potential for displacement and alienation of local communities. For example, the case of eco-tourism on Komodo Island, NTT, shows that eco-tourism tends to cause dispossession and displacement wherever the government-private sector partnership introduces capitalist accumulation at the expense of society.[89] In Rote Ndao, NTT, the local community's land, currently controlled by business privatisation, has closed the people's access by prohibiting entry to the hotel area.[90] For women in the local Rote Ndao community, tourism has no economic impact and disrupts the livelihoods of those who depend on seaweed.[91]

The traditions of the Rendu community were also affected because natural resources and the environment began to decrease and change. The illustrations related to traditional woven cloth show that access and depletion of natural resources due to conflict are slowly breaking the Rendu community's relationship with the Creator, their ancestors and their traditional territory.[92] As a comparison, this condition is also apparent from the condition of weavers in Sumba, NTT, where out of 476, only 19 remain.[93] The need of the hour for the Rendu community is to survive, at least so that they can slow down the damage and simultaneously protect tradition and nature.

Though it is difficult to find stories of the Rendu community being directly affected by climate change, information gathered from surrounding areas, particularly from farmers and dry cultivators on small islands from East Flores NTT shows that the impact is being felt. Farmers and dry cultivators experience disasters: dry seasons are getting longer and longer causing droughts; rainy seasons are getting shorter but the amount of rainfall received is much higher and is always accompanied by loud lightning, strong winds, tidal waves and storms, destroying possessions, property, and the sources of livelihood of residents.[94] The adaptation efforts carried out range from the use and consumption of local Sorghum plants to adapting dry farming based on the principles of reducing human intervention, reduction in carbon and the use of organic materials and environmentally friendly technology; use of more labour intensive than capital intensive methods; and orienting one's self to meet one's own needs rather than market demands.[95]

In closing this section, neglecting the rights of indigenous people does reflect that the Government is more inclined to support economic politics with greater emphasis on development of state and private elites. Eve Warburton called this developmentalism, in short, a pragmatic and growth-oriented approach in its policies, statist-nationalist in its ideological position, and conservative in its approach to issues of transparency and governance as well as human rights and justice.[96] The government has the backing and the strength of an oligarchic network that extends to the state bureaucracy and manifests through the interests of capital owners who utilise democracy.[97] Political power will thus be pervading every layer, so resistance never reaches the top. This condition is fairly apparent from the fragmentation of laws and policies, achieving nothing but a dead end in the fulfilment of human rights.

[88] Aja [67].

[89] Dale and Afioma [68].

[90] Neef [69].

[91] Ibid.

[92] Dalidjo [70].

[93] Ato [71].

[94] Baran [72].

[95] Ibid., 108.

[96] Warburton [73].

[97] Wiratraman [74].

4 Fragmentation of Law and Policy: Vulnerabilities for Indigenous Peoples in Indonesia

Human rights violations, land grabbing, and environmental degradation, which result from the lack of state support and effective laws, impact the survival and continuity of indigenous people. Regulations impacting indigenous people are currently spread across sectoral laws, such as the Forestry Law, the Plantation Law, and the Coastal Area and Small Island Management Law. This sectoral approach becomes an obstacle in protecting the rights of indigenous peoples.[98] For instance, of the 30 laws governing indigenous people in Indonesia, 12 were changed by Law No. 11 of 2020 concerning Job Creation.[99] These changes in laws make the existence and rights of indigenous peoples depend solely on whether they get is recognition from the state without any opportunity for self-identification. This position needs to be revised since self-identification constitutes a core element in preserving and protecting the rights of indigenous peoples.[100]

First, the absence of special laws poses a dilemma regarding the State's position and the existence of indigenous people as socio-political entities. Indigenous people or *Masyarakat Hukum Adat* are recognised under Article 18 B paragraph (2) of the 1945 Constitution of the Republic of Indonesia, 'as long as they are alive, following the development of society; following the principles of the Unitary State of the Republic of Indonesia; and regulated in law'.[101] From the above, it is clear that even under the Constitution, self-identification is lacking. Lack of recognition of the principle of self-identification is hazardous for indigenous peoples because it hinders the possibility of indigenous peoples to register collective rights to land, and the lack of monitoring mechanisms has made it easier for state and private parties to take over land, territory, and natural resources without free, prior and informed consent, by indigenous people.[102] In the absence of self-identification to state-centric approaches, it is necessary to interpret the indigenous people's rights through human rights, which are regulated relatively more fully in Chapter X of the 1945 Constitution.[103] It should be noted that demands related to human rights to obtain rights to land and other resources can be seen from the national movement of indigenous peoples.[104] The movement aims to recognise indigenous peoples and customary forests. Simultaneously, the national movement of the Indigenous Peoples Alliance of the Archipelago (national movement of indigenous peoples)/AMAN is transforming from being a non-governmental organisation.[105]

Before Article 18 of the 1945 Constitution was amended to become Articles 18, 18 A, and 18 B, the restrictions on indigenous communities were more relaxed because they provided direct protection for their rights by considering the rights of origin and particular territories.[106] In the explanation of this article,

[98] Arizona and Cahyadi [75].

[99] Zakaria [76]. Currently, the Job Creation Law is recognised with stipulation of government regulations, the replacement for Law Number 2 of 2022 concerning job creation.

[100] Hauser-Schäublin [77].

[101] Arizona [78].

[102] U.N. Committee on the Elimination of Racial Discrimination, Formal Communication to the Government of Indonesia, 30 April 2021, U.N. Doc CERD/EWUAP/103rd session/2021/MJ/CS/ks.

[103] Herlambang P. Wiratraman, ' Masyarakat Adat, Konstitusi dan HAM' ('Mendorong Proses Legislasi Daerah Untuk Mewujudkan Pembangunan Sumba Berbasis Masyarakat Adat Demi Pemenuhan Hak Tenurial Bagi Masyarakat Adat Sebagai Subjek Dari Pembangunan Berkelanjutan Sesuai Dengan NAWACITA Presiden Jokowi', kerjasama Aliansi Masyarakat Adat Nusantara (Sumba) dan Universitas Kristen Wira Wacana, Waingapu, 22 Mei 2018)<https://herlambangperdana.files.wordpress.com/2018/05/herlambang-2018-masyarakat-adat-konstitusi-dan-ham.pdf>accessed on 2 October 2023 5.

[104] van der Muur et al. [79].

[105] Arizona et al. [80].

[106] Bedner and Van Huis [81]

even the autonomy of indigenous communities is guaranteed inasmuch as to say that they are given rights to have their own government, autonomous communities, and customary rights as long as they are not a state within a State.[107] Currently, however, the process of recognition as an indigenous people or based on the Indonesian Constitution as Adat Law community can be carried out in the following manner: (a) legal recognition by the communities and/or the government; (b) data and information related to indigenous peoples is verified by the Indigenous Peoples Committee (district/city or provincial or national); (c) The committee announces the results of the verification to the public before further submitting it to the Regent/Mayor or the Governor or the President according to their authority to be designated through a decree.[108]

Formalism coupled with a state-centric approach to recognising indigenous communities is also manifested in other regulations, which negate self-identification and give rise to administrative and bureaucratic problems. The registration of the rights of indigenous peoples is currently regulated through:

a. Indigenous village recognition under Articles 103, 104, and 105 of Law No. 6 of 2014 concerning villages;
b. Indigenous people as subjects of rights under Article 1 Number 2 and Article 6 paragraph (2) Minister of Home Affairs Regulation No. 52 of 2014 concerning Guidelines for the Recognition and Protection of Indigenous Peoples; and
c. Indigenous people's land rights under Article 3 paragraph (3), Article 17, Article 29, Article 30 paragraph (3), Article 32, and Article 33 paragraph (2) of the Regulation of the Minister of Agrarian and Spatial Planning 18/2021 Concerning Procedures for Determining Management Rights and Land Rights.

This recognition through various regulations reflects that the administration and the bureaucracy also burden the indigenous community. To gain recognition, the indigenous community needs to study the typology's suitability, which requires the gathering of experts, assistants, and networks. This bureaucratic and administrative complexity is also accompanied by the emergence of informal/political close relationships that trap indigenous communities in obtaining land rights.[109] At the same time, there needs to be more consistency in the administration of indigenous communities, which is lacking because of the absence of institutions and registration of indigenous territories. The indigenous community could then use the network to advocate and facilitate the registration of customary territories through non-governmental organisations. So far, the Indigenous Area Registration Agency (BRWA) has recorded that the total land area of the Indigenous community is around 26,970,220.82 ha.[110] Registration of customary territories is intended to reorganise the relationship between the indigenous community and the state, especially regarding ownership and preservation of natural resources in customary territories, as well as resolve thousands of agrarian conflicts.[111]

However, the existing laws and policies threaten the existence of indigenous communities. Firstly, through the Stipulation of Government Regulations, the replacement for Law Number 2 of 2022 concerning job creation, also known as the job creation law, threatens the indigenous community because of its private and capitalistic characteristics and orientation. As environmental rights defenders, indigenous

[107] Ibid.

[108] Sumardjono [82]. The problems of state recognition start from the myth of homogeneity-harmony in indigenous communities. Local governments also need help in determining indigenous communities' origins; recognising indigenous peoples is complex and expensive because it requires lobbying and research. See Arizona [83].

[109] van der Muur [84].

[110] The reason for establishing BRWA can be seen on the link: https://brwa.or.id/pages/about.

[111] Widodo et al. [85].

peoples are vulnerable to physical violence and criminalisation due to the strong-arm tactics of corporations who enjoy impunity. However, the job creation law dilutes the role of the public and indigenous people in the conduct of environmental impact analysis (AMDAL), and thus, there is greater need for more adequate protection for activists.[112]

In addition, the Presidential Regulation Number 98 of 2021 concerning structuring the economic value of carbon in the context of achieving national contribution targets and controlling greenhouse gas emissions in national development is an instrument that leads to the usurpation of the rights of indigenous peoples. Implementing carbon trading is detrimental because it does not open up space for participation by indigenous communities and or respect generations for utilising natural resources.[113] For this reason, the Association of Defenders of Indigenous Peoples of the Archipelago (PPAMAN) submitted a judicial review to the Supreme Court with the argument that the participation of indigenous peoples was not reflected and there was a threat to confiscate land in the area without measures for sustainability; no carbon diversity was taken into account; and that the community would be faced with the burden to pay for the carbon footprint caused for no fault of itself.[114] This request for judicial review was granted by the Supreme Court, granting the participation of the indigenous community in the carbon context through the Supreme Court of the Republic of Indonesia—Decision No.61/P/HUM /2022[115] The Supreme Court stated that Indigenous Peoples are providers of Economic Carbon Value as per Article 46 paragraph (2) letter d of the Presidential Decree Number 98 of 2021 because of their contribution to increasing carbon reserves through forest utilisation in business activities that utilise environmental services.[116]

Laws and policies related to climate migration also add to the problem of vulnerability of indigenous peoples. In the context of the climate crisis, Indonesia has a National Action Plan–Climate Change Adaptation (RAN-API) and a National Target Roadmap for Climate Change Adaptation (NDC Roadmap). In the RAN-API document, relocation is only mentioned in the infrastructure context, for example, in settlement relocation planning, relocation of strategic roads, and urban infrastructure.[117] The NDC Roadmap only focused on infrastructure resilience, such as risk standards, increasing disaster preparedness through infrastructure, climate-resistant housing, and revitalising climate-resilient cities.[118] The absence of a relocation policy leads to dependence on local policies, lack of proper focus, and lack of empowerment of communities.[119] The limitations of the local government also resulted in difficulties for the indigenous peoples; for example, in the case of Simonet Hamlet, Pekalongan, there was no availability of relocation land, and ultimately the residents lost their jobs as jasmine farmers and fishermen.[120] To resist eviction, the Sikep community urged a policy to preserve the Kendeng mountains by rejecting the licenced and illegal cement industry and pushing for a policy plan for protecting and managing the karst ecosystem.[121] Meanwhile, in the Rendu community, there has been no plan or response to relocation because of climate change. In other cases, such as the displacement of Dayak communities whose voices have been neglected by the new capital mega-infrastructure there are consequences that threaten the sustainability of their environment and culture.[122] In cases with

[112] Wiratraman [86].

[113] AMAN [87].

[114] Hariandja [88].

[115] Perhimpunan Pembela Masyarakat Adat Nusantara (PPAMAN) [89].

[116] Ibid.

[117] Badan Perencanaan Pembangunan Nasional [90].

[118] KLHK [91].

[119] Gemilang [92].

[120] Ibid.

[121] Walhi (n 59).

[122] See Yumagulova et al. [93].

a socio-economic dimension, people in West Kalimantan are even threatened with inter-generational displacement because palm plantation results in land being taken, giving rise to new structural poverty in the community.[123]

Lastly, the fragmentation of laws and policies that are detrimental to indigenous people needs to be addressed with clarity and firm protection of basic rights through the ratification of the Indigenous Peoples Bill. Proposals for a draft law on indigenous peoples have been made since 2012, but they have yet to succeed. This draft has provided legal education to indigenous peoples to show that the problems they face at the local level are related to national regulations.[124] In future, the ratification of the Indigenous Peoples Law must also be reviewed in conjunction with the legal politics of recognition and implementation of the decentralised rights of indigenous peoples.

5 Conclusion

The adaptation and mitigation of environment and climate change based on the knowledge of indigenous people sounds like an exciting premise but lacks the force to bring about the emergence of practical solutions. Indigenous peoples' experiences and way of life are essential to why these communities have survived hundreds of years. This chapter uses the case of the Sikep and Rendu communities to also reflect on marginalisation due to the intersection of land grabbing, human rights violations, and dilemmatic migration choices. The increasing expansion of the cement industry and dam construction means that the Sedulur Sikep community and the Rendu community are caught at a crossroads between the uncertainty of protecting a degrading environment and the wisdom of their ancestors and their traditional knowledge. Migration and relocation of the Sedulur Sikep community and the Rendu community will raise the question of how to move communities whose culture and identity are one with nature. Additionally, the fragmentation of law and policy creates multiple vulnerabilities due to the absence of indigenous laws, limited legal recognition, the bureaucratic-administrative nature of land registration, and the loss of meaningful participation in environmental and climate policy.

[123] Li [94].

[124] Arizona [95].

References

1. Moggridge BJ et al (2022) IPCC reports Still exclude indigenous voices. Join us at our sacred fires to find answers to climate change. Conversat 1. https://theconversation.com/ipcc-reports-still-exclude-indigenous-voices-come-join-us-at-our-sacred-fires-to-find-answers-to-climate-change-178045. Accessed 3 May 2023
2. IPCC (2022) Climate change 2022: impacts, adaptation and vulnerability. In: Pörtner H-O et al (eds) Contribution of working group II to the sixth assessment report of the intergovernmental panel on climate change. Cambridge University Press
3. Locke A, Childress M (2022) New IPCC climate report stresses indigenous & local land rights 58 times : let's respond with a concrete tenure plan (commentary). Mongabay 1. https://news.mongabay.com/2022/03/after-ipcc-climate-report-stresses-indigenous-local-land-rights-58-times-lets-respond-with-a-concrete-tenure-plan-commentary/. Accessed on 3 May 2023
4. Indonesia Government (2022) Enhanced nationally determined contribution Republic Indonesia, 35–40
5. United Nations Environment Programme (2022) Emissions gap report 2022: The closing window-climate crisis calls for rapid transformation of societies. https://www.unep.org/emissions-gap-report-2022. XVI
6. World Economic Forum (2023) The global risks report 2023—18th Edition. World Economic Forum 6. https://www.weforum.org/reports/global-risks-report-2023/. Accessed 27 July 3 2023
7. Solomon S et al (2009) Irreversible climate change due to carbon dioxide emissions. Proc Natl Acad Sci USA 106:1709
8. IPCC (2022) Summary for policymakers. In: Pörtner HO et al (eds) Climate change 2022: impacts, adaptation and vulnerability. Contribution of working group II to the sixth assessment report of the intergovernmental panel on climate change. Cambridge University Press, 9
9. International Labour Organization (ILO) (1989) Indigenous and tribal peoples convention, C169, 27 June 1989, C169, Article 27. https://www.refworld.org/docid/3ddb6d514.html. Accessed 20 Oct 2023
10. UN General Assembly (2007) United Nations Declaration on the Rights of Indigenous Peoples:

Resolution/adopted by the General Assembly, 2 October 2007, A/RES/61/295, Article 31
11. Molnár ZS, Babai D (2021) Inviting ecologists to delve deeper into traditional ecological knowledge. Trends Ecol Evol 36:679
12. Berkes F (1993) Traditional ecological knowledge in perspective. In: Inglis JT (ed) Traditional ecological knowledge concept and case. International Program on Traditional Ecological Knowledge and International Development Research Center, 5
13. Van Eijck M, Roth W-M (2007) Keeping the local local: recalibrating the status of science and traditional ecological knowledge (TEK) in education. Sci Educ 91:938
14. Ballard M (2012) Flooding sustainable livelihoods of the Lake St Martin First Nation: the need to enhance the role of gender and language in Anishinaabe knowledge systems. University of Manitoba, 5. https://mspace.lib.umanitoba.ca/xmlui/bitstream/handle/1993/5312/myrle_ballard.pdf?sequence=1&isAllowed=y. Accessed 13 June 2023
15. Cajete GA (2020) Indigenous science, climate change, and indigenous community building: a framework of foundational perspectives for indigenous community resilience and revitalization. Sustainability 12:2
16. Seftyono C (2011) Pengetahuan Ekologi Tradisional Masyarakat Orang Asli Jakun Dalam Menilai Ekosistem Servis Di Tasik Chini, Malaysia. J Ilmu Sos Dan Ilmu Polit 15:55 (indonesian)
17. Sumarwati S (2022) Traditional ecological knowledge on the slope of Mount Lawu, Indonesia: all about non-rice food security. J Ethnic Foods 9:1
18. Siahaya ME et al (2016) Traditional ecological knowledge on shifting cultivation and forest management in East Borneo, Indonesia. Int J Biodivers Sci, Ecosyst Serv Manag 12:14
19. Ariando W (2022) Developing a model for the integration of Bajau traditional ecological knowledge in the management of locally managed marine area: a case study of Wakatobi Regency, Indonesia developing a model for the integration of Bajau traditional ecological knowledge. Chulalongkorn
20. Yuliani EL et al (2018) The roles of traditional knowledge systems in Orang-Utan Pongo Spp. and forest conservation: a case study of Danau Sentarum, West Kalimantan, Indonesia. Oryx 52:156
21. Buchori I et al (2018) Adaptation to coastal flooding and inundation: mitigations and migration pattern in Semarang City, Indonesia. Ocean Coast Manag 163:20
22. Zickgraf C (2021) Climate change, slow onset events and human mobility: reviewing the evidence. Curr Opin Environ Sustain 50:27
23. Jacobson C (110) When is migration a maladaptive response to climate change? RegNal Environ Chang 19:110
24. Sakdapolrack P et al (2023) The limits of migration as adaptation. A conceptual approach towards the role of immobility, disconnectedness and simultaneous exposure in translocal livelihoods systems. Clim Dev 0:5
25. Miller MA et al (2021) Hydrosocial rupture: causes and consequences for transboundary governance. Ecol Soc 26:3
26. Middleton C (2022) The political ecology of large hydropower dams in the Mekong Basin: a comprehensive review. Water Altern 15(2):273
27. Nadasdy P (1999) The politics of TEK: power and the "integration" of knowledge. Arct Anthropol 36:9
28. Berkes F (2008) Sacred ecology, 2nd edn. Routledge, 258
29. Shiva V (2020) Reclaiming the commons biodiversity, indigenous knowledge, and the rights of mother earth. Synergetic Press
30. Shiraishi T (1990) Dangir's testimony: Saminism reconsidered. Indonesia. Cornell University, vol 50, p 113
31. Korver APE (1976) The Samin movement and millenarism. Bijdragen tot de Taal-, Land- en Volkenkunde 132:250
32. AMAN (2016) Kabar Nusantara. Aliansi Masyarakat Adat Nusantara, 2
33. Hendro EP (2021) Komodifikasi Budaya Dalam Pengembangan Wisata Di Kabupaten Nagekeo Nusa Tenggara Timur. ENDOGAMI: J Ilm Kaji Antropol 4:71. (Indonesian)
34. Safitri MA (2018) Social movement in Indonesian Mining Law Enforcement: The case of peasants-scholars nexus in karst mining dispute in Java. In: Advances in economics, business and management research. Atlantis Press, 228
35. Peluso NL (1992) Rich forests, poor people. University California Press, 70
36. Ford JD et al (2020) The resilience of indigenous peoples to environmental change. One Earth 2(6):539
37. Ngongo Y et al (2022) The Roles of Cassava in Marginal Semi-Arid Farming in East Nusa Tenggara—Indonesia. Sustain (Switz) 14:16
38. Miranda, Eghenter C (2021) Transformation of food systems needs more inclusion. The Jakarta Post 5. https://www.thejakartapost.com/academia/2021/10/16/transformation-of-food-systems-needs-more-inclusion-.html. Accessed 7 May 2023
39. Lake RC, Purbadi YD, Harmans HF (2020) Identification and orientation on spatial arrangement of Wajo traditional village, Keo Tengah, Nagekeo Regency. Local Wisdom : Jurnal Ilmiah Kajian Kearifan Lokal 12:28
40. Mojo E, Hadi SP, Purnaweni H (2015) Saminist's Indigenous Knowledge in Water Conservation in North Karts Kendeng Sukolilo. Komunitas 7(2):241

41. Putri PS (2017) Re-claiming lost possessions: a study of the Javanese Samin (Sedulur Sikep) movement to maintain their peasant identity and access to resources. University of Oslo, 80
42. Ngongo et al (2022) Local wisdom of West Timorese farmers in land management. Sustainability 14(10):6023
43. Arif A (2020) Sorgum: Benih Leluhur untuk Masa Depan. Kepustakaan Populer Gramedia
44. BPS Kabupaten Rembang (2023) Kabupaten Rembang Dalam Angka 2023. BPS Kabupaten Rembang, 273–274
45. Kresna M (2022) Menambang Nestapa di Pegunungan Kendeng Utara. Project Multatuli. https://projectmultatuli.org/menambang-nestapa-di-pegunungan-kendeng-utara/. Accessed on 26 Sept 2023
46. Ramsden K (2020) Cement and concrete: the environmental impact. Princeton Education. https://psci.princeton.edu/tips/2020/11/3/cement-and-concrete-the-environmental-impact. Accessed on 16 Dec 2023
47. Mora C et al (2018) Broad threat to humanity from cumulative climate hazards intensified by greenhouse gas emissions. Nat Clim Chang 8(12):1062
48. Putri PS (2017) The meaning making of an environmental movement: a perspective on Sedulur Sikep's narrative in anti-cement movement. PCD J 5(2):318
49. WALHI (2023) Bencana Banjir Meluas, JM-PPK Datangi Istana Tagih Pelaksanaan Rekomendasi KLHS Pegunungan Kendeng dan Pengesahan RPP Karst. Siaran Pers WALHI. https://www.walhi.or.id/bencana-banjir-meluas-jm-ppk-datangi-istana-tagih-pelaksanaan-rekomendasi-klhs-pegunungan-kendeng-dan-pengesahan-rpp-karst. Accessed on 26 Sept 2023
50. Wagner I (2021) Pandemi COVID-19, Proyek Semen dan Komunitas Rakyat Pegunungan Kendeng Utara. Heinrich Böll Stiftung. https://th.boell.org/en/2021/10/10/pandemi-covid-19-kendeng
51. PRAKARSA (2022) Pelanggaran Hak Warga Dan Tanggung Jawab Bank Dalam Pembiayaan Industri Semen Di Pegunungan Kendeng Utara, 60–62
52. Sumarlan Y, Rumpia JR (2021) The Wong Sikep Movement of Central Java's Longue Dureé in its paradoxical nature: a portent form of non-violence human rights struggles through different names. 43. https://th.boell.org/sites/default/files/2021-08/Samin-Sikep-Central-Java-final.pdf
53. Utami KD (2023) Kisah Sedulur Sikep Melawan Perusakan Lingkungan. Kompas id. https://www.kompas.id/baca/nusantara/2023/03/23/kisah-sedulur-sikep-melawan-perusakan-lingkungan. Accessed 1 Oct 2023
54. Kurniasari D, Cahyono E, Yuliati Y (2018) Kearifan Lokal Petani Tradisional Samin Di Desa Klopoduwur, Kecamatan Banjarejo, Kabupaten Blora. Habitat 29:35–36
55. de Rosary E (2016) Mengapa Pembangunan Waduk Lambo Ditolak Masyarakat Di Tiga Desa Adat. Apa Masalahnya? Mongabay 2:2. https://www.mongabay.co.id/2016/10/31/mengapa-pembangunan-waduk-lambo-ditolak-tiga-desa-adat-apa-masalahnya/. Accessed 5 May 2023.
56. Kementerian PUPR (2023) Menteri Basuki: Pembangunan Bendungan Upaya Nyata Atasi Ancaman Perubahan Iklim. Kementerian PUPR. https://pu.go.id/berita/menteri-basuki-pembangunan-bendungan-upaya-nyata-atasi-ancaman-perubahan-iklim
57. Fairhead J, Leach M, Scoones I (2012) Green grabbing: a new appropriation of nature. J Peasant Stud 39(2):254
58. Konsorsium Pembaruan Agraria (2021) Catatan Akhir Tahun 2021 Konsorsium Pembaruan Agraria "Penggusuran Skala Nasional (PSN)", 21. http://kpa.or.id/publikasi/catatan-akhir-tahun/2/catatan-akhir-tahun-2021-penggusuran-skala-nasional-psn. Accessed 24 Sept 2023
59. Dalidjo N (2019) Women are heroes stories of resistance behind the beauty of Kain. Kain Kita (Kain by Indonesia), 13–14. https://medium.com/kain-kita/women-are-heroes-7e3ead436825. Accessed 6 May 2023
60. Kurniasih Y (2021) Masyarakat Adat: Dibicarakan Hanya Saat Konflik atau Bajunya Dipakai Pejabat. Project Multatuli. https://projectmultatuli.org/masyarakat-adat-dibicarakan-hanya-saat-konflik-atau-bajunya-dipakai-pejabat/. Accessed 29 Sept 2023
61. Arianto K, Al Imran H (2023) Memahami Pembangunan Sosial Dibalik Pembangunan Waduk Lambo Di Nusa Tenggara Timur: Tinjauan Sosiologis. Responsive: Jurnal Pemikiran dan Penelitian Bidang Administrasi, Sosial, Humaniora Dan Kebijakan Publik 6(2):122
62. Nurita D (2022) Terusir dari Kampung Sendiri: Perampasan Tanah Ulayat dan Satu Dekade RUU Masyarakat Adat Mangkrak. Tempo.co. http://interaktif.tempo.co/proyek/rencana-undang-undang-masyarakat-adat/. Accessed 20 Nov 2023.
63. Kementerian PUPR (2023) Penggalian Terowongan Pengelak Bendungan Mbay Lambo. Kementerian PUPR. https://sda.pu.go.id/berita/view/penggalian_terowongan_pengelak_bendungan_mbay_lambo#:~:text=Bendungan%20Mbay%2FLambo%20merupakan%20Proyek,genangan%20mencapai%20499%2C5%20Ha. Accessed 25 Nov 2023
64. de Rosary E (2022) Apa Masalah Pembangunan Waduk Lambo. Mongabay 2022. https://www.mongabay.co.id/2022/04/23/apa-masalah-pembangunan-waduk-lambo/. Accessed 25 Nov 2023
65. Kementerian PUPR (2021) Kementerian PUPR Bangun Jaringan Irigasi, Jamin Ketahanan Pangan di NTT. Kementerian PUPR. https://pu.go.id/berita/kementerian-pupr-bangun-jaringan-irigasi-jamin-ketahanan-pangan-di-ntt. Accessed 28 Sept 2023

66. Badan Pusat Statistik Provinsi Nusa Tenggara Timur (2023) Jumlah Akomodasi Hotel Menurut Kabupaten/Kota (Unit), 2020–2022. BPS Provinsi Nusa Tenggara Timur. https://ntt.bps.go.id/indicator/16/387/1/jumlah-akomodasi-hotel-menurut-kabupaten-kota.html. Accessed 29 Sept 2023
67. Aja AF (2020) Analisis Potensi Ekowisata dan Kesiapan Masyarakat Desa Rendu Tutubadha dalam Pengembangan Ekowisata. J Destin Pariwisata 8(2):231
68. Dale CJP, Afioma G (2020) Puzzling Confluence of Conservation and Ecotourism in Komodo National Park, Indonesia. Japan-ASEAN Transdisciplinary Studies Working Paper Series 10:15
69. Neef A (2019) Tourism, land grabs and displacement a study with particular focus on the global South, 36–37. https://researchspace.auckland.ac.nz/bitstream/handle/2292/49710/Tourism-Land-Grab-Study-Complete-Final.pdf?sequence=2
70. Dalidjo N (2019) Women are heroes stories of resistance behind the beauty of Kain. Kain Kita (Kain by Indonesia), 13–14. https://medium.com/kain-kita/women-are-heroes-7e3ead436825
71. Ato S (2023) Maestro Tenun Sumba Tergerus Zaman. Kompas id. https://www.kompas.id/baca/nusantara/2023/05/25/maestro-tenun-sumba-tergerus-zaman. Accessed 30 Sept 2023
72. Baran MK (2022) Ura Timu Etnografi Iklim Mikro Flores. INSIST Press, 24
73. Warburton E (2016) Jokowi and the New Developmentalism. Bull Indones Econ Stud 52:297
74. Wiratraman HP (2021) Political Cartels and the Judicialization of Authoritarian Politics in Indonesia. Int'l J. Const. L. Blog, 1. http://www.iconnectblog.com/2021/02/symposium-constitutional-struggles-in-asia-part-ii-political-cartels-and-the-judicialization-of-authoritarian-politics-in-indonesia/%0A5/5. Accessed 7 May 2023
75. Arizona Y, Cahyadi E (2013) The Revival of indigenous peoples: contestations over a special legislation on Masyarakat Adat. In: Hauser-Schäublin B (ed) Adat and Indigeneity in Indonesia. Culture and entitlements between heteronomy and self-ascription. Göttingen University Press, 51
76. Zakaria Y (2021) Nasib Masyarakat Adat dalam UU Cipta Kerja. Forest Digest. https://www.forestdigest.com/detail/988/omnibus-law-uu-cipta-kerja-masyarakat-adat. Accessed 2 Oct 2023
77. Hauser-Schäublin B (2013) Introduction. The power of indigeneity: reparation, readjustments and repositioning. In: Hauser-Schäublin B (ed) Adat and indigeneity in Indonesia. Culture and entitlements between heteronomy and self-ascription. Göttingen University Press, 9
78. Arizona Y (2014) Konstitusionalisme Agraria. STPN Press, 50
79. van der Muur W et al (2019) Changing indigeneity politics in Indonesia: from Revival to projects. Asia Pac J Anthropol 20:385
80. Arizona Y, Wicaksono MT, Vel J (2019) The role of indigeneity NGOs in the legal recognition of Adat communities and customary forests in Indonesia. Asia Pac J Anthropol 20:501–502
81. Bedner A, Van Huis S (2008) The return of the native in Indonesian law: Indigenous communities in Indonesian legislation. Bijdragen tot de taal-, land-en volkenkunde/J Humlties Soc Sci Southeast Asia 164(2):171
82. Sumardjono MSW (2018) Regulasi Pertanahan Dan Semangat Keadilan Agraria. STPN Press
83. Arizona Y (2019) Adat in Indonesian land law: a promise for the future or a dead end? Asia Pac J Anthropol 20(5):425
84. van der Muur W (2018) Forest conflicts and the informal nature of realising indigenous land rights in Indonesia. Citizsh Stud 171–172:160
85. Widodo K et al (2015) Pedoman Registrasi Wilayah Adat Badan Registrasi Wilayah Adat (BRWA). BRWA, 2
86. Wiratraman HP (2020) 3 Ancaman UU Ciptaker Bagi Para Pembela Lingkungan Dan HAM. The Conversation, 2–3. https://theconversation.com/3-ancaman-uu-ciptaker-bagi-para-pembela-lingkungan-dan-ham-148988. Accessed 5 May 2023
87. AMAN (2022) Catatan Tahun 2022 Melawan Penundukan, 12. https://www.aman.or.id/publication-documentation/catatan-tahun-2022-aman:-melawan-penundukan
88. Hariandja R (2022) Kala Masyarakat Adat Gugat Aturan Nilai Ekonomi Karbon. Mongabay, 3–5. https://www.mongabay.co.id/2022/10/15/kala-masyarakat-adat-gugat-aturan-nilai-ekonomi-karbon/. Accessed 4 May 2023
89. Perhimpunan Pembela Masyarakat Adat Nusantara (PPAMAN) (2023) Putusan Mahkamah Agung Menyatakan Masyarakat Adat Sebagai Penerima Manfaat Nilai Ekonomi Karbon. PPAMAN. https://ppman.org/putusan-mahkamah-agung-menyatakan-masyarakat-adat-sebagai-penerima-manfaat-nilai-ekonomi-karbon/. Accessed 2 Oct 2023
90. Badan Perencanaan Pembangunan Nasional (2014) Rencana Aksi Nasional Perubahan Iklim (RAN-API). Bappenas, 112–142
91. KLHK (2020) Roadmap Nationally Determined Contribution (NDC) Adaptasi Perubahan Iklim. Kementerian Lingkungan Hidup dan Kehutanan Republik Indonesia, 122–132
92. Gemilang MSC (2023) Perubahan iklim sebabkan banyak 'bedol desa' di Pantura, tapi kebijakan perlindungan warga belum tersedia. Conversation. https://theconversation.com/perubahan-iklim-sebabkan-banyak-bedol-desa-di-pantura-tapi-kebijakan-perlindungan-warga-belum-tersedia-203407. Accessed 1 Oct 2023
93. Yumagulova L et al (2023) Indigenous perspectives on climate mobility justice and displacement-mobility-immobility continuum. Clim Dev 0:12

94. Li T (2017) Intergenerational displacement in Indonesia's oil palm plantation zone. J Peasant Stud 44(6):16
95. Arizona Y (2023) Adat as strategy for legal struggle and legal mobilization (Adat Sebagai Strategi Perjuangan dan Mobilisasi Hukum). Indones J Socio-Leg Stud 2(2):15–16

James Reinaldo Rumpia is a Researcher at the Center for Public Policy and Human Rights Studies, Faculty of Law, University of Lampung. James is an alumnus of the Master of Arts in Human Rights, Institute of Human Rights and Peace Studies, Mahidol University, Thailand. The focus of expertise other than human rights is political economy, environmental studies, indigenous peoples, and marginalized communities. In 2021 he was involved as a project researcher at the Henrich Boll Stiftung, Southeast Asia Regional Office for a study related to the "Environmental Movement of the Sedulur Sikep Indigenous People in Central Java, Indonesia" together with a team from the Institute of Human Rights and Peace Studies, Mahidol University. In 2023 he is doing ongoing research on "Climate Change, Indigenous Knowledge, Land Rights, and Human Rights" with the Raoul Wallenberg Institute, Jakarta. Apart from being active as a researcher, James is also active as the Organizational Development and Research Coordinator of the Perkumpulan Damar, Lampung, Indonesia, as well as an adviser at the Legal Aid and Human Rights Association (PBHI) in the Lampung region, Indonesia which focuses on legal aid for people experiencing poverty and victims of human rights violations.

Climate Change and Socio-economic Issues in Sri Lanka: An Assessment of Landslide-Induced Relocations in Aranayake

Dinushika M. Yapa Abeywardhana

Abstract

Being a tropical island with a developing economy in the South Asian region, Sri Lanka is exposed to the adverse impacts of climate change. Extreme weather events and their slow-onset effects have threatened well-functioning communities, serving as a core factor for climate mobility. This paper explores the socio-economic impacts of landslide-induced relocations in Aranayake, Sri Lanka. Communities in Aranayake were exposed to a massive landslide in 2017, which led to a wide range of relocations. The relocation of the affected community to peripheral areas intensified the economic and social vulnerabilities, leaving their well-being at risk. The lack of consideration for the societal implications of the relocation and the rebuilding process has further enhanced their vulnerability by disrupting social networks and reducing individual livelihoods. Though it has been five years since the landslide relocation, the insecurities of the communities persist, and access to fundamental rights, including food, development, and freedom, are negatively impacted. Proper identification of the community's needs before the relocation and proper post-monitoring is a must in order for policy formation in Sri Lanka to be more effective and community sensitive.

1 Introduction

Climate change is exacerbating the sudden onset of disaster events worldwide. Most developing countries are at risk of these disasters due to the uncertain nature of hazard events, high levels of exposure and vulnerability, and low coping mechanisms and infrastructure available to them compared to developed countries.[1] Moreover, disasters mainly affect people with low income and aggravate their vulnerability.[2] The countries in the South Asian region are passing through a 'new climate normal' in which intensifying heat waves, cyclones, droughts, and floods are testing the limits of government, businesses, and citizens to adapt.[3] One of the significant consequences of climate change is the rise in the need for communities to relocate as their homes become increasingly exposed to climate-related hazards.

D. M. Y. Abeywardhana (✉)
Department of Sociology, University of Peradeniya, Peradeniya, Sri Lanka
e-mail: dinushikay@arts.pdn.ac.lk

[1] United Nations Office for Disaster Risk Reduction [1].

[2] Lal et al. [2].

[3] The World Bank [3].

S. Jolly et al. (eds.), *Climate-Related Human Mobility in Asia and the Pacific*, Sustainable Development Goals Series, https://doi.org/10.1007/978-981-97-3234-0_14

Despite the expectation that relocation will enable individuals to improve their well-being, the process is not always voluntary. Authorities may require displaced communities to permanently migrate even if they are not willing to, due to prevailing risks. Relocation is a strategy that has been adopted in many parts of the world, as reflected in a number of case studies from the United States of America (36), the Philippines (29), India (22), Sri Lanka (19), China (17), Indonesia (17), Vietnam (17), Fiji (15), Japan (15) and Colombia (8), among others.[4] Apart from the United States, Fiji and Colombia, the rest of the countries having the highest identified relocation projects are found in Asia. Involuntary relocation poses several key challenges in the form of livelihood disruptions and fragmentation of social ties for the displaced communities. Hence, policymakers face mammoth challenges in terms of envisaging a mechanism that protects people from natural hazards, while also respecting other fundamental rights.

As far as the policy frameworks are concerned, Sri Lanka has over the years established a robust catalogue of laws and policies to deal with environmental degradation and disaster management. However, it was only in 2011 that climate change became part of the explicit policy framework under the 2011 amendment to the Sri Lanka Disaster Management Act No 13 of 2005.[5] Even then climate-related human mobility was not explicitly addressed.

Even under the National Climate Change Policy in Sri Lanka 2012 there is no reference to climate-related human mobility.[6] However, from 2013, the Sri Lanka National Disaster Management Plan (2013–2017) and Sri Lanka Comprehensive Disaster Management Programme (2014–2018) link disasters, climate change, and displacement and highlight the need for relocation of people out of disaster-prone areas.[7] It should be noted that the National Involuntary Relocation Policy (NIRP, 2001), which emphasises the issues associated with involuntary relocation induced by development initiatives and emergencies, is the policy framework that addresses relocation in Sri Lanka, though it has no thorough focus on disaster- or climate-induced involuntary relocation.[8] A Resettlement Authority (2007) was established as an authority vested with the power to formulate a national policy and to plan, implement, monitor, and coordinate the relocation of internally displaced persons, and to provide for matters connected therewith or incidental thereto.[9]

This chapter explores the socio-economic impacts of post-landslide-induced relocations in Aranayake, Sri Lanka. This study has two sub-objectives, namely identifying livelihood issues experienced by the relocated communities and identifying new forms of challenges experienced by them. This empirical study was conducted in two selected relocated sites, China Friendship Village and Panapurewatta. These villages were established to cater to the displaced community of landslide events that occurred in the 'Samasara' Mountain, which is located in the Aranayake divisional secretariat in the Kegalle district, Sri Lanka, in May 2017.[10] Qualitative data was gathered through twenty semi-structured interviews, five in-depth interviews, and observation in the two relocated communities, across 2022 and 2023.

The chapter is divided into four sections, starting from the introduction. Section two provides a brief introduction to the climate-induced disaster profile in Sri Lanka. The third section deals with landslides as a disaster in Sri Lanka with specific reference to Aranayake, displacement and relocation, assessing two selected sites. The fourth section provides conclusions.

4 Bower and Weerasinghe [4].

5 Disaster Management Act No 13 of 2005, Sri Lanka.

6 National Climate Change Policy in Sri Lanka 2012.

7 Sri Lanka Comprehensive Disaster Management Program (2014–2018).

8 Mombauer and Wijenayake [5].

9 Sri Lanka, Resettlement Authority Act 2007 [No. 09] 1.

10 Disaster Service Centre [6].

2 Sri Lanka and Climate Change-Induced Disasters

Sri Lanka is an island in the Indian Ocean with a land area of 65,610 km^2 and a population of 22.1 million.[11] Topographically, the island consists of a south-central mountainous region and is surrounded by broad lowland plains. The frequency of disasters in Sri Lanka has increased in the recent past, leading to an elevated risk of exposure to such incidents.[12] The devastating Indian Ocean Tsunami that occurred on 26 December 2004 had a profound impact on the social fabric and economic landscape of Sri Lanka.[13] In a span of fifteen minutes, the tsunami resulted in the loss of over 30,000 lives, inflicted injuries upon more than 15,000 individuals, and caused displacement for over 800,000 individuals, while causing the destruction of their properties.[14]

Beyond the tsunami and its socio-environmental impacts, the climate crisis has been exacerbating in the recent past.[15] Major floods hit Sri Lanka in 2010, 2011, 2014, 2016, and 2018, and these events caused more than USD 2 billion in damages between 1990 and 2018 alone.[16]

While the wet zone of the country is exposed to flooding, the dry zone has been continuously impacted by widespread drought. Landslides are uncommon, but when the rainy season arrives, they cause enormous human and physical harm and necessitate relocation programmes. These hazards and associated displacement become a huge burden on the country's economy[17] and rob people of their lives and livelihoods.[18] As per the latest data available for 2017, the series of floods and landslides had a significant impact on fifteen districts located in the southern region of the country, and the financial impact of those events was estimated to be over 1,262,884 USD; the projected expenses for recovery were around 782,350 USD.[19]

Thus, the country is suffering from the complex impacts of such climate change-induced disasters, which are further enhanced by the drowning economy of the country. The following section evaluates the threat of landslides as a disaster in the Sri Lankan context to later assess the relocation strategy for the landslide-affected population at Aranayake in Kegalle district, Sri Lanka.

3 Landslide as a Disaster in Sri Lanka with Specific Reference to Aranayake

Landslide events have traditionally been isolated incidents caused by local and natural factors in Sri Lanka, with around 20 per cent of the country's land area being mountainous or hilly. However, the areas prone to landslides are now predominantly occupied by tea and cinnamon plantations, leading to about 35per cent of the Sri Lankan population residing in these areas.[20] Landslides occurring in Sri Lanka are

[11] BBC News [7].

[12] Abeywardhana [8].

[13] Bower and Weerasinghe (n 4) 30–32.

[14] Vaes and Goddeeris [9].

[15] Disaster Management Centre [10].

[16] UNDRR [11]

[17] World Bank [12].

[18] Climate Change Secretariat, Ministry of Mahaweli Development and Environment [13].

[19] The Disaster Management Centre, State Ministry of National Security and Disaster Management [14].

[20] Konagai et al. [15]

not as dominant as other disaster events such as floods and drought but their impact has risen in the recent past. Based on the available data from the National Building Research Organisation (NBRO),[21] it is reported that there has been a significant increase in landslides during the years 2003, 2007, 2011, 2014, 2016, and 2017. Particularly between 2011 and 2018, there was a notable surge in landslides, with significantly higher mortality rates observed in 2003, 2014, 2016, and 2017. Despite the landslide rates reaching their peaks in 2007, 2011, and 2018, the mortality rates were not as high as those experienced in 2003, 2006, 2016, and 2017 in Sri Lanka.[22]

Human-induced interventions such as rapid urbanisation, population growth towards the foot of mountains, inappropriate land management, deforestation in steep slopes, etc., have influenced nearly 70 per cent of the landslides in Sri Lanka.[23] These factors are further explained in detail in the next paragraph.

One of the most common causes in the Sri Lankan context is construction activity carried out on steep slopes, which can destabilise the soil and increase the risk of slope failure. The composition of the soil itself can also be a contributing factor, with loose and poorly compacted soils being more susceptible to landslides. Other factors, such as deforestation or urbanisation, can also have a significant impact on the stability of slopes. In addition, waterlogging—which occurs when the soil becomes saturated with water—is another factor that can contribute to landslides. Furthermore, human intervention, such as through mining or excavation, can alter the natural stability of slopes and trigger landslides in Sri Lanka.[24] The unpredictable rainfall patterns and the land-use-related practices seem highly impactful in this rise in landslides compared to the past. The districts in Sri Lanka namely, Kandy, Matale, Nuwara Eliya, Badulla, Kegalle, Ratnapura, Kalutara, Galle, Matara, and Hambantota have been declared as landslide-prone districts by the National Building Research Organisation (NBRO).[25] Though these landslide threats are visible, people tend to settle in these regions, which have favourable climatic and environmental conditions for living.[26] Therefore, the risk of landslides is further increased by settlements, rapid urbanisation, and modification of land in mountainous and sloped areas.

Landslides cause damage to life, property, infrastructure, and the natural surroundings. Due to floods and landslides in 2016, five million people were affected.[27] During May 2017, 35 landslide events alone were reported and those events accounted for 176 deaths out of a total of 219 fatalities caused by floods and landslides.[28] The estimated consolidated damages and losses from the floods and landslides in May 2016 were over 275,813,280 USD, and in the aftermath of the events in May 2017, it was estimated at 215,134,358 USD.[29] Hence, both years

[21] Following the catastrophic landslides of the mid-1980s, the Sri Lankan government embarked on an initiative to study and understand the phenomenon and its effects. Accordingly, the National Building Research Organisation (NBRO) has been designated as the National Focal Point for landslide risk management by the Government of Sri Lanka (GoSL). Landslide Research and Risk Management Division (LRRMD) established in 1993 has been the powerhouse and driving force behind NBRO in managing the landslide risk in the country to date. The NBRO performs multiple tasks such as landslide hazard zonation maps, issuance of landslide clearance for buildings, constructions, land development projects, landslide early warning, landslide mitigation projects, slope stability research, etc.,
'Landslide Hazard Zonation Mapping' (National Building Research Organisation), https://www.nbro.gov.lk/index.php?option=com_content&view=article&id=48&Itemid=264&lang=en#kegalle. Accessed on: 10 November 2023.

[22] Konagai and others (n 20) 219.

[23] Ibid, 218.

[24] Jayasingha [16].

[25] Rathnasiri and Wijegunarathne [17].

[26] Wickramasooriya and Dilini [18].

[27] Ministry of National Policies and Economic Affairs Ministry of Disaster Management [19].

[28] Ministry of National Policies and Economic Affairs, Ministry of Disaster Management [20]

[29] 'Post-Disaster Recovery Plan: Sri Lanka Floods and Landslides' (n 27) 15.

represent huge economic loss which makes the community and physical rebuilding process hard. The most severe landslides occurred in the Kegalle district, and the following table shows the impact of the said landslides.

There does seem to be a gap between legal frameworks in relocation policies in Sri Lanka. Though the Disaster Management Act of 2005 and the National Policy on Disaster Management (2010) are available as legal policy frameworks to address disaster-related issues, an explicitly focused legal framework addressing climate and disaster-related relocation is not available. The concept of relocation can be found in the NIRP (2001) and the Land Acquisition Act (LAA) 1950, yet both of those are specially formed to facilitate development projects. Disasters and war are two of the many situations where the NIRP has been put into practice since its inception.[31] The official bodies design relocation planning while being inspired by the NIRP and due to the unavailability of a particular law that governs climate- and disaster-related relocation. Accordingly, the communities that were impacted by the Aranayake landslide were also considered under this policy regulation in terms of relocation.

Aranayake, a village area located in the Kegalle district of Sri Lanka, 70 km east of Colombo, was one of the sites of a devastating landslide disaster that left a lasting impact in 2016.[32] As previously explained, this district is in the country's central highlands and is highly prone to landslide risks. Furthermore, as given in Table 1, the Aranayake landslide depicts the highest missing rate and a considerable death rate. The map in Fig. 1 shows the landslide hazard zones in the Kegalle district and the location of Aranayake is marked in red. The NBRO has classified the areas as portrayed in the map into four main categories: safe slopes, landslides not likely to occur, modest level landslide hazard, and landslides are to be expected. Accordingly, many areas in the Kegalle district including Aranayake have a high risk towards landslides. The next paragraph will briefly explain the landslide event that occurred in Aranayake in May 2016.

On 17 May 2016, a massive landslide of large magnitude occurred on a mountain named 'Samasara' in Aranayake, leading to two additional landslide cases in the following days. This was triggered by heavy rainfall and slow-moving cyclone occurrences in those days.

In Sri Lanka, there are four rainfall seasons during the year, namely the southwest monsoon (May to September), the inter-monsoon period following the southwest monsoon (October to November), the northeast monsoon period (December to February), and the inter-monsoon period following the northeast monsoon (March to April). An annual average rainfall of 5000 mm is expected in the Kegalle and Nuwara Eliya districts.[34] The Aranayake landslide occurred during the rainy season of the southwest monsoon in 2016. The precipitation reached 446.5 mm in the three days, from 14 to 17 May, and this caused a landslip mass that travelled about 2 km, demolishing 75 buildings.[35]

As reported by the Japan International Cooperation Agency's (JICA), Sri Lanka Office, the Aranayake landslide is a very complex incident, and the anticipation of such landslides remains hard. However, they report that the Aranayake landslide was triggered by a failure of the crown area,[36] which was subjected to deforestation and rising human settlements.[37] Hence, the accumulation of climate-induced impacts and more direct anthropogenic factors have created the ground for this landslide.

The Aranayake landslide has impacted not only the community that settled in the mountain for the time being but also the peripheral communities settled in the low-flow channel of the mountain valley. The areas of the Aranayake

31 Fernando et al. [22].

32 Konagai and other (n 20) 217.

34 Department of Census and Statistics [23].

35 Tan et al. [24].

36 Petley [25].

37 Froude and Petley [26].

Table 1 Affected people by Landslides in Kegalle District in May 2016

Date	District	Divisional secretariat	Affected families	Affected people	Deaths	Missing
15 May 2016	Kegalle	Dehiovita	832	3342	4	1
17 May 2016	Kegalle	Aranayake	996	2756	31	96
17 May 2016	Kegalle	Bulathkohupitiya	758	2756	15	2
18 May 2016	Kegalle	Yatiyantota	997	3810	0	0
Total			**3583**	**12,664**	**50**	**99**

Source Adapted from Sri Lanka Post-disaster Needs Assessment, 2016[30]

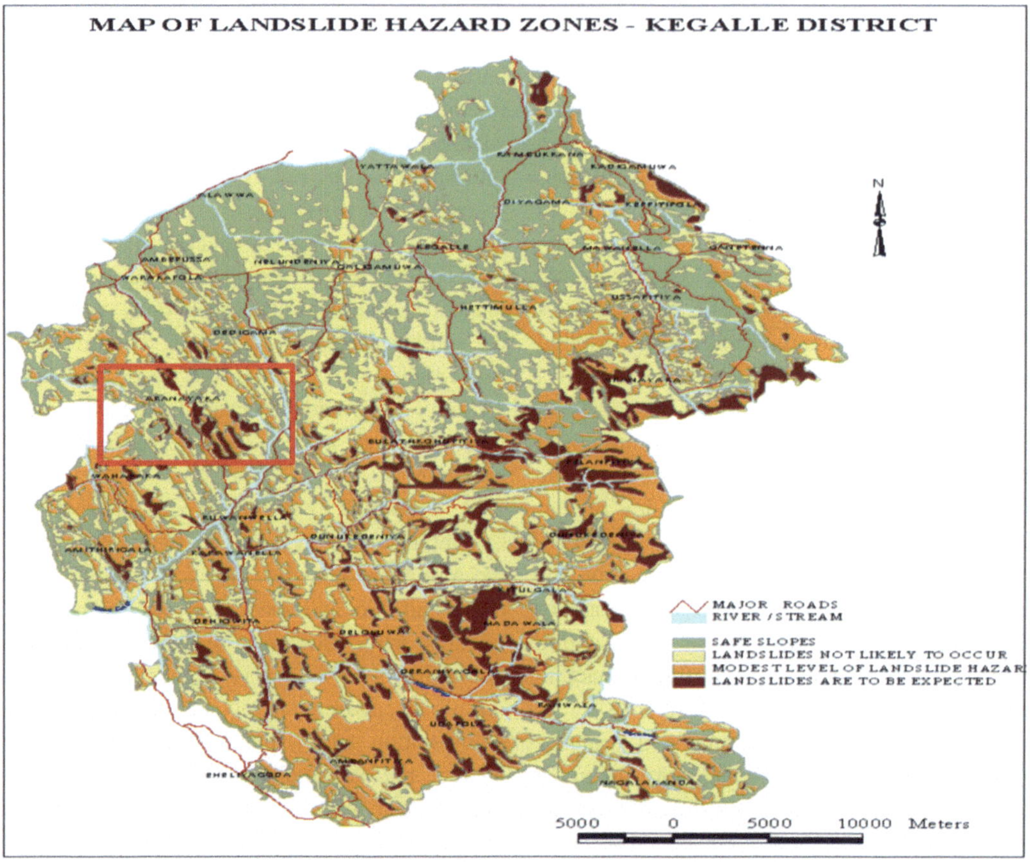

Fig. 1 Map of landslide hazard zones—Kegalle district, Source: adapted from NBRO, n.d[33]

divisional secretariat that the landslide abruptly impacted include Siripura, Elangapitiya, and Pallebage.[38] The landslide made the lower channel areas of the mountain risky for future landslides. Many houses and cultivation lands located on the hill were either highly damaged or buried due to this massive landslide. The infrastructural facilities in the area, such as electricity, telecommunication, water, and school facilities, were disrupted for the time being by the landslide occurrence. As Table 1 shows, 2756 people were affected by the Aranayake event. The landslide affected more than 600 hectares of land, with the mud level covering up to

[30] Ministry of Disaster Management [21].
[33] NBRO (n 21).
[38] Sangasumana [27].

30–40 feet in some areas.[39] Immediately following the landslide, the Government of Sri Lanka (GoSL) attempted to relocate the victims who lost their houses and/or families, as well as the population at risk of being victimised by future landslides. The following section elaborates on the process of relocation and its challenges.

3.1 Landslide and Relocation of the Victims in Aranayake

The displaced communities of the Aranayake landslide were relocated to permanent shelters after two years of stay in the temporary shelters. Some of the displaced lost their physical and non-physical assets such as family members and neighbourhoods. The others were considered at high risk of being impacted by the landslide though their physical and non-physical assets were not harmed. Those who were at risk were also considered victims here. The reason being, that although their houses and family members were not lost due to the landslide, their houses were not in a liveable state and the landslide made them leave their place of origin. Therefore, it is fair to consider both those directly influenced by the damage and those who were lucky enough to save their lives and houses, yet under severe risk, together as displaced. The quote below from one of the respondents proves the nature of their displacement:

> I witnessed the landslide; it was so close to destroying my house too. Luckily my house was not destroyed. I escaped with my family as soon as possible. If another landslide happened our house would be wiped off. We were relocated to this village years after staying in the shelters. Though our house was not fully harmed, our land was. We were not allowed to live in there after the landslide. (Respondent 4, Panapurewatta, 47 years old)The above quote shows the nature of the risk that some of the community members were under. Their houses were either damaged or partially damaged, yet they too were at a high risk of harm. It was prohibited to live in the evacuated areas and the community was moved to a transitional zone where temporary shelters were provided for nearly two years. A total of 512 families displaced due to this event were relocated within the Kegalle district.[40]

The areas where the displaced were relocated were within the same province, yet away from the original location. The Kegalle divisional secretariat has 11 divisions, and a total number of 1729 houses were constructed to facilitate all landslide displaced in the said district, including the Aranayake displaced. The donor-driven approach[41] was used to build 211 houses, and the owner-driven approach[42] was used to build 949 houses in Government Relocation Sites (GRS). The government relocation sites were the lands owned by the government and dedicated to relocating people who were displaced by the landslide. A further 569 owner-driven houses were built at individual relocation sites which the community members had selected themselves.[43]

The GoSL was able to relocate communities in the housing schemes that were built by the donors specifically for the landslide victims. The China Friendship Village (CFV) is one of the housing schemes built under the financial support of China, which became a relocated village itself. The Panapurewatta relocated site is one of the owner-driven sites in Aranayake. Though the communities were relocated to new spheres, the issues related to their well-being are notable in the empirical data as well as in the literature.

---

[39] Disaster Service Centre (n 10).

[40] Jagath [28].

[41] Owner-driven approach has two aspects. One is, that the displaced persons are given an opportunity to select a land they would like to live in. However, the selected land's value should be compatible with the government compensation's allocated amount for the land. The other aspect is the government finds and pays for the land; the settler should build his own house with the compensation provided by the government. When that is the case, the settler receives compensation after deducting the land's value.

[42] Donor-driven approach is providing already built houses for the IDPs. A housing scheme with houses of equal characteristics will be provided to the community.

[43] Senanayake et al. [29].

3.2 Research Methods and Techniques

The methods and techniques used for the study are elaborated here. Firstly, the conceptual foundation of the study relied on the definition of Internally Displaced Persons (IDPs) as given by the United Nations Guiding Principles on Internal Displacement, 1998 (guiding principles). The guiding principles define IDPs as 'persons or groups of persons who have been forced or obliged to flee or leave their homes or places of habitual residence, in particular as a result of or to avoid the effects of armed conflict, situations of expanded violence, violations of human rights, or natural or man-made disasters, and who have not crossed an internationally recognised border'.[44] Further, the IASC Framework on Durable Solutions[45] was also majorly used to strengthen the basis of the study. Along with that, relevant key principles of the Guidance on Planned Relocation[46] were also used in the analysis where necessary. The study employed an exploratory research design with the purpose of formulating a problem for more accurate investigation.[47]

A qualitative approach was used for the study and semi-structured and in-depth interviews, together with empirical observations were the primary data collection methods. The selected relocated sites were China Friendship Village and Panapurewatta. The convenient sampling method was used to select the sample of 20 people (10 from each village) to conduct semi-structured interviews and these respondents were heads of the household of each family.

Given that females were predominantly not represented in the sample of heads of households, five female respondents were selected through the purposive sampling method from the same villages, to conduct in-depth interviews. Age was considered in selecting these female respondents, and they were ages 19, 27, 35, 54, and 70. Furthermore, five in-depth interviews were conducted with officials who were aware of the relocation process of the selected sites. In keeping with the ethical considerations, all the names of the participants are anonymised in the analysis.

[44] UNHCR [30].

[45] Inter-Agency Standing Committee [31].

[46] UNHCR [32].

[47] Akhtar [33].

3.3 Assessment and Analysis of the Relocation at Aranayake

The nature and the issues experienced by the communities due to relocation caused by the Aranayake landslide are evaluated in this section with special reference to the selected sites mentioned previously. As the Guidance on Planned Relocation affirms, governments may undertake relocation as an anticipatory measure where hazards threaten to render certain areas uninhabitable. This is further affirmed through the Sendai Framework for Disaster Risk Reduction 2015–2030, which highlights the need for formulating policies to address the issues generated through relocation.[48] When appropriate policies are formulated, the relocation process will eventually contribute to reducing disaster risk.

However, relocation initiatives around the world illustrate that relocating people is a complex endeavour with a high risk of deprivation of basic rights leaving individuals significantly worse off. States, when confronted with scenarios requiring deliberate relocation, often lack clarity on the fundamental principles that apply to this demanding scenario. Even though Sri Lanka has previous disaster-induced relocation and resettlement experiences, the planning and success of those attempts have not been examined in depth, either by academic or policy actors. The lack of community-sensitive planning, influence of political agendas, improper use of funding, and limited holistic understanding of the events and the community's unachieved expectations may have been

[48] UNISDR [34].

the reasons behind the infant nature of relocation attempts in Sri Lanka. The most prominent example is this Aranayake event, which shows the incompatibility between community wellbeing and what the relocation has addressed.

3.3.1 Assessing the Relocation Sites at China Friendship Village and Panapurewatta

One of the key areas of post-landslide relocation planning was relocating communities from where they were temporarily accommodated to permanent settings. The preliminary evaluation of the Aranayake landslide-induced relocations seems to be a durable solution to address the extended stay in temporary shelters. Arranging their settlement or facilitating their settling in a safe, permanent environment has taken away their insecurities as they now have a place to permanently mobilise from the temporary shelters. The IDPs had no choice but to agree with relocation as moving back to the mountain was prohibited. They were supposed to apply for a village that they preferred for relocation. However, though some of their requests were approved, not everyone has received the village or the type of relocation assistance that they requested. The government officials decided on the allocation of some IDPs in relocated villages. This was mostly found in the Panapurewatta relocation site and the following quote of one of the respondents further proves that

> When we were in temporary shelters, we had an opportunity to request one of the options the government gave us: to find a land value about 3 lacks, request for an already built house, request land to be built the house ourselves with the compensated money, were those options. Finding a nearby land for 3 lacks was very hard as those who had lands increased the prices of them. We did not want to take the burden of building a new house from the beginning. Therefore, many of us requested a built house. But including me, many of our wishes was not granted... (Respondent 11, Panapurewatta, 36 years old)The above quote clarifies the nature of facilitation in relocation and the impact and practical status of the relocation measures. As per the findings of the study, a majority of the respondents in the Panapurewatta had a similar issue while the respondents in the China Friendship Village (CFV) were satisfied that their wish of having a built house was granted, though they had to face different kinds of issues after relocating there.

As mentioned previously, the CFV was a relocation project comprised of 60 housing units that were built and given to the IDPs. The housing units are two-story which were initially planned by the Chinese government and later built by the Sri Lankan government. As per the data received from an in-depth interview with an official, the initial plan was to build 100 housing units for the village, but it ended up building only 60 units in 2018. As previously mentioned not all the vulnerable were relocated and those who were in high-risk areas were relocated under the first phase due to the high demand which was driven by the higher number of landslide displaced communities. According to the information given by the respondents, another 100 housing units will be built in the CFV in 2023, for those who live in the areas which are classified as 'moderate risk areas' in the mountain, as they were not counted in the first phase of relocation.

The location of the CFV was originally privately owned land, and as per the respondents, the relocated communities were not given the deeds yet. The reason was that the land ownership-related clearance was yet to be received. The land selected for the relocation was compiled from two privately owned lands named Weediya Mankada Hena and Udu Wihara Thanna, which were initially a cemetery; though the government's land acquisition regulations were followed in acquiring this land for this purpose, the multiple ownership demands made the process complicated and time-consuming.[49] Accordingly, the right to the land has not yet been given to the current settlers.

In contrast, the Panapurewatta relocation site is owner-driven, and the settlers were supposed to build their own houses by using the money given by the GoSL. The settlers were given 4,903 USD of which 1,225 USD were for the land and the rest was to be used to build the

[49] Fernando and others (n 31) 136–137.

house. The said balance has been given to the community as instalments based on their progress in utilising the money provided. The government suggested a housing plan of 650 square metres with an estimation of 4,290 USD as the construction cost. Even if the relocated family wanted to build their house according to their plan, still the government would only provide the above-mentioned amount. The rest of the expenses would have to be covered by the members themselves. The lands provided for relocated communities affected by the Aranayake landslide span from a minimum of 250 square metres to a maximum of 650 square metres, and this is because of the limited lands available for relocation purposes.[50]

While Panapurewatta is located around 4.4 kms away from the main road, the CFV is located 1.3 kms away. Comparatively, the CFV is in a community-friendly location which is accessible to the town area far more easily than Panapurewatta. The location of Panapurewatta is distant and can be defined as a more isolated relocated village. With reference to religion-wise composition, the CFV is religiously homogenous with a Buddhist population, while Panapurewatta is mixed with Buddhist and Catholic populations making it religiously heterogeneous.

The issues that occurred after permanently settling in the above two locations deeply affected their well-being particularly in economic and social spheres. This will be discussed in the following section with special reference to the identified dimensions of rights violations and vulnerability.

[50] Norris et al. [35]. Norris and others have focussed on four sites and investigated livelihood issues, which is not the only focus of the current study. I was also a part of that study which was done in 2018. That experience inspired me to investigate the current situation (2023) of the relocated community in terms of livelihoods as well as social aspects from a sociological perspective.

3.3.2 Assessing the Landslide Relocation: Post-Relocation and Issues.

The issues experienced by both the communities in the post-relocation phase are discussed here. The housing units in both villages are different as mentioned in the previous section. The residents of the CFV were given a built house while the Panapurewatta's residents were supposed to construct their own houses. Despite the difference in the nature of construction, social and economic challenges were identified in both locations even after five years of their relocation. These issues will be discussed specifically within the framework of IASC Durable Solutions Guidelines. These guidelines aim to foster a better understanding of the concept of durable solutions for IDPs which makes the mechanisms implemented for IDPs effective and durable while ensuring the need to protect their rights.[51] Therefore, following these guidelines would be helpful in planning and initiating relocation in a way that people are experiencing satisfactory conditions.

As per the IASC Durable Solutions Guidelines, eight criteria determine the extent to which a durable solution has been achieved: (1) safety and security, (2) an adequate standard of living, (3) access to livelihoods, (4) restoration of housing, land and property, (5) access to documentation, (6) family reunification, (7) participation in public affairs, and (8) access to effective remedies and justice.[52] The satisfactory facilitation in these components can make the relocation process a success. However, within the relocation sites, there were issues that showed that the relocation process in Aranayake did not encompass these essential measures. This has resulted in multi-faceted issues and those are discussed below with relevant examples from the selected sites.

The first aspect to consider is the housing unit-related issues. The housing units in the

[51] Ibid, 27–43. See also Scott [36].

[52] Ibid, 51.

Image 1 and 2 Housing Units in CFV. *Source* By the researcher, 2023

CFV comprised two bedrooms, a living room, a kitchen, and a bathroom. These all-housing units were of the same size, i.e. 600 m^2. However, as per the respondent's perspective, there were issues with the new setting of the house, which was typically different from what they had in the mountain. Herein the issues they are experiencing are that the housing provided is not compatible with their needs or with the criteria on safety and security, adequate standard of living, access to livelihoods, and restoration of housing.

As shown in Image 1, the CFV community was given a house in a two-storey struct*ure*. However, the major concerns regarding the housing unit pertained to the limited space, the unavailability of an outside kitchen and toilet (they were given in-house kitchen and bathroom facilities), unsafe windows (windows without grills creating unsafe situations), and architectural concerns (wherein the architecture doesn't facilitate any ventilation). These concerns will be further explained in the latter part of this section. The second image shows a stove built within the house to cook using firewood.

The following quote from a respondent further explains the issues mentioned above, as experienced by them:

We had 4 acres of land in the mountain with a large house of four bedrooms. But here we only have a very small space for six members. We had an outside kitchen to cook in and we used free firewood to cook, but here we didn't have a kitchen that could use firewood and it was hard to spend on gas monthly. When gas was not available during the time of economic crisis, we had no solution but to build a small stove for cooking with firewood. Since our house is upstairs, it was not easy. We covered our small balcony and made a stove, but when it is cooking on it, the whole house is covered by smoke. (Respondent 2, CFV, 45 years old)The relocated community had bigger houses in their previous location as compared to the houses that they were given under the relocation project. As per the above quote, it was not easy for them to adapt to the new setting. Being an extended family, they found it hard to fit into the limited space. This is further worsened by the changes in conditions that they were used to before, such as livelihoods, methods of cooking, structure of the house, and access to essential services and goods. Overall, the issues in these conditions limit their fulfilment of adequate standards of living which is what they were striving for.

The economic status of most of the respondents of the CFV was informal as most were engaged in tea and cinnamon cultivations in the mountains. This was their only income source. They had limited expenses in terms of food and

Image 3 Unsafe windows. *Source* By the researcher, 2023

gas, as they were used to growing their own food for household consumption and used the firewood collected from their own garden as fuel for cooking. Using gas and buying every food item was a novelty and not a part of their life in the previous locations. The relocation has compelled them to shift from their frugal living to entailing higher expenses on both food and fuel, as they do not have a substance garden space to grow their vegetables[53] or a stove to cook using firewood.

Given that the economic crisis in the country exacerbated in 2022 up until early 2023, using or finding gas was challenging for them. Cooking gas was unavailable at the time, and it necessitated the need to go back to the use of firewood. The problem was the unavailability of space built to use for the said purpose. Since they were not allowed to upgrade the housing unit, they started covering the small balconies in their houses and made a stove to use firewood (shown in Image 2). The use of firewood aggravated the situation as the entire house would get filled up with smoke due to poor ventilation. This impacted their overall well-being and was violative of their right to live in a healthy environment and their right to access essential services.

This leads us to the next set of issues experienced by the community in the CFV with respect to safety. Though safety and security are priorities of the IASC durable solutions framework, the physical structure of the houses and the nature of the facilities provided do little to allay fears with regard to safety. It is apparent from Image 3 that the windows in the houses in the CFV do not have grills thereby compromising on safety. This has been a common concern for the majority of the respondents and especially for families living upstairs and having little children. Having window grills is important as window grills unnerve burglars, and discourage them from targeting property; further, families with children do not need to worry about their children's safety, thus upholding overall safety.

The above windows were referred to by respondents as unsafe because they did not have grills. Many families with children who were occupying housing units upstairs complained about this saying that they could not keep the windows open as they did not have any cover. This limits the ventilation in the house that is

[53] The substance garden is a part of rural communities in Aranayake. This means growing food to consume at the household level. The concept of substance economy can be found in Lillian Norris and others' writing, ibid, 50.

further worsened by the smoke of the stove. Respondents living both upstairs and downstairs shared their fear of being robbed and the lack of privacy due to the nature of the windows. It is a fact that these windows are big enough for burglars to enter and houses across can see right through these large windows if they are not curtained. As a result, this has become a safety and privacy concern for them. Since the community has no experience of living in condominium housing-type structures, the structure itself makes the 'restoration of housing' in an environment with satisfactory conditions, problematic.

The IASC durable solutions framework adopts a rights-based approach encompassing key principles including among others, voluntary and informed choice on what durable solutions to pursue, and active participation in planning and management of durable solutions.[54] However, according to the respondents of the CFV, they did not participate in the planning stages and were only moved into built up houses after applying for them. The emergence of issues has a direct link with non-community inclusive planning. Accordingly, the CFV's issues show that the relocation attempts were not durable as they did not fully address the needs of communities.

When considering the IASC Durable Solutions Framework, the 'restoration of housing and land and property' clearly relates to the current analysis as the relocation itself is the restoration of housing. However, the seven other criteria, which were mentioned previously, were more or less not covered in the CFV. The access to livelihoods can be elaborated as it is the most visible issue in this context. The following quote from a female respondent in the CFV explains this status.

> I was used to helping my husband in tea cultivation. Since we are relocated away from our cultivated sites, both of us can't travel there. I need to take care of my children and there is no trustworthy person to keep them here. Many thieves are making money off our lands there. Now my husband is working as an informal labourer. I wrap beedi to earn some; since the companies are sending us essential stuff to our doorstep, I can engage in that work while working at home. (Respondent 12, CFV, 35 years old)According to the Guidelines on Planned Relocation (15a), 'planned relocation should provide opportunities and conditions to enable relocated persons to improve, or, at a minimum restore, their living standards'.[55] This is also highlighted in the criteria on durable solutions as 'access to livelihoods'. In contrast, though the community was relocated physically, their livelihood restoration has not been done satisfactorily. Therefore, the females who used to support their family's livelihood in their previous locations were no longer doing the same. The reasons were the hardships of travelling a long way from where they live now, and the safety of their children which is highly dependent on women in the Sri Lankan context. This was mentioned as a concern by many females interviewed, as previously they heavily relied on neighbours and relatives for childcare but this was not possible in the new locations. The relocated community belongs to the same mountain, but their previous neighbours and relatives are no longer with them in their new locations. Accordingly, the lack of childcare systems has compelled women to stay at home to take care of their children. It is clear that the above-mentioned durable solutions for 'family reunification' and 'safety and security' have not been properly addressed, leading to concerns about experiencing a quality life.

A new means of livelihood that of wrapping *beedi* was seen among many women in the CFV.[56] This is a locally used cigarette, and the companies supplied all the essentials to the village and women were able to wrap them while staying home. The following images show their new livelihood, Image 4 shows the beedi leaves, and Image 5 shows the final production.

Women were thus able to contribute to the household income through this as the sole breadwinner could not fulfil household needs anymore. They earn approximately 3 USD (1100 LKR) for wrapping 1000 beedis and the time that they spend on this task depends on

[54] 'IASC Framework on Durable Solutions for Internally Displaced Persons' (n 45) 11–24.

[55] Guidance on Planned Relocation (n 46) 12.

[56] Beedi is a mini cigarette filled with tobacco flake and usually wrapped in a Maila leaf (Scientific name: bauhinia racemose) tied with a string at one end.

Image 4 and 5 Beedi wrapping, a common female livelihood in CFV. *Source* By the researcher

their engagement in it while doing household chores. Accordingly, this has become a community-invented alternative livelihood which can be identified as one of their attempts in restoration of household income. Though the above-mentioned guidance on planned relocation highlights the need for the restoration of livelihoods, the community was not happy with the support of the GoSL. According to them, the Chinese Government Representatives were coming annually to give them a yearly parcel of essentials, but the GoSL has no such care for them. As highlighted by the respondents, the livelihood support systems for them were not sufficient. This shows the importance of identifying community needs deeply, before planning livelihood restoration methods in terms of uplifting the community directly.

Various housing unit-related issues are visible among communities in Panapurewatta as well. Unlike in the CFV, Panapurewatta's communities were supposed to build their own houses by using the monetary compensation provided by the government. As noted in the current study, the majority of the housing units were partially completed and they didn't have much equipment at their houses such as chairs, tables, beds, and electric items. According to the respondents, the reason behind the incompleteness of the houses was that the amount of money they were given was not sufficient to meet their home construction.

> We could not complete my house because the provided money was not enough. One of the residents left the village for the mountain after laying the foundation, and now we only have 26 houses here. Like me, many of the others have not completed their houses yet. I have even applied for a bank loan to complete my house. (Respondent 9, Panapurewatta 51 years old)The insufficiency of compensated money to build houses was a common concern for many respondents. The above respondent is planning to complete the house by using the money that he will receive through the loan he applied for. However, an in-depth interview with a professional involved in the relocation, highlighted that the community's poor management of monetary support was the reason for the incomplete nature of houses. The contradiction between the community's perspective on compensated money and the official's perspective shows the complexity of the use and provision of money for this purpose. Still, what was identified was that the use of money was not properly managed by the people as they had multiple expenses other than building their house, such as transportation, food and water, and electricity. Furthermore, the surge in the prices of construction-related material prices, which was out of the control to the community, was what aggravated this issue.

Image 6 and 7 Housing units in Panapurewatta. *Source* By the researcher

Image 6 shows a partially completed housing unit, and Image 7 shows a house with a poorly constructed pathway that residents are required to take in order to access the house. Infrastructural facilities, specifically with respect to the quality of roads and access to water, are poor in this village. According to the Guidance on Planned Relocation, to address and reduce the risks of poverty, caused by landlessness, unemployment, homelessness, marginalisation, food insecurity, morbidity and mortality, loss of access to common property and services, social and cultural displacement, the states should offer adequate support, resources, and services to those who have been relocated.[57] Though the GoSL has relocated them and helped them become free from homelessness and landlessness, proper awareness and continuous community support system are needed to make the community use what they were given to build back better. The exceeding costs of building a new house beyond the GoSL estimate seemed to be the major reason behind the incompleteness of the houses in the Panapurewatta region. The poor management of money in the community has also impacted on the worsening of this concern.

This can be further identified in terms of livelihood issues that led them to move back to their original lands which was allowed by the government only for livelihood purposes (not for residential purposes).

> I am a tea, pepper, and cloves cultivator. I'm not good at any other job. I travel daily to my old land where my cultivations are going on. I must take two buses every day and come back in the evening. I feel like going back and living in my former land is better than travelling every day by spending what I earn so hard. (Respondent 14, Panapurewatta 41 years old)The trend of moving back to their original location to fulfil their economic needs is visible among both communities relocated, especially among the Panapurewatta's community. As most of them were cultivating minor export crops, their life was bound to the mountain with those economic investments. However, the community had to spend a lot of time and money on daily travelling to their lands to cultivate, use economically grown plants, etc. The total area of the Aranayake landslide area has been cultivated with minor export crops such as cloves, cardamom, pepper, and tea.[58] The respondents were also used to cultivating some of those products, which was their usual livelihood, but this was hard to engage in while living in new locations. This shows how their lives are tied to the lands where they live, and their livelihood needs motivate them to move back either to

[57] Guidance on Planned Relocation (n 46) 11–22.

[58] Perera et al. [37].

live illegally or even under risk as travelling daily costs a lot. The non-restoration of livelihoods has led them to economic impoverishment drawing them to move back to their previous locations, where the risk is still detectable. Therefore, livelihood restoration has an unavoidable impact on the effectiveness and durability of the relocation which has been highly neglected in both selected sites.

Another major concern highlighted by the community was the mixed religious composition of the community and locating them near a village comprised of different religious groups. As reported, the Panapurewatta site was specifically reserved for the Catholic population and the community still has concerns about this decision taken by the government.[59] As per the current study, the community had both Buddhists and Catholics and the nearest small-town area named Thalgaspitiya comprises a majority of the Muslim population. Since the host community holds a different religious identity than the relocated community, this creates issues in terms of their cultural rights as well as overall well-being. For example, the unavailability of religious places and schools specifically reserved for the Muslim community makes this environment unsupportive to their life order. This compels the community to travel a long way to their religious places and schools. The community doesn't have a cemetery and they are having issues regarding that as well.

As proposed by the Guidance on Planned Relocation, the rights to self-determination, preservation of identity and culture, and control of land and resources are important in planned relocation.[60] However, the mixed relocation of communities in Panapurewatta has negatively impacted the people by creating a non-cultural and conflicting setting which ultimately affects the fulfilment of their cultural rights. In this background, they have become a culturally isolated community while being physically distinct from the host community.

According to the planned relocation guidelines, the planned relocation ends when the relocated persons no longer have needs or vulnerabilities related to the planned relocation.[61] However, the communities in Panapurewatta are still experiencing issues specifically analysed before and specifically related to the poor infrastructure support provided. This can be specifically noted in terms of water and transportation. The drinking water facility is available, but for all other needs such as bathing and washing clothes, they are required to use the common well located away from their houses. This was challenging to them as they were used to having free water access at their previous locations and were now required to pay for drinking water. Furthermore, the poor transportation facilities have only added to their woes. The limited public transportation facility includes only two small buses that travel from the town to their village and those are not frequent, and they usually end up travelling from their village to the town, a distance of about 4.4 kms, using a hired vehicle or by foot. Since their economic status is not so strong in the community, they're reluctant to spend on a taxi and mostly travel by foot, which is hard. Accordingly, the non-fulfilment of their needs makes them further vulnerable after the relocation. The realisation of their human rights needs to be fulfilled in the infrastructural as well as the social, economic, water, education, and cultural needs which are not being properly met, and which harms the realisation of the said rights. Moreover, the Panapurewatta's relocated communities keep having to sacrifice their basic rights including the right to bury their dead, the right to water and sanitation, and the right to freedom of religion per se.

One of the major issues commonly visible in both the CFV and Panapurewatta was the unavailability of the deeds pertaining to the lands. It's been almost five years since their relocation, yet the legal rights to their lands and houses remain pending. As previously mentioned, the IASC framework on durable solutions highlights

59 Jagath (n 38) 61.

60 Guidance on Planned Relocation (n 46) 11.

61 Ibid, 22.

the need for access to documentation, which is a right that a relocated community should have.[62] The document that proves their land tenure should be provided to them to legally confirm their rights to the land. As found in a previous study by Fernando and others, the possibility of selling lands by the relocated community has delayed this legal transfer process.[63] However, living on those lands without having legal rights creates uncertainty for the relocated community and aggravates insecurity in land tenure that can eventually trigger future issues in terms of unlawful acquisitions and even the inability to prove their lawful entitlement to the land or house.

Though the relocation of the Aranayake landslide victims has saved the lives of community members who were at risk of future landslides, this move has negatively impacted their social and economic well-being. It has led some to consider returning to their previous locations, despite being prohibited from doing so. If the community's needs had been adequately addressed and their participation included in the relocation planning process, these issues could have been avoided. The challenges faced by the community, which directly or indirectly infringe upon their rights, serve as a worthwhile lesson for future relocation attempts.

When comparing the two relocation sites, the CFV has relatively better conditions as the community was handed over built up houses and their right to shelter was granted through the relocation measures; however, this is unfortunately not the case with the community at Panapurewatta where a majority of the people continue to face multiple obstacles in their attempt to settle in the relocated site. Considering that Panapurewatta, followed the owner-driven approach, the actual issues on the ground have made it difficult for most members of the community to realise their right to shelter. The location of the CFV is closer to the city as compared to Panapurewatta and this helps the community to easily fulfil their needs from the city. However, Panapurewatta is in a more remote area, isolated from the city and the community faces daily issues specifically in terms of transportation. In terms of infrastructural facilities also, the CFV has better conditions which were fully taken care of by the Chinese Government directed funds. Panapurewatta, on the other hand, has limited infrastructural facilities including water, poor roads, limited transportation facilities, limited accessibility to religious places, lack of burial facilities, and so on.

Though the CFV has better conditions as mentioned above, they are also experiencing issues associated with the physical structure of the houses which are not compatible with their needs. In contrast, Panapurewatta's community has had comparative freedom in the building of the house based on their requirements. Both the CFV and Panapurewatta have pros and cons with respect to their relocation and most of these issues have emerged due to the non-participation of the community in the exercise of relocation planning. Had both the communities been consulted and the planning process made participatory, the issues with respect to limited livelihood restoration and legal rights would not have emerged. Therefore, it is clear that for the relocation to be durable and effective, the processes should be community sensitive and participatory.

4 Conclusion

The chapter has discussed the vulnerability and disasters which have been exacerbated by climate change in Sri Lanka. Among the disasters affecting Sri Lanka, floods and droughts occupy prominent places. However, since 2003 landslides are also emerging as a prominent disaster affecting people and communities.

The chapter has analysed the pattern and model of relocation following the landslide at Aranayake, Kegalle. The analysis reveals that relocation measures were taken to build the lives of IDPs at the China Friendship Village and Panapurewatta relocation sites. However, even

[62] IASC Framework on Durable Solutions' (n 45) 4, 40.

[63] Fernando and others (n 31) 137.

after seven years since the Aranayake landslide, and five years since the relocation, the communities at both the China Friendship Village as well as Panapurewatta are still experiencing issues associated with the relocation. There is a mismatch in the community's housing expectations and the reality of what was provided in terms of architecture and ventilation-related issues, diminishing of traditional childcare systems, the inability to livelihood restoration and absence of legal rights to the house, were all identified as problems in the donor-driven China Friendship Village. The same concerns of livelihood restoration and legal rights to the land were apparent in owner-driven Panapurewatta as well. The insufficient compensation in building houses, and limited access to basic, essential rights such as water, burial grounds, sustainable transportation, and non-availability of a religious place to access were found to be the major issues faced by the Panapurewatta community. The issues mentioned above prove that the relocation has threatened their social and economic sustainability, as the relocation has taken their basic rights away.

The lack of a direct policy to address relocation caused by climate-induced relocation essentially leads to involuntary relocation. A policy directly addressing these relocation needs and applicable to the Sri Lankan context must be formulated. That will help create a shift from involuntary relocation to an acceptable and voluntary move which will prove to be more stable and sustainable. It is apparent that the IASC durable solutions framework's conditions have been largely incompatible with the relocation projects discussed here. In fact, some of the guidelines for planned relocation that were deliberated upon in this chapter were also not reflected in the relocation process of these communities. The failure of these attempts shows the need to address such events with a rights-based community-sensitive approach.

The current study highlights that the community should not be left out of relocation planning, and they should be considered authorised actors. At the end of the day, they are the ultimate beneficiaries of the planned outcome. Hence, a bottom-up approach which starts from the community and ends with the policymakers and the government would contribute to the non-violation of essential rights of the communities in terms of relocation. This will aid in shifting the paradigm from error correction to low-mistake facilitation in disaster-induced relocations.

References

1. United Nations Office for Disaster Risk Reduction. Disaster risk reduction in least developed countries. https://www.undrr.org/disaster-risk-reduction-least-developed-countries. Accessed on 12 June 2023.
2. Lal PN, Singh R, Holland P (2009) Relationship between natural disasters and poverty: SOPAC miscellaneous report 678 (international strategy for disaster reduction) 17
3. The World Bank. Climate and development in South Asia. https://www.worldbank.org/en/region/sar/brief/integrating-climate-and-development-in-south-asia/integrating-climate-and-development-in-south-asia-region. Accessed on 01 March 2023.
4. Bower E, Weerasinghe S (2021) Leaving place, restoring home. In: Andrew, Kaldor R (eds) Platform on disaster displacement (PDD). Centre for International Refugee Law 2021, p 29.
5. Mombauer D, Wijenayake V (2020) Policy brief climate change and human mobility in Sri Lanka. SLYCAN Trust. p 2
6. Disaster Service Centre (2020). Aranayake landslide. https://disaster.lk/aranayaka/. Accessed on 10 Aug 2023
7. BBC News (2023) Sri Lanka country profile. https://www.bbc.com/news/world-south-asia-11999611. Accessed on 10 Aug 2023
8. Abeywardhana DMY (2020) Environmental sociology of floods in the Colombo District of Sri Lanka. In: Singh A, Lalitha R, Fernando S, Haran NP (eds) Development in coastal zones and disaster management. Springer, Singapore, pp 271–272
9. Vaes B, Goddeeris M (2012) Sri Lanka tsunami-lessons learned: a donor and owner-driven reconstruction approach construction delegates. Belgian Red Cross Finders, p 4
10. Disaster Management Centre (2009) Sri Lanka national report on disaster risk, poverty and human development relationship, pp 33–34
11. UNDRR (2019) Disaster risk reduction in Sri Lanka: status report, p 8.
12. World Bank (2021) 'Historical hazards' in Sri Lanka. World Bank Climate Change Knowledge Portal. https://climateknowledgeportal.worldbank.org/country/sri-lanka/vulnerability. Accessed on 4 Apr 2023

13. Climate Change Secretariat, Ministry of Mahaweli Development and Environment (2016) National adaptation plan for climate change impacts in Sri Lanka 2016–2025, p 5.
14. The Disaster Management Centre, State Ministry of National Security and Disaster Management (2022) National Disaster Management Plan 2022–2030 (draft version), p 1
15. Konagai K et al (2022) Early warning system against rainfall-induced landslide in Sri Lanka. In: Progress in landslide research and technology, vol 1, issue 1, pp 217–218.
16. Jayasingha P (2016) Social geology and landslide disaster risk reduction in Sri Lanka. J Trop Forest Environ 6:5–6
17. Rathnasiri PHCS, Wijegunarathne EE (2015) An evaluation of landslide risk assessment process of construction activities in landslide prone areas'. In: NBRO Symposium, pp 1–4
18. Wickramasooriya A, Dilini S (2022) Geoinformatics approach to classifying landslide hazard in the Hantana Mountain and its vicinity in Sri Lanka. IOP conference series: earth and environmental science 1109:1–17
19. Ministry of National Policies and Economic Affairs Ministry of Disaster Management (2017) Post-disaster recovery plan: Sri Lanka floods and landslides, p 15.
20. Ministry of National Policies and Economic Affairs, Ministry of Disaster Management (2017) Sri Lanka rapid post disaster needs assessment, p 168.
21. Ministry of Disaster Management (2016) Sri Lanka post disaster needs assessment: floods and landslides-May 2016:22
22. Fernando N et al (2019) Disaster, displacement and relocation: an analysis of the needs and policy implications on a displaced community in Sri Lanka. In: 10th International conference on structural engineering and construction management—ICSECM, pp 133–134
23. Department of Census and Statistics (2019) Statistical abstract-2019: chapter 1—Area and climate. Department of Census and Statistics, Sri Lanka. http://www.statistics.gov.lk/abstract2019/CHAP1. Accessed on 12 Sept 2023
24. Tan Q et al (2020) Estimation of the past and future landslide hazards in the neighbouring slopes of the 2016 Aranayake Landslide, Sri Lanka. Landslides 17:1727–1738
25. Petley D (2016) Landslide Glossary defines the crown area as the practically nondisplaced material that is still there and near the highest areas of the main edges. The Aranayake landslide disaster in Sri Lanka—A JICA report. Advancing Earth and Space Sciences Blogsphere. https://blogs.agu.org/landslideblog/2016/07/19/aranayake-landslide-1/. Accessed on 25 May 2023
26. Froude MJ, Petley DN (2018) Global fatal landslide occurrence from 2004 to 2016. Nat Hazards Earth Syst Sci 18:2161–2181
27. Sangasumana P (2018) Post disaster relocation issues: a case study of Samasarakanda Landslide in Sri Lanka. Eur Sci J 14(32):1–17
28. Jagath M (2021) Strengths and weaknesses of the resettlement process in the post disaster era: a case study of Aranayake, Kegalle, Sri Lanka. Colombo Arts J Soc Sci Hum 6:57
29. Senanayake A et al (2022) Landslide induced displacement and relocation options: a case study of owner-driven settings in Sri Lanka. 1906 Sustainability 14(3):15
30. UNHCR (2023) Emergency handbook: IDP definition, p 1
31. Inter-Agency Standing Committee (2010) IASC framework on durable solutions
32. UNHCR (2015) Guidance on planned relocation: guidance on protecting people from disasters and environmental change through planned relocation
33. Akhtar I (2016) Research design. Res Soc Sci: Interdisc Perspect 1:73–74
34. UNISDR (2015) Sendai framework for disaster risk reduction 2015–2030 A/CONF.224/CRP.1, p 18
35. Norris L et al (2019) Resettlement stories of loss and barriers to recovery following the 2016 Aranayaka Landslide, Sri Lanka. In: NBRO research symposium, pp 197–206
36. Scott M (2019) Background brief: key international standards and guidelines relating to displacement in the context of disasters and climate change. Raoul Wallenberg Institute of Human Rights and Humanitarian Law, p 21
37. Perera ENC et al (2018) Direct impacts of landslides on socio-economic systems: a case study from Aranayake. Sri Lanka. Geoenviron Disasters 5(11):1–12

Dinushika M. Yapa Abeywardhana is a Lecturer in the Department of Sociology, Faculty of Arts, University of Peradeniya, Sri Lanka. She holds a BA in Sociology from the University of Colombo in 2018 with first-class honours. Dinushika has achieved two gold medals for the best performance in the Sociology of Development and Sociological Theory courses. Her research interests lie in Climate Change and Disaster Management, Environmental Sociology, Sociology of Development, and Urban Sociology. She has published research on climate change, disaster management, urban development, ecological education, and human-animal relations. She is reading for her MA in Sociology at the University of Colombo, Sri Lanka.

Conclusion

Stellina Jolly, Nafees Ahmad, and Matthew Scott

A legal research and policy agenda for addressing climate-related human mobility in Asia and the Pacific in the context of interdisciplinary rights-based approaches.

International academic and policy circles have continued to give significant attention to human mobility in the context of climate change. Human populations would suffer if extreme weather events like heatwaves, typhoons, droughts, and floods occur more frequently and with greater severity. Extreme weather will likely lead to more water scarcity, food insecurity, and illness, as well as new vulnerabilities that might result in displacement and the creation of new migration patterns. This volume addresses climate-related human mobility in Asia and the Pacific using an interdisciplinary rights-based approach. A solid qualitative groundwork has been created by reviewing existing legal and policy frameworks and researching how these frameworks are applied in different climate-related human mobility situations. This groundwork has made it possible to identify areas that, from law and policy standpoint, require special attention. Some recognisable paths and possibilities seem very conducive to additional research.

Climate change-related human mobility will be a significant challenge for Asia and the Pacific's sustainable prosperity and stability, and early coordinated action across sectors and governance levels is required to prevent, prepare for, and mitigate anticipated adverse impacts on the enjoyment of human rights by people who move, as well as people who are unable to move or otherwise choose to remain in areas exposed to hazards. Acknowledging the instant threat's importance, the RWI and FLS-SAU have collaborated to produce this edited volume '*Climate-related Human Mobility in Asia and the Pacific in the Context of Interdisciplinary Rights-based Approaches*' to enhance knowledge and understanding of the varieties of mobility patterns and the associated human rights impacts.

This edited volume compiles empirical and field-level research conducted by home grown researchers to provide the most current and thorough analysis of the phenomena in one of the most populated and disaster-prone regions of the world. This book places climate-related human mobility in the larger context of migratory dynamics focusing mostly on issues of droughts, floods, coastal erosion, cyclones, etc. The chapters do not argue that climate change creates a new class of migrants; instead, they consider how it interacts with existing social-ecological

S. Jolly (✉) · N. Ahmad · M. Scott
South Asian University, New Delhi, India
e-mail: stellinajolly@sau.ac.in

N. Ahmad
e-mail: drnafeesahmad@sau.ac.in

M. Scott
e-mail: matthew.scott@rwi.lu.se

S. Jolly et al. (eds.), *Climate-Related Human Mobility in Asia and the Pacific*,
Sustainable Development Goals Series, https://doi.org/10.1007/978-981-97-3234-0_15

dynamics. Similarly, it acknowledges that mobility encompasses a broad range of responses to social-ecological dynamics unfolding in the context of climate change, including migration, displacement, planned relocation, as well as voluntary and involuntary immobility. This last chapter summarises the learnings gathered and offers recommendations for future legal research and policy propositions.

In the context of the Section on 'Dimensions of Migration and Displacement: Experiences, Policies, Laws and State Practice' the edited volume provides a conceptual background and covers a range of topics such as '*Tropical Cyclones and the Mobility of Older Persons: Insights from Coastal Bangladesh*' by Abdul Malak and Nazia Khan Lin that examines how elderly people move across Bangladesh after major storms, emphasising the substantial effects on coastal populations; '*To Move or Not to Move? Drought-related Internal Migration and Voluntary Immobility among the Hmong Forest Community in Chiang Rai, Northern Thailand*' by Reni Juwitasari discussing how the Hmong ethnic minority population in Chiang Rai is being impacted by the drought in Northern Thailand; '*Sensitisation of Disaster Relief Operations towards Persons with Disabilities*' by Ishari Gunarathna and Priyadarshani Premarathne, which highlights how people with disabilities are sometimes marginalised in society, particularly in the wake of climate-related disasters; '*Rising Waters, Stagnant Paths: Gendered Experiences of Flooding and Restricted Mobility in Can Tho City, Viet Nam*' by Danang Aditya Nizar and Ly Quoc Dang that describes how climate change has been included into Vietnamese legislation, acknowledging the connection between gender equality, mobility, and climate change; '*Establishing a Human Rights-Based Approach to Climate Change-Induced Internal Displacement in the Regime of Bangladesh: Challenges and Way Forward*' by Md Abdul Awal Khan dealing with the impact and implications of climate change leading to internal displacement in Bangladesh; 'Climate Change, Agriculture and Internal Mobility in the Bhutan Himalayas' by Om Katel, Anooja Nair, Ugyen Yangchen, and Chogyel Wangmo, which traces the impact of climate change on agricultural livelihoods and associated rural–urban migration; '*Climate Change-induced Disaster Displacement and Law in India: Positioning the Operationalization of Artificial Intelligence for Protecting Human Rights*' by Nafees Ahmad elucidating CiDD in India and other countries are causing human rights concerns requiring AI applications for anticipatory actions; and '*Humanitarian Aid Distribution in the Context of Human Rights-Based Approach among Vulnerable Communities: Flash Floods and Climate Change in North Luwu, Indonesia*' by Dina Ruslanjari, Cahyadi Ramadhan, Inayah Bastin Al Hakim, and Feby Aulia Marsida dealing with the multi-dimensional consequences of a catastrophic flash flood that occurred in 2020 in South Sulawesi Province, Indonesia's North Luwu Regency.

In the context of the Section on 'Planned Relocation, Resettlement, and State Responsibility', the volume addresses topics such as 'Climate Change-Related Displacement: Inter-Island and Rural–Urban Migration in the Solomon Islands: Options for a Viable National Resettlement Plan' by Lovelyn K.M. Otoiasi, which deals with increased forced migration that is being brought on by climate change, especially from low-lying atolls in the Solomon Islands; '*Impact of Climate Change and Accessing Services in Papua New Guinea*' by Dora Kuir-Ayius concerning Papua New Guinea's susceptibility to climate change as a result of causes like increasing sea levels, changed rainfall patterns, and occurrences similar to El Niño; '*State Responsibilities and International Obligations in Responding to Climate Mobilities: What should International Assistance Look Like?* by Liam Moore demonstrating how creative policies have been devised in Fiji and Vanuatu, acknowledging the rights of those who are forced to relocate and the obligations of nations to protect these rights; '*Marginalisation of Adat Communities: Intersectionality of Land Grabbing, Human Rights, Climate Adaptation and Human Mobility in Indonesia*' by James Reinaldo Rumpia discussing the vulnerabilities and marginalisation of indigenous peoples in Indonesia; and '*Climate Change and Socio-Economic Issues in Sri Lanka: An Assessment of the Landslide-Induced Relocations in Aranayake*' by Dinushika M. Yapa Abeywardhana investigating the socio-economic ramifications of the landslides that caused major community relocations and displacement in Aranayake, Sri Lanka.

These chapters, focusing on a wide range of topics, shed light on the intricate link that exists between various types of human mobility and the consequences of climate change. The chapters acknowledge the effects of climate change, forcing individuals to relocate and seek new opportunities, and try to address human rights concerns and associated environmental planning and humanitarian operations issues to tackle climate change effects better. It looks for essential players, pertinent legal and regulatory frameworks and tools, and policy gaps relevant to climate-related human mobility, climate change, and displacement. According to the edited volume, paradigms and tools currently in use frequently overlook many of the interrelated aspects of human mobility and protection of people who move inside and across borders due to the effects of climate change. Although there are some procedures and instruments for displacement, planned relocation, and migration, they are not yet entirely in operation and are developing slowly. However, the few tools, currently in use, concentrate more on disaster risk reduction and adaptation strategies than on promoting human mobility, which is increasingly viewed by many nations as a security danger or, at the very least, as politically undesirable. The chapters, through case studies, underpin the local-level impact and climate mobility. Local communities that are impacted also have significant challenges in obtaining information, accessing international policy processes, and finding suitable financing sources. The above steps regarding collection of information and access to policies with fiscal backup are essential to creating localised, people-centred solutions. The volume's focus on the lived experience of the affected populations is consistent with the research into climate change and mobility which has most typically been local, with in-depth cases using qualitative methods.[1]

In summary, the book provides ground-level insights into lived experiences and international standards, highlighting the need for a more rights-based law, policy, and practice from local to international levels.

1 Law and Policy Propositions Emerging from the Research

The various chapters in the volume highlight that climate mobility is a complex issue influenced by various factors, including economic, social, political, historical, and cultural circumstances. Factors such as attachments to one's place, social networks, political forces, shifting economic conditions, and gender norms complicate these patterns. Similarly, the consequences of climate mobility vary across populations based on their context and conditions highlighting 'vulnerability' as a core element. To address how climate change and mobility impact people, policymakers must decrease forced relocation, safeguard relocated individuals, and provide new opportunities for those who wish to leave dangerous situations behind. In this regard, empirical studies from the chapters have highlighted instances where communities did not prefer to migrate. For instance, the elderly Hmong population is endeavouring to adapt their traditional forest management strategies in response to climate change despite the prevalence of drought. This underscores the point that while migration may be a strong coping or adaptive strategy which needs to be respected, it is equally important to respect the people's right to stay and facilitate that discussion.

One of the underlying themes of the chapters is the recognition that an effective strategy addressing climate-related human mobility needs to address the inequality and vulnerability that already exists and makes

[1] Oakes et al. [1].

individuals susceptible to the consequences of climate change. Because human mobility related to climate change is multi-causal, policy responses should take a holistic approach rather than concentrating on specific factors that influence migration. The more important concerns of what makes individuals susceptible and why some groups of people are more vulnerable than others should be taken into account by decision-makers when making plans for climate adaptation. The word 'vulnerability' is often used in development discourse, particularly for the impoverished, who are often portrayed as the most marginalised population. It involves exposure to stress and unforeseen circumstances and challenges in managing them. Chapters in this volume have highlighted the vulnerability of different groups of people including women, older persons, and persons with disabilities. It was seen that people who are disadvantaged due to their gender, age, class, mobility, indigeneity, or disability face distinctive challenges, both in terms of being able to move and particular adversity associated with being in situations of migration, displacement, or planned relocation. Assessing vulnerability from an interdisciplinary and multi-dimensional approach is beneficial and encouraging and can lead to a deeper understanding of mobility dynamics. The edited volume emphasises combining socio-economic and biophysical aspects to provide a more thorough evaluation of vulnerability.

Though there is still more to be explored, the gendered effects of climate change are well-explained by the available data. Vulnerabilities are gendered and appear differently for different groups of men and women. Gender largely determines an individual's vulnerability level, but it is also shaped by the interaction of gender with other social segregations. Preconceived notions about gender as a homogenous group obscure disparities. The approach highlights the various social contexts and exogenous variables intersecting at a particular place and moment to shape gendered vulnerabilities. Gender simplification can overlook important details essential for successful adaptation policies. Asia and the Pacific nations must acknowledge that biophysical shifts determine the impact of climate change not only on the region but also on existing social, economic, and political systems. In short, endeavours to mitigate and adapt to climate change must also focus on lowering socio-economic and gender inequality. The chapters in the volume underpin that differentiated vulnerabilities and discrimination based on gender, class, ethnicity, citizenship, immigration status, age, disability, and other factors restrict people's capacity to benefit from climate action. Hence, planning and implementation efforts must prioritise a human security-centred HRBA that acknowledges the intersectional implications of climate-related human mobility. This is important because while mobility can be a mechanism for people to reduce vulnerability and diversify livelihoods, such a right to move cannot and should not obliterate the complex interplay of power, gender, race, etc.[2]

The edited volume does not explore the vulnerability only from a victimhood perspective, but it highlights the resilience shown by the vulnerable communities including the indigenous population and women and the need to include their perspective in the climate mobility discourse and while planning relocation. The chapters showcase that in increasing the resilience of communities that are already at risk, communities ought to be directed by the concepts of human rights and dignity. At the same time, the volume also highlights case studies which point to the overall satisfaction of people affected by disasters to the government responses. For example, in Indonesia, flash floods in Masamba and Baebunta sub-districts in July 2020 caused fatalities, damaged infrastructure, and destroyed homes. Despite a comprehensive humanitarian aid and logistics policy, challenges arose due to limited accessibility to affected areas. However, the point to be highlighted is that the community was mostly satisfied with the government's performance in providing aid, with good coordination and no discrimination in aid distribution.

---

[2] Bergmann and Martin [2].

The book highlights that governments and subnational authorities have started addressing the intricate connections between climate change and human mobility in planning and policy procedures, such as NDCs and NAPs, as awareness of these connections has grown. According to the volume, many countries see migration as a risk, difficulty, or issue, but some have embraced its advantages (e.g. migration as an adaptive technique). The volume highlights best practices that can improve the integration of human mobility in various priority policy sectors for adaptation, loss, and damage through case examples. The authors have stressed the necessity of sufficient funding, institutional capabilities, and data to advance human mobility integration into policy framework. At the local, national, and regional levels, there is a need to recognise and comprehend possible policy interventions, evaluate their effects, and map out potential collaborations. In short, in light of the difficulties presented by climate-related human mobility and disasters in the region, the edited volume intends to provide the ground for safe, orderly, and regular migration to craft an adaptation strategy to confront climate change and envision aid and protection for the impacted people and look for durable solutions.

The volume also points that climate change poses a danger to regional and international security, and it needs to be tackled as such, utilising intersectional analysis and resources that consider the situated expertise of those most impacted. The contributors to this volume are scholars from Asia and the Pacific who have witnessed the damage caused by climate-related human mobility, displacement, and relocation during the past decades. The learnings from these studies may aid in framing pre-emptive steps to speed up action to prevent another disaster. The foundation of such global and regional action on climate change must be cooperation, solidarity, and shared responsibility for a disaster that will have catastrophic effects on all of us.

Despite the focus on solidarity, cooperation, and shared responsibility, international conventions on development and climate change have taken quite varied approaches to and considerations of climate-related mobility caused by the climate.[3] The national governments attempting to adopt these frameworks under their respective development objectives are misled by their ambiguity. Addressing the needs of climate-related mobility and displaced populations on the move presents a significant opportunity as pertinent national and international policy contexts change. However, there is a growing integration of mobility and climate change into various legal and policy frameworks across multiple spheres, such as disaster risk reduction (Sendai Framework for Disaster Risk Reduction), migration (Global Compact for Safe, Orderly, and Regular Migration (GCM)), and climate change (United Nations Framework Convention on Climate Change Task Force for Displacement). The Inter-Agency Standing Committee Operational Guidelines for Protection in Disasters, the Guiding Principles on Internal Displacement, and the Guidance on Protecting People from Disasters and Environmental Change Through Planned Relocation are other extremely pertinent tools. Similarly at the regional level, the UN Pacific Strategy, the Framework for Pacific Regionalism, and the Framework for Resilient Development in the Pacific: An Integrated Approach to Address Climate Change and Disaster Risk Management are some regional frameworks that direct the development of climate mobility policies in the Pacific Islands. The Pacific Climate Change Migration and Human Security Programme (PCCMHS) and the recently completed Pacific Climate Change and Migration Project are two examples of regional activities in the Pacific Islands.[4] These are based on the protection of displaced persons and human security principles, but they also emphasise voluntary movement as a means of adapting to climate change.[5] However, there is a need for more ground-level research to implement

[3] See generally Scott and Salamanca [3].

[4] Environmental Migration Portal [4].

[5] Vanhala and Calliari [5].

the framework in local stettings. Ground-level studies can be instrumental in developing toolkits which ensure community participation in promoting an inclusive, gender-responsive strategy for involving all relevant actors, including the public, private, and third sectors, as well as indigenous peoples, migrants, displaced populations, and communities affected by displacement for addressing human mobility concerns in the context of climate change at all levels, from local to global.

2 Considerations for Further Research

Despite the three decades of international climate law developments, a significant deficiency in international law is the absence of 'hard' law, as there are currently no specific laws that apply to migrants and displaced individuals affected by climate change. The Paris Agreement, however, expressly recognises the rights of all people who are vulnerable, including migrants, and it requires nations to uphold, advance, and consider human rights while implementing climate change policies. In addition, it is noteworthy how much attention and visibility this topic has received at the UNFCCC's COP28 in Dubai, United Arab Emirates. During COP28 negotiations, human mobility has been recognised as one of the most extensive effects of climate change on people and communities.[6] A good compromise between the interests of so many different nations can be found in the decision text language of the first global stocktake agreed in Dubai. It is an extensive evaluation of the world's progress on climate action and an updated roadmap to raise ambition to fulfil the goals of the Paris Agreement. It is encouraging that the global stocktake explicitly mentions displacement from the standpoint of human mobility. In addition to acknowledging the efforts already made to respond to displacement and the unique vulnerability of displaced populations, it calls on governments and pertinent institutions to move forward in preventing, minimising, and addressing loss and damage, including through measures pertaining to migration, relocation, and displacement.

The global stocktake's use of terminology related to displacement establishes a 'hook' to motivate and assist nations in reporting on their progress in responding to human mobility. It also encourages states to make more robust promises and measures to prevent, reduce, and deal with displacement and associated losses and damages in the upcoming round of NDCs due by 2025. On the first day of COP 28, a monumental agreement was reached to operationalise the Loss and Damage Fund and include displacement and displaced people within its purview.[7]This would enable the Fund to offer funding to assist vulnerable developing nations in implementing human mobility-related policies and initiatives, particularly by strengthening evidence and filling data gaps. The implementation of these initiatives at the ground level and its challenges is worth exploring.

Numerous subjects have been addressed in the volume in relation to climate-related mobility. Nevertheless, it was not possible to thoroughly examine all of the nuanced intersectionality and complicated interplay of concerns in the context of climate mobility. For example, problems related to heat waves and their generated mobility could not be investigated. Similarly, the issue of children and their vulnerability concerning climate mobility was not investigated. Climate change and its related problems are among the most urgent issues endangering the rights of children. Children who are forced to move in under these circumstances are particularly vulnerable. It is only in recent times, the policies pertaining to migration and children have started incorporating environmental and climate change considerations. International organisations, civil society organisations, and local and national governments

[6] Donoghoe and Perry [6].

[7] United Nations Framework Convention on Climate Change [7].

who engage with children on the move in the context of climate change are encouraged to utilise the Guiding Principles for Children on the Move in the Context of Climate Change.[8] These Principles are predicated on operational rules or frameworks that have been formulated by several parties, in addition to extant international law. The Principles are not compelling and do not establish any new international legal duties; rather, they are taken from the 1989 UN Convention on the Rights of the Child.

Challenges of climate-related human mobility in response to seasonal labour movement, drought and food poverty, and various health possibilities and dangers need diversifying livelihoods across climate-vulnerable locales. These difficulties add to the scant empirical evidence of health risks associated with climate-related mobility, such as changed eating habits, impaired access to healthcare, income loss, hazardous employment, unhygienic urban environments, changed exposure to infectious diseases, disrupted social networks, and loss of place attachment. Another aspect is the emotional and physical well-being of both stationary and mobile populations. These challenges require more in-depth research, with robust conceptual frameworks, methodological innovations, and comparative approaches for understanding a multitude of factors, including family and individual characteristics, socio-political and economic circumstances, and environmental dangers. One way to focus on these research areas is through analysing the linkage of climate change and SDGs, informed by normative guidance on the right to the highest attainable standard of physical and mental health.

Similarly, there is a need for disparate research focusing on country-specific volume or thematic volume in Asia and the Pacific region, for instance, floods or droughts, for pragmatic comparison or similarity among Bangladesh, India, Pakistan, etc. Other areas of research focus could be food production in the region that will be threatened by the adverse effects of global warming on the present scale of increasing temperatures, which will lower human and land productivity. Considerable harm will be caused by erratic rainfall and an escalation in the incidence of extreme weather occurrences. Land pressure, habitat degradation and biodiversity loss, water shortages and pollution, air pollution, global warming, and climate change require country-specific research volumes.

International policy largely ignores cross-cutting areas like water, food, energy, seas, ecosystems, and interlinkage of social issues in the context of human mobility and climate change. For instance, the inability of financing agreements and mechanisms such as the Grand Bargain or the Green Climate Fund (GCF) to specifically include human mobility about climate change in their overarching goals impede efforts to finance large-scale action. Nonetheless, there are positive indications of some existing GCF projects showing some project-level integration of human mobility features.[9] This development needs to be further explored. Though, several avenues for cross linking issues connected with climate-related human mobility are provided by the Agenda 2030 for Sustainable Development, these cross-connections may need to be more clearly stated and need to be researched to influence global policy and can be taken forward in future research.

Assistance and protection policies that grant the same rights to individuals who are displaced or compelled to move, regardless of their location and based on need rather than status, must be agreed upon to address climate mobility. The policies should support long-term solutions, primarily local integration and relocation, comparable to those required for refugees and individuals displaced permanently due to irreversible environmental change. It should also support policies that encourage voluntary migration as a tactic for adaptation. This

[8] UNICEF and IOM [8].

[9] GCF Project Portfolio [9], Where it is most needed, GCF is providing significant climate action via more than 200 projects and initiatives in developing nations.

would involve actions to promote local integration, such as easy access to public services that benefit the community's public sector and way of life. This is possible if there is an integration of the language and commitments already made in the SDGs and climate and disaster policies, bringing these together under an HRBA for climate-induced human mobility. Consistent with the political sensitivities, bringing all of this together in a single global agreement is a tall order and will probably take some time. The way out could be through regional cooperative mechanisms.

The coordination and networking among the Pacific countries under the PCCMHS could be an example to inspire states in Southeast Asia and South Asia to develop regional frameworks for addressing climate-related human mobility. In this context, during the 52nd Meeting of Pacific Islands Forum Leaders in the Cook Islands in November 2023, the Pacific Governments approved a Regional Framework on Climate Mobility.[10] The framework acknowledges the pressing need to improve knowledge, coordination, and collaboration regarding human mobility related to climate change and to set up sufficient mechanisms for responding to and supporting people at risk based for addressing climate mobility and enhancing resilience and well-being in the Pacific. One of its distinctive features is the framework's categorical recognition of the Pacific people's right to remain in their homes as a paramount urgency. It attempts to guide governments on how to deal with particular legal, policy, and practical concerns related to staying put, displacement, migration, planned relocations, and stranded migrants that would come from increased climatic mobility with a regional initiative on PCCMHS that would also provide support. Similar strategies can be explored at other regional levels like South Asia and Southeast Asia.

Climate change impacts individuals and groups differently. Due to societal variables and different risk perceptions, migration may not be permanent calling for different adaptation approaches such as permanent, temporary, or no migration. This is challenging as the local effects of climate change are uncertain, and their extent, magnitude, and social reactions are complex. Human mobility as a coping strategy requires significant expenditure and the creation of bilateral and multilateral institutions to promote organised planning, reduce socio-economic effects, and prevent conflicts. Many emerging countries are unprepared to handle demographic shifts. The potential function of migration as an adaptation mechanism will also be influenced by gender considerations, with marginalised populations and impoverished women often facing the danger of climate effects in their current environment. Therefore, addressing the impacts of climate change requires a comprehensive approach that considers local and global factors.

Human mobility can be a valuable tool for adaptation and risk reduction, reducing vulnerability. However, barriers to adequate protection pose risks to migrants, including exploitation and human rights breaches. The international community must protect those negatively impacted by climate change, support adaptation efforts, and deal with loss and damage. Climate change has led to climate hazards, displacement, and migration, but anticipatory action in the context of human mobility remains relatively unexplored. Frameworks capable of understanding and predicting climate hazards, anticipating climate change-related human mobility, disaster-induced displacement, and migration, and its implications for vulnerable populations in Asia and the Pacific have gained new focus.

Therefore, responses to human mobility should be implemented within development frameworks and integrated into efforts for mitigating climate change. Migration choices should be made available to the most vulnerable groups, allowing them to relocate to safer environs or stay where they are and adapt. More public funding for social protection, urban investment, disaster risk reduction, and livelihood development is needed, while the private sector may also play a significant role in

[10] Pacific Islands Forum [10, 11].

lowering risk. Supporting governments and at-risk populations in averting future displacement and better managing the consequences of those who have already moved is crucial to addressing the needs of displaced people and communities hosting them in an age of climate change. All tiers of government and regional and international organisations must take immediate action to influence human mobility development in future, promoting sustainable solutions, resilience, and protection of communities affected by climate change in the Asia–Pacific region and beyond.

References

1. Oakes R et al (2023) A future agenda for research on climate change and human mobility. Int Migration 61(5):116–125
2. Bergmann J, Martin SF (2023) Addressing climate change-related human immobilities. In: KNOMAD paper, p 54. https://www.knomad.org/sites/default/files/publication-doc/knomad_paper_54_addressing_climate_changerelated_human_immobilities_bergmann_martin_nov_2023.pdf. Accessed 10 Jan 2024
3. Scott M, Salamanca A (2022) Climate change, disasters, and internal displacement in Asia and the Pacific: a human rights-based approach. Routledge
4. Environmental Migration Portal. PCCMHS enhancing protection and empowerment of migrants and Environmental Migration Portal. https://environmentalmigration.iom.int/pccmhs-enhancing-protection-and-empowerment-migrants-and-communities-affected-climate-change-and-disasters-pacific-region> accessed 10 January 2024.
5. Vanhala L, Calliari E (2022) Governing people on the move in a warming world: framing climate change migration and the UNFCCC task force on displacement. Glob Environ Change 76:102578
6. Donoghoe M, Perry AM (2023) The successes and failures of COP28. Brookings. https://www.brookings.edu/articles/the-successes-and-failures-of-cop28/. Accessed 21 Jan 2024
7. United Nations Framework Convention on Climate Change (2023) In: Conference of the Parties serving as the meeting of the Parties to the Paris Agreement, Decision 1/CMA.4, para 15 (2023) (establishing the global stocktake and outlining its scope). https://unfccc.int/sites/default/files/resource/cma2023_L17_adv.pdf. Accessed 10 Jan 2024
8. UNICEF and IOM (2022) Guiding principles for children on the move in the context of climate change, pp 23–52
9. GCF Project Portfolio. https://www.greenclimate.fund/projects?f[]=field_theme:236. Accessed 21 Jan 2024
10. Pacific Islands Forum (2023) 52nd Pacific Islands forum leaders meeting—Cook Islands. https://piflm52.com/
11. Pacific Islands Forum (2023) Pacific regional framework on climate mobility. https://www.forumsec.org/wp-content/uploads/2023/11/Annex-C-Pacific-Regional-Framework-on-Climate-Mobility-1.pdf. Accessed 10 Jan 2024

Stellina Jolly is a Lead Editor and an Associate Professor at the Faculty of Legal Studies, South Asian University (SAU). She is a Visiting Senior Research Associate with the Centre for Emerging Countries in Private international law at the University of Johannesburg. Dr. Jolly is a Fulbright Scholar with the University of San Francisco and a recipient of the International Visitors Leadership Program (IVLP). She researches international environmental law and conflict of laws. In environmental law, her research papers have explored the north-south dimensions of environmental justice. Her book, 'Climate Refugees in South Asia' published by Springer, explored the legal and policy framework of climate refugees in South Asia; she has also elaborately written on the environmental justice and governance concerns and intersectionalities of climate change, water, gender, disaster management, and sustainable development goals focusing on South Asian jurisdictions. Her recent researches also focus on the emerging global trend of the rights of nature. In her research, she advocates an approach based on just sustainability, hybrid law mechanism, and environmental justice to tide over the ecological crisis. Dr. Jolly has published in several leading international journals like the *Transnational Environmental Law, Washington Journal of Environmental Law and Policy, Texas Environmental Law Journal*, and *Chinese Journal of Environmental Law*, to name a few. Hart published her book on Indian Private International law. She has also co-edited a book collection published by Springer on Private International Law of South Asian Jurisdictions. She is a member of the *IUCN World Commission on Environmental Law and part of the Network of Environmental Law Champions, Asian Development Bank.* She has undertaken projects and consultancies with various organizations, including the *Rajiv Gandhi Foundation, European Union, International Renewable Energy* Agency (IRENA), and *Asia Europe Foundation.* She was awarded an educational grant on Civil Society Law from International Centre for Not-for-Profit Law (ICNL) and USAID.

Nafees Ahmad is an Editor, and an Associate Professor at the Faculty of Legal Studies, South Asian University (SAU)-New Delhi. He holds a doctorate in International Refugee Law and Human Rights. His scholarship focuses on RAMS (Refugees, Asylum-seekers, Migrants,

Stateless) and the role of Artificial Intelligence (AI) in their protection, Global Forced Displacement, Global Circumstantial Migration (GCM) Governance, and Climate Refugees in South Asia. He also addresses the International Politics of Asylum, Refugee Policy Paradigms, Invisible Frames of Asylum, Disconnects of Durable Solutions, and SAARC connects and contexts of refugee protection. He conceived and introduced a new SAARC-specific Program in 2011 called *Comparative Constitutional Law of SAARC Nations* (CCLSAARCN) at the LLM level. He publishes *inter-alia* in the *International Handbook on Disaster Research* (Springer 2023), *The Asian Yearbook of Human Rights and Huma*nitarian Law (Brill 2023), *Groningen Journal of International Law* (The Netherlands), *Harvard International Law Journal* (Harvard) *Asia-Pacific Journal of Human Rights and Law* (Brill), *International Journal of Environment and Waste* Management indexed in SCOPUS, *International Journal on Minority and Group Rights*, *Kings' Student Law Review* (KCL-London), *ISIL Year Book on International Humanitarian Law* and Refugee Law, *ELCOP Year Book of Human Rights* and *NUJS International Journal of Legal Studies and* Research (IJLSR) etc. Dr. Ahmad has co-authored a book on Climate Refugees in South Asia (Springer 2019). He is a member of Editorial Advisory Board of *Iranian Journal of International and Comparative* Law. Dr. Ahmad has been a Resource Person and External Reviewer for the Ministry of Law; Government of India-sponsored Research Project on "*Judicial Reforms since June 2016*" at the Indian Institute of Management (IIM-Kashipur) and accomplished an ICSSR Project on "*The Muncipal Solid Waste Management in Delhi: A Socio-Legal Study of Okhla Lanfill.*" Further, he has been a visiting professor at the Indian Society of International Law (ISIL), Jamia Milia Islamia-New Delhi, and Judicial Academies of various states of India.

Matthew Scott is an Editor and co-leads the Human Rights and the Environment thematic area at the Raoul Wallenberg Institute of Human Rights and Humanitarian Law in Lund, Sweden. His scholarship integrates social science perspectives with international legal standards to promote context-sensitive, human rights-based law, policy, and practice relating to disaster risk reduction and climate change adaptation. Dr. Scotts' primary area of expertise concerns migration and displacement in the context of disasters and climate change, on which he published a monograph entitled *Climate Change, Disasters and the Refugee Con*vention (CUP 2020), an edited volume entitled *Climate Change, Disasters and Internal Displacem*ent in Asia and the Pacific: A Human Rights-Based Approach (Routledge 2021), and a range of book chapters and academic articles in, amongst others, the *International Journal of Refugee Law*, the *Nordic Journal of International Law* and the *Yearbook of International Disaster Law*. The edited volume was a major output of a regional thematic study he coordinated as part of RWI's Asia-Pacific programme on human rights and sustainable development. Dr. Scott works on the role of local authorities in addressing climate- and disaster-related migration and displacement and how human rights law can contribute to building resilience to pandemic risk. He holds a Ph.D. in Public International Law from Lund University (2018) and a MA in Social Anthropology of Development from SOAS (1998). He practiced immigration and asylum law before entering academia. He is a member of the advisory committee of the Platform on Disaster Displacement, a member of the editorial board of the *Yearbook of International Disaster Law*, and a founding member of the *Nordic Network on Climate-Related Displacement* and Mobility. At Lund University, he convenes the LLM course on human rights law, the environment, and climate change and lectures on international refugee and human rights law at the Faculty of Law. He also lectures on the M.Sc. programme in Disaster Risk Reduction and Climate Change Adaptation at the Department of Risk Management and Societal Safety. He is also actively engaged in international collaboration initiatives and currently contributing technical expertise on human rights-based approaches to disaster risk reduction across eight countries in Asia in collaboration with the Asia Disaster Preparedness Center, Stockholm Environment Institute, and the Swedish Civil Contingencies Agency.

The manufacturer's authorised representative in the EU is Springer Nature Customer Service Centre GmbH, Europaplatz 3, 69115 Heidelberg, Germany. If you have any concerns regarding our products, please contact ProductSafety@springernature.com

Printed and bound by CPI Group (UK) Ltd, Croydon, CR0 4YY
01/10/2026
02225862-0020